Powerful techniques have been developed in recent years for the analysis of digital data, especially the manipulation of images. This book provides an in-depth introduction to a range of these innovative, avant-garde data-processing techniques. It develops the reader's understanding of each technique and then shows with practical examples how they can be applied to improve the skills of graduate students and researchers in astronomy, electrical engineering, physics, geophysics and medical imaging.

What sets this book apart from others on the subject is the complementary blend of theory and practical application. Throughout, the book is copiously illustrated with real-world examples from astronomy, electrical engineering, remote sensing and medicine. It also shows how many, more traditional, methods can be enhanced by incorporating the new wavelet and multiscale methods into the processing.

For graduate students and researchers already experienced in image processing and data analysis, this book provides an indispensable guide to a wide range of exciting and original data-analysis techniques.

Image Processing and Data Analysis

Image Processing and Data Analysis

The Multiscale Approach

J.-L. STARCK

Centre d'Etudes de Saclay

F. MURTAGH

University of Ulster

A. BIJAOUI

Observatoire de la Côte d'Azur

PUBLISHED BY THE PRESS SYNDICATE OF THE UNIVERSITY OF CAMBRIDGE
The Pitt Building, Trumpington Street, Cambridge CB2 1RP, United Kingdom

CAMBRIDGE UNIVERSITY PRESS
The Edinburgh Building, Cambridge CB2 2RU, United Kingdom
40 West 20th Street, New York, NY 10011-4211, USA
10 Stamford Road, Oakleigh, Melbourne 3166, Australia

First published 1998

Printed in the United Kingdom at the University Press, Cambridge

Typeset in Monotype Times

A catalogue record of this book is available from the British Library

Library of Congress Cataloguing in Publication data

Starck, J.-L. (Jean-Luc), 1965–
Image processing and data analysis: the multiscale approach / J.-L. Starck,
F. Murtagh, A. Bijaoui.
p. cm.
Includes bibliographical references and index.
ISBN 0 521 59084 1. – ISBN 0 521 59914 8 (pbk.)
1. Physical sciences–Data processing. 2. Wavelets (Mathematics)
3. Signal processing–Mathematics. I. Murtagh, Fionn. II. Bijaoui, A. III. Title.
Q183.9.S83 1998
621.36′7′0245–dc21 97-17393 CIP

ISBN 0 521 59084 1 hardback
ISBN 0 521 59914 8 paperback

Contents

Preface

There is a very large literature on the theoretical underpinnings of the wavelet transform. However, theory must be complemented with a significant amount of practical work. Selection of method, implementation, validation of results, comparison with alternatives – these are all centrally important for the applied scientist or engineer. Turning theory into practice is the theme of this book. Various applications have benefited from the wavelet and other multiscale transforms. In this book, we describe many such applications, and in this way illustrate the theory and practice of such transforms. We describe an 'embedded systems' approach to wavelets and multiscale transforms in this book, in that we introduce and appraise appropriate multiscale methods for use in a wide range of application areas.

Astronomy provides an illustrative background for many of the examples used in this book. Chapters 5 and 6 cover problems in remote sensing. Chapter 3, dealing with noise in images, includes discussion on problems of wide relevance. At the time of writing, the authors are applying these methods to other fields: medical image analysis (radiology, for mammography; echocardiology), plasma physics response signals, and others.

Chapter 1 provides an extensive review of the theory and practice of the wavelet transform. This chapter then considers other multiscale transforms, offering possible advantages in regard to robustness. The reader wishing early 'action' may wish to read parts of Chapter 1 at first, and dip into it again later, for discussion of particular methods.

In Chapter 2, an important property of images – noise – is investigated. Application of the lessons learned in regard to noise is then illustrated. Chapter 3 describes deconvolution, or image sharpening and/or restoration. This includes drawing various links with entropy-based smoothness criteria. Chapter 4 covers (i) spectral analysis and

(ii) general themes in multivariate data analysis. It is shown how the wavelet transform can be integrated seamlessly into various multivariate data analysis methods. Chapter 5 covers image registration, in remote sensing and in astronomy. Chapter 6 deals with stereo image processing in remote sensing. Chapter 7 describes highly effective image compression procedures based on multiscale transforms. Chapter 8 deals with object detection in images and also with point pattern clustering. The concluding chapter, Chapter 9, covers object recognition in images. This chapter is oriented towards image interpretation and understanding. A number of appendices round off the book, together with a bibliography and index.

Chapters 5 and 6 are due mainly to Jean-Pierre Djamdji (Observatoire de la Côte d'Azur, Nice).

Other colleagues we would like to acknowledge include: Marguerite Pierre (CEA, Service Astrophysique) – Chapter 2; Eric Pantin (CEA, Saclay), Bruno Lopez (Observatoire de la Côte d'Azur, Nice), and Christian Perrier (Observatoire de Grenoble) – Chapter 3; Ralf Siebenmorgen (European Space Agency, Vilspa), Roland Gredel (European Southern Observatory, Chile), and Alexandre Aussem (Université Blaise Pascal – Clermont-Ferrand II) – Chapter 4; Roger Manière – Chapter 5; Pierre-François Honoré (CEA, Saclay) – Chapter 7; Frédéric Rué (Observatoire de la Côte d'Azur, Nice) – Chapter 9.

We would like to acknowledge the European Science Foundation scientific network on 'Converging Computing Methodologies in Astronomy', through which support for a number of visits between the authors was obtained. Part of the work presented here was carried out at Observatoire de la Côte d'Azur, Nice, and at CEA-Saclay. Fionn Murtagh acknowledges the European Space Agency's Hubble Space Telescope support group at the European Southern Observatory, where part of this work was carried out. Fionn Murtagh also acknowledges partial support from Office of Naval Research Grant N-00014-91-J-1074 ('Model-based clustering for marked spatial point processes') and discussions within the associated Working Group, in regard to Chapter 8.

We would also like to acknowledge the considerable assistance provided by the referees of a number of papers, on which part of the work presented here is based. These include Rick White and Robert J. Hanisch of the Space Telescope Science Institute, Baltimore. The mammographic image used in Chapter 2 was provided by Dr Matthew Freedman, Georgetown University, and our attention was brought to it initially by Robert Hanisch.

Jean-Luc Starck, Fionn Murtagh and Albert Bijaoui

1 The wavelet transform

1.1 Multiscale methods

A range of very different ideas have been used to formulate the human ability to view and comprehend phenomena on different scales. Wavelet and other multiscale transforms are treated in this book. Just a few alternative approaches are surveyed in this section. In this section also, various terms are introduced in passing, which will reappear many times in later chapters.

Data classification may be carried out using hierarchical classification (Murtagh, 1985). A sequence of agglomerative steps is used, merging data objects into a new cluster at each step. There is a fixed number of total operations in this case. Such an approach is bottom-up, starting with the set of data objects, considered independently. Spatial and other constraints may be incorporated, to provide segmentation or regionalization methods. This approach is combinatorial since neither continuous data values, nor stochasticity, are presupposed. For alternative combinatorial methods, see Breiman *et al.* (1984) and Preparata and Shamos (1985). For image data, split-and-merge approaches are introduced in Schalkoff (1989).

Let us now consider two-dimensional (or other) images. An image represents an important class of data structures. Data objects may be taken as pixels, but it is more meaningful for image interpretation if we try, in some appropriate way, to take regions of the image as the data objects. Such regions may be approximate. One approach is to recursively divide the image into smaller regions. Such regions may be square or rectangular, to facilitate general implementation. Decomposition halts whenever a node meets a homogeneity criterion, based on the pixel values or gray-levels within the corresponding image region. This tree data structure, associated with an image, is a quadtree (Samet, 1984).

A pyramid is a set of successively smoothed and downsampled versions of the original image. Usually the amount of data decreases two-fold at each successive step. One sees then that the total storage required is bounded by the storage required for the original data.

A wavelet is a localized function of mean zero. Wavelet transforms often incorporate a pyramidal representation of the result. We will also see examples later of cases where a set of successively smoother versions of an image are not downsampled. Wavelet transforms are computationally efficient, and part of the reason for this is that the scaling or wavelet function used is often of compact support, i.e. defined on a limited and finite domain. Wavelets also usually allow exact reconstitution of the original data. A sufficient condition for this in the case of the continuous wavelet transform is that the wavelet coefficients, which allow reconstitution, are of zero mean. Wavelet functions are often wave-like but clipped to a finite domain, which is why they are so named.

A different idea is that of scale-space filtering (Lindeberg, 1994). In this method, the image is smoothed by convolving with a Gaussian kernel (usually), of successively increasing width at successive scales. The Gaussian function has been shown to be of most interest for this purpose, since it fulfils the conditions necessary for no new structure to be introduced at any scale. The idea is that all structure should be present in the input signal, and structure should not be added by the convolutions. Zero-crossings are examined in the context of this approach. These are extrema, defined using the second derivative of the signal or its increasingly smoothed versions.

Compared to other methods described here, the wavelet transform can be determined very efficiently. Unlike scale-space filtering, it can introduce artifacts. To limit the retrograde impact of these, we may wish to develop other similar multiscale methods, with specific desirable properties. The choice of method to apply in practice is a function of the problem, and quite often of the properties of the signal.

Some expository introductions to the wavelet transform include: Graps (1995), Nason and Silverman (1994), Vidaković and Müller (1995), Bentley and McDonnnell (1994), and Stollnitz, DeRose and Salesin (1995). Later chapters in many instances offer new reworkings of these concepts and methods, and focus strongly on applications. Our aim is to describe the theoretical underpinnings and to illustrate the broad utility and importance of wavelets and other related multiscale transforms.

1.1.1 Some perspectives on the wavelet transform

Wavelets can be introduced in different ways. In the following we can think of our input data as a time-varying signal. If discretely sampled, this amounts to considering an input vector of values. The input data may be sampled at discrete wavelength values, yielding a spectrum, or one-dimensional image. A two-dimensional, or more complicated input image, can be fed to the analysis engine as a rasterized data stream. Analysis of such a two-dimensional image may be carried out independently on each dimension. Without undue loss of generality, we will now consider each input to be a continuous signal or a discrete vector of values.

- In the continuous wavelet transform, the input signal is correlated with an analyzing continuous wavelet. The latter is a function of two parameters, scale and position. An admissibility condition is required, so that the original signal can be reconstituted from its wavelet transform. In practice, some discrete version of this continuous transform will be used. A later section will give definitions and will examine the continuous wavelet transform in more detail.
- The widely-used Fourier transform maps the input data into a new space, the basis functions of which are sines and cosines. Such basis functions extend to $+\infty$ and $-\infty$, which suggests that Fourier analysis is appropriate for signals which are similarly defined on this infinite range, or which can be assumed to be periodic. The wavelet transform maps the input data into a new space, the basis functions of which are quite localized in space. They are usually of compact support.

  The term 'wavelet' arose as a localized wave-like function. Wavelets are localized in frequency as well as space, i.e. their rate of variation is restricted. Fourier analysis is not local in space, but is local in frequency.

  Fourier analysis is unique, but wavelet analysis is not: there are many possible sets of wavelets which one can choose. One trade-off between different wavelet sets is between their compactness versus their smoothness.

  Compactness has implications for computational complexity: while the Fast Fourier Transform (FFT) has computational complexity $O(n \log n)$ for n-valued inputs, the wavelet transform is often more efficient, $O(n)$.
- Another point of view on the wavelet transform is by means of filter banks. The filtering of the input signal is some transformation of it, e.g. a low-pass filter, or convolution with a smoothing function.

Low-pass and high-pass filters are both considered in the wavelet transform, and their complementary use provides signal analysis and synthesis.

We will continue with a short account of the wavelet transform, as described from the point of view of filtering the data using a cascade of filters.

The following example uses a Haar wavelet transform. Basis functions of a space indicated by V_j are defined from a *scaling function* ϕ as follows:

$$\phi_{j,i}(x) = \phi(2^{-j}x - i) \quad i = 0, \ldots, n-1 \text{ with } \phi(x) = \begin{cases} 1 & \text{for } 0 \leq x < 1 \\ 0 & \text{otherwise} \end{cases} \tag{1.1}$$

Note the dimensionality of space V_j, which directly leads to what is called a *dyadic* analysis. The functions ϕ are all box functions, defined on the interval $[0, 1)$ and are piecewise constant on 2^j subintervals. We can approximate any function in spaces V_j associated with basis functions ϕ_j, clearly in a very fine manner for V_0 (all values!), more crudely for V_{j+1} and so on. We consider the nesting of spaces, $\ldots V_{j+1} \subset V_j \subset V_{j-1} \ldots \subset V_0$.

Next we consider the orthogonal complement of V_{j+1} in V_j, and call it W_{j+1}. The basis functions for W_j are derived from the *Haar wavelet*. We find

$$\psi_{j,i}(x) = \psi(2^{-j}x - i) \quad i = 0, \ldots, n-1 \text{ with } \psi(x) = \begin{cases} 1 & 0 \leq x < \frac{1}{2} \\ -1 & \frac{1}{2} \leq x < 1 \\ 0 & \text{otherwise} \end{cases} \tag{1.2}$$

This leads to the basis for V_j as being equal to the basis for V_{j+1} together with the basis for W_{j+1}. In practice we use this finding like this: we write a given function in terms of basis functions in V_j; then we rewrite in terms of basis functions in V_{j+1} and W_{j+1}; and then we rewrite the former to yield, overall, an expression in terms of basis functions in V_{j+2}, W_{j+2} and W_{j+1}. The wavelet parts provide the detail part, and the space V_{j+2} provides the smooth part.

For the definitions of scaling function and wavelet function in the case of the Haar wavelet transform, proceeding from the given signal, the spaces V_j are formed by averaging of pairs of adjacent values, and the spaces W_j are formed by differencing of pairs of adjacent values. Proceeding in this direction, from the given signal, we see that application of the scaling or wavelet functions involves downsampling of the data.

The low-pass filter is a moving average. The high-pass filter is a moving difference. Other low- and high-pass filters could alternatively be used, to yield other wavelet transforms. We see that an input signal has been split into frequency bands, by means of application of low-pass and high-pass filters. Signal splitting of this type is termed subband coding. The collection of filters is termed an analysis bank or a filter bank. The subsignals thus constructed can be compressed more efficiently, compared to the original signal. They have storage and transfer advantages.

A filter is a linear, time-invariant operator. We can therefore write the low-pass filtering as the matrix product Hx, and the high-pass filtering as Gx. The analysis bank can make use of the following matrix:

$$\begin{bmatrix} H \\ G \end{bmatrix} \tag{1.3}$$

Reconstituting the original data involves inversion. Subject to orthogonality, we have:

$$\begin{bmatrix} H \\ G \end{bmatrix}^{-1} = \begin{bmatrix} H^T G^T \end{bmatrix} \tag{1.4}$$

where T is the transpose. For exact reconstitution with such an orthogonal filter bank, we have

$$\begin{bmatrix} H^T G^T \end{bmatrix} \begin{bmatrix} H \\ G \end{bmatrix} = H^T H + G^T G = I \tag{1.5}$$

where I is the identity matrix. The term 'conjugate mirror filters' is also used for H and G above. We have taken the filter bank as orthogonal here, but other properties related to other wavelet transforms have also been studied: biorthogonality (H orthogonal to G, H and G independently orthogonal), semi-orthogonality (H and G orthogonal but spaces associated with H and G are not individually orthogonal), non-orthogonal schemes (e.g. for G). An example of a non-orthogonal wavelet transform is the *à trous* wavelet transform which will be used extensively in this book. With regard to orthogonal filter banks, it was once thought that the Haar wavelet transform was the sole compact representative. However Daubechies found what has become a well-known family of orthogonal, compact filters satisfying certain regularity assumptions.

A very readable introductory text on the wavelet transform from the filter bank perspective is Strang and Nguyen (1996). Other books include Chui (1992), Daubechies (1988), Meyer (1993), Meyer, Jaffard and Rioul (1987), and Ruskai *et al.* (1992).

1.1.2 The wavelet transform and the Fourier transform

In the early 1980s, the wavelet transform was studied theoretically in geophysics and mathematics by Morlet, Grossman and Meyer. In the late 1980s, links with digital signal processing were pursued by Daubechies and Mallat, thereby putting wavelets firmly into the application domain.

The Fourier transform is a tool widely used for many scientific purposes, and it will serve as a basis for another introduction to the wavelet transform. For the present, we assume a time-varying signal. Generalization to any x as independent variable, or image pixels (x, y), in the place of time t, is immediate. The Fourier transform is well suited only to the study of stationary signals where all frequencies have an infinite coherence time, or – otherwise expressed – the signal's statistical properties do not change over time. Fourier analysis is based on global information which is not adequate for the study of compact or local patterns.

As is well-known, Fourier analysis uses basis functions consisting of sine and cosine functions. These are time-independent. Hence the description of the signal provided by Fourier analysis is purely in the frequency domain. Music, or the voice, however, impart information in both the time and the frequency domain. The windowed Fourier transform, and the wavelet transform, aim at an analysis of both time and frequency. A short, informal introduction to these different methods can be found in Bentley and McDonnell (1994) and further material is covered in Chui (1992).

For non-stationary analysis, a windowed Fourier transform (STFT, short-time Fourier transform) can be used. Gabor (1946) introduced a local Fourier analysis, taking into account a sliding Gaussian window. Such approaches provide tools for investigating time as well as frequency. Stationarity is assumed within the window. The smaller the window size, the better the time-resolution. However the smaller the window size also, the more the number of discrete frequencies which can be represented in the frequency domain will be reduced, and therefore the more weakened will be the discrimination-potential among frequencies. The choice of window thus leads to an uncertainty trade-off.

The STFT transform, for a signal $s(t)$, a window g around time-point

τ, and frequency ω, is

$$\mathrm{STFT}(\tau,\omega) = \int_{-\infty}^{+\infty} s(t)g(t-\tau)e^{-j\omega t}dt \tag{1.6}$$

Considering

$$k_{\tau,\omega}(t) = g(t-\tau)e^{-j\omega t} \tag{1.7}$$

as a new basis, and rewriting this with window size a, inversely proportional to the frequency ω, and with positional parameter b replacing τ, as follows:

$$k_{b,a}(t) = \frac{1}{\sqrt{a}}\psi^*\left(\frac{t-b}{a}\right) \tag{1.8}$$

yields the continuous wavelet transform (CWT). In the STFT, the basis functions are windowed sinusoids, whereas in the CWT, they are scaled versions of a so-called mother function (ψ, where ψ^* is the conjugate).

A wavelet mother function can take many forms, subject to some admissibility constraints: see Freeman (1993) for an informal discussion. The best choice of mother function for a particular application is not given a priori.

From the basic wavelet formulation, one can distinguish (see Daubechies, 1992) between: (i) the continuous wavelet transform, described above; (ii) the discrete wavelet transform, which discretizes the continuous transform, but which does not in general have an exact analytical reconstruction formula; and within discrete transforms, distinction can be made between (iii) redundant versus non-redundant (e.g. pyramidal) transforms; and (iv) orthonormal versus other bases of wavelets.

Among points made in Graps (1995) in favor of the wavelet transform are the following. 'Choppy' data are better handled by the wavelet transform, and periodic or other non-local phenomena by the Fourier transform. To 'choppy' data we could add edge-related phenomena in two-dimensional imagery, or local scale-related features. Many additional application-oriented examples will be considered in the following chapters. The wavelet transform provides a decomposition of the original data, allowing operations to be performed on the wavelet coefficients and then the data reconstituted.

1.1.3 Applications of the wavelet transform

We briefly introduce the varied applications which will be discussed in the following chapters.

The human visual interpretation system does a good job at taking scales of a phenomenon or scene into account simultaneously. A wavelet or other multiscale transform may help us with visualizing image or other data. A decomposition into different resolution scales may open up, or lay bare, faint phenomena which are part of what is under investigation.

In capturing a view of multilayered reality in an image, we are also picking up noise at different levels. Therefore, in trying to specify what is noise in an image, we may find it effective to look for noise in a range of resolution levels. Such a strategy has proven quite successful in practice.

Noise of course is pivotal for the effective operation of, or even selection of, analysis methods. Image deblurring, or deconvolution or restoration, would be trivially solved, were it not for the difficulties posed by noise. Image compression would also be easy, were it not for the presence of what is by definition non-compressible, i.e. noise.

Image or data filtering may take different forms. For instance, we may wish to prioritize the high-frequency (rapidly-varying) parts of the image, and de-emphasize the low-frequency (smoother) parts of the image. Or, alternatively, we may wish to separate noise as far as possible from real image signal. In the latter case, we may wish to 'protect' important parts of the image from the slightest alteration.

An image may contain smooth and sharp features. We may need to consider a trade-off in quality between results obtained for such types of features. Introducing an entropy constraint in the image analysis procedure is one way to do this. This comes under the general heading of regularization.

An image analysis often is directed towards particular objects, or object classes, contained in the image. Template matching is the seeking of patterns which match a query pattern. A catalog or inventory of all objects may be used to facilitate later querying. Content-based queries may need to be supported, based on an image database.

Image registration involves matching parts of images taken with different detectors, or taken at different times. A top-down approach to this problem is offered by a multiscale approach: the crudest, most evident, features are matched first; followed by increasingly better resolved features.

In the analysis of multivariate data, we integrate the wavelet transform with methods such as cluster analysis, neural networks and (supervised and unsupervised) pattern recognition.

In all of these applications, efficiency and effectiveness (or quality of the result) are important. Varied application fields come immediately to mind: astronomy, remote sensing, medicine, industrial vision, and so on.

All told, there are many and varied applications for the methods described in this book. Based on the description of many applications, we aim to arm the reader well for tackling other similar applications. Clearly this objective holds too for tackling new and challenging applications.

We proceed now to look at the main features of various wavelet transforms, and also at closely related strategies for applying them.

1.2 The continuous wavelet transform

1.2.1 Definition

The Morlet-Grossmann definition (Grossmann and Morlet, 1984) of the continuous wavelet transform for a 1-dimensional signal $f(x) \in L^2(R)$, the space of all square integrable functions, is:

$$W(a,b) = \frac{1}{\sqrt{a}} \int_{-\infty}^{+\infty} f(x)\psi^*\left(\frac{x-b}{a}\right) dx \tag{1.9}$$

where:

- $W(a,b)$ is the wavelet coefficient of the function $f(x)$
- $\psi(x)$ is the analyzing wavelet
- a (> 0) is the scale parameter
- b is the position parameter

In Fourier space, we have:

$$\hat{W}(a,\nu) = \sqrt{a}\hat{f}(\nu)\hat{\psi}^*(a\nu) \tag{1.10}$$

When the scale a varies, the filter $\hat{\psi}^*(a\nu)$ is only reduced or dilated while keeping the same pattern.

1.2.2 Properties

The continuous wavelet transform (CWT) is characterized by the following three properties:

1 CWT is a linear transformation:

- if $f(x) = f_1(x) + f_2(x)$ then $W_f(a,b) = W_{f_1}(a,b) + W_{f_2}(a,b)$
- if $f(x) = kf_1(x))$ then $W_f(a,b) = kW_{f_1}(a,b)$

2 CWT is covariant under translation:

$$\text{if } f_0(x) = f(x - x_0) \text{ then } W_{f_0}(a,b) = W_f(a, b - x_0)$$

3 CWT is covariant under dilation:

$$\text{if } f_s(x) = f(sx) \text{ then } W_{f_s}(a,b) = \frac{1}{s^{\frac{1}{2}}} W_f(sa, sb)$$

The last property makes the wavelet transform very suitable for analyzing hierarchical structures. It is like a mathematical microscope with properties that do not depend on the magnification.

1.2.3 The inverse transform

Consider now a function $W(a,b)$ which is the wavelet transform of a given function $f(x)$. It has been shown (Grossmann and Morlet, 1984; Holschneider *et al.*, 1989) that $f(x)$ can be restored using the formula:

$$f(x) = \frac{1}{C_\chi} \int_0^{+\infty} \int_{-\infty}^{+\infty} \frac{1}{\sqrt{a}} W(a,b) \chi\left(\frac{x-b}{a}\right) \frac{da\, db}{a^2} \tag{1.11}$$

where:

$$C_\chi = \int_0^{+\infty} \frac{\hat{\psi}^*(v)\hat{\chi}(v)}{v} dv = \int_{-\infty}^{0} \frac{\hat{\psi}^*(v)\hat{\chi}(v)}{v} dv \tag{1.12}$$

Generally $\chi(x) = \psi(x)$, but other choices can enhance certain features for some applications.

Reconstruction is only possible if C_χ is defined (admissibility condition). In the case of $\chi(x) = \psi(x)$, this condition implies $\hat{\psi}(0) = 0$, i.e. the mean of the wavelet function is 0.

1.3 Examples of wavelet functions

1.3.1 Morlet's wavelet

The wavelet defined by Morlet (Coupinot *et al.*, 1992; Goupillaud, Grossmann and Morlet, 1985) is:

$$\hat{g}(v) = e^{-2\pi^2(v-v_0)^2} \tag{1.13}$$

This is a complex wavelet which can be decomposed into two parts, one for the real part, and the other for the imaginary part:

$$g_r(x) = \frac{1}{\sqrt{2\pi}} e^{-x^2/2} \cos(2\pi v_0 x)$$
$$g_i(x) = \frac{1}{\sqrt{2\pi}} e^{-x^2/2} \sin(2\pi v_0 x)$$

where v_0 is a constant. Morlet's transform is not admissible. For v_0 greater than approximately 0.8 the mean of the wavelet function is very small, so that approximate reconstruction is satisfactory. Figure 1.1 shows these two functions.

1.3.2 Mexican hat

The Mexican hat used e.g. Murenzi (1988) or Slezak, Bijaoui and Mars (1990) is in one dimension:

$$g(x) = (1 - x^2)e^{-x^2/2} \tag{1.14}$$

This is the second derivative of a Gaussian (see Fig. 1.2).

1.3.3 Haar wavelet

Parametrizing the continuous wavelet transform by scale and location, and relating the choice of a and b to fixed a_0 and b_0 (and requiring b

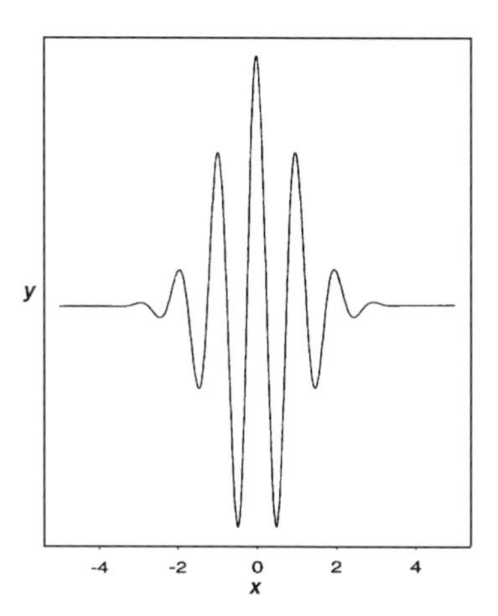

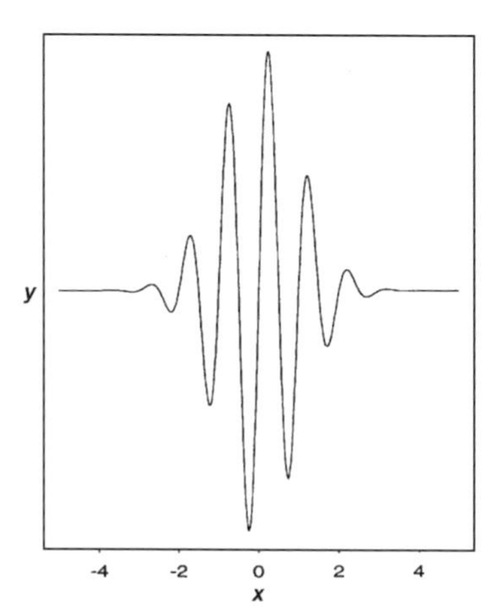

Figure 1.1 Morlet's wavelet: real part on left and imaginary part on right.

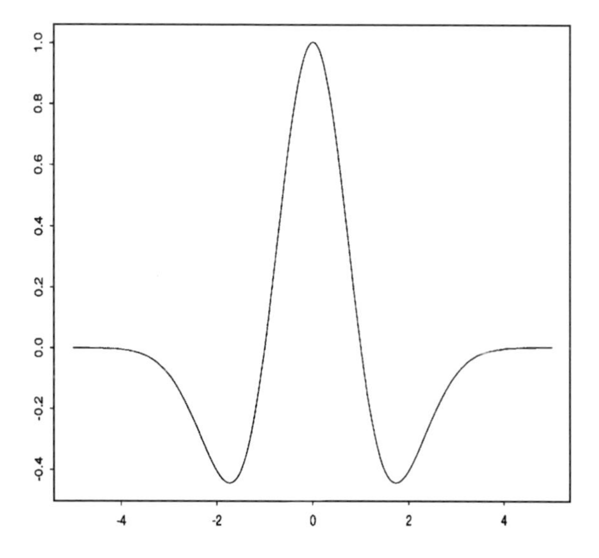

Figure 1.2 Mexican hat function.

to be proportional to a), we have (Daubechies, 1992):

$$\psi_{m,n}(x) = a_0^{-m/2}\psi(a_0^{-m}(x - nb_0a_0^m)) \quad (1.15)$$

The Haar wavelet transform is given by $a_0 = 2$ and $b_0 = 1$. The compact support of $\psi_{m,n}$ is then $[2^m n, 2^m(n+1)]$.

As far back as 1910, Haar described the following function as providing an orthonormal basis. The analyzing wavelet of a continuous variable is a step function (Fig. 1.3).

$$\begin{aligned} \psi(x) &= 1 && \text{if } 0 \leq x < \tfrac{1}{2} \\ \psi(x) &= -1 && \text{if } \tfrac{1}{2} \leq x < 1 \\ \psi(x) &= 0 && \text{otherwise} \end{aligned}$$

The Haar wavelet constitutes an orthonormal basis. Two Haar wavelets of the same scale (i.e. value of m) never overlap, so we have scalar product $\langle \psi_{m,n}, \psi_{m,n'} \rangle = \delta_{n,n'}$. Overlapping supports are possible if the two wavelets have different scales, e.g. $\psi_{1,1}$ and $\psi_{3,0}$ (see Daubechies, 1992, pp. 10–11). However, if $m < m'$, then the support of $\psi_{m,n}$ lies wholly in the region where $\psi_{m',n'}$ is constant. It follows that $\langle \psi_{m,n}, \psi_{m',n'} \rangle$ is proportional to the integral of $\psi_{m,n}$, i.e. zero.

Application of this transform to data smoothing and periodicity detection is considered in Scargle (1993), and to turbulence in fluid mechanics in Meneveau (1991). A clear introduction to the Haar wavelet transform is provided in particular in the first part of the two-part survey in Stollnitz *et al.* (1995).

Relative to other orthonormal wavelet transforms, the Haar basis

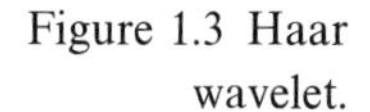
Figure 1.3 Haar wavelet.

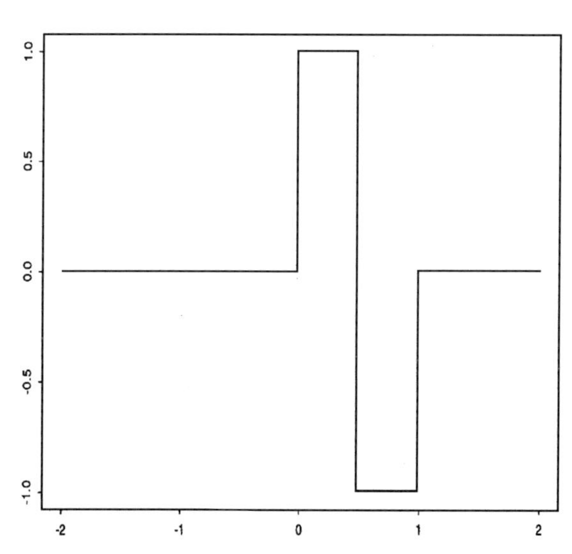

lacks smoothness; and although the Haar basis is compact in physical space, it decays slowly in Fourier space.

1.4 The discrete wavelet transform

In the discrete case, the wavelet function is sampled at discrete mesh-points using not δ functions but rather a smoothing function, ϕ. Correlations can be performed in physical space or in Fourier space, the former in preference when the support of the wavelet function is small (i.e. it is non-zero on a limited number of grid-points).

For processing classical (regularly pixelated) signals, sampling is carried out in accordance with Shannon's (1948) well-known theorem. The discrete wavelet transform (DWT) can be derived from this theorem if we process a signal which has a cut-off frequency. If we are considering images, we can note that the frequency band is always limited by the size of the camera aperture.

A digital analysis is made possible by the discretization of eqn. (1.9), with some simple considerations given to the modification of the wavelet pattern due to dilation. Usually the wavelet function $\psi(x)$ has no cut-off frequency and it is necessary to suppress the values outside the frequency band in order to avoid aliasing effects. It is possible to work in Fourier space, computing the transform scale-by-scale. The number of elements for a scale can be reduced, if the frequency bandwidth is also reduced. This is possible only for a wavelet which also has a cut-off frequency. The decomposition proposed by Littlewood and Paley (1931) provides a very nice illustration of the scale-by-scale reduction of elements. This decomposition is based on an iterative dichotomy of the frequency band. The associated wavelet is well localized in Fourier space where a reasonable analysis is possible. This is not the case, however, in the original space. The search for a discrete transform which is well localized in both spaces leads to multiresolution analysis.

1.4.1 Multiresolution analysis

Multiresolution analysis (Mallat, 1989) results from the embedded subsets generated by interpolations at different scales.

In formula (1.9), $a = 2^j$ for increasing integer values of j. From the function, $f(x)$, a ladder of approximation spaces is constructed with

$$\ldots \subset V_3 \subset V_2 \subset V_1 \subset V_0 \ldots \qquad (1.16)$$

such that, if $f(x) \in V_j$ then $f(2x) \in V_{j+1}$.

The function $f(x)$ is projected at each step j onto the subset V_j. This projection is defined by $c_j(k)$, the scalar product of $f(x)$ with the scaling function $\phi(x)$ which is dilated and translated:

$$c_j(k) = \langle f(x), 2^{-j}\phi(2^{-j}x - k)\rangle \tag{1.17}$$

As $\phi(x)$ is a scaling function which has the property:

$$\frac{1}{2}\phi\left(\frac{x}{2}\right) = \sum_n h(n)\phi(x-n) \tag{1.18}$$

or

$$\hat{\phi}(2v) = \hat{h}(v)\hat{\phi}(v) \tag{1.19}$$

where $\hat{h}(v)$ is the Fourier transform of the function $\sum_n h(n)\delta(x-n)$, we get:

$$\hat{h}(v) = \sum_n h(n)e^{-2\pi inv} \tag{1.20}$$

Equation (1.18) permits the direct computation of the set $c_{j+1}(k)$ from $c_j(k)$. If we start from the set $c_0(k)$ we compute all the sets $c_j(k)$, with $j > 0$, without directly computing any other scalar product:

$$c_{j+1}(k) = \sum_n h(n-2k)c_j(n) \tag{1.21}$$

At each step, the number of scalar products is divided by 2. Step-by-step the signal is smoothed and information is lost. The remaining information can be restored using the complementary orthogonal subspace W_{j+1} of V_{j+1} in V_j. This subspace can be generated by a suitable wavelet function $\psi(x)$ with translation and dilation.

$$\frac{1}{2}\psi\left(\frac{x}{2}\right) = \sum_n g(n)\phi(x-n) \tag{1.22}$$

or

$$\hat{\psi}(2v) = \hat{g}(v)\hat{\phi}(v) \tag{1.23}$$

The scalar products $\langle f(x), 2^{-(j+1)}\psi(2^{-(j+1)}x - k)\rangle$ are computed with:

$$w_{j+1}(k) = \sum_n g(n-2k)c_j(n) \tag{1.24}$$

With this analysis, we have built the first part of a filter bank (Smith and Barnwell, 1988). In order to restore the original data, Mallat uses the properties of orthogonal wavelets, but the theory has been generalized to a large class of filters (Cohen, Daubechies and

Feauveau, 1992) by introducing two other filters $\tilde{h}$ and $\tilde{g}$, defined to be conjugate to h and g. The reconstruction of the signal is performed with:

$$c_j(k) = 2\sum_l [c_{j+1}(l)\tilde{h}(k+2l) + w_{j+1}(l)\tilde{g}(k+2l)] \tag{1.25}$$

In order to get an exact reconstruction, two conditions are required for the conjugate filters:

– *Dealiasing condition*:

$$\hat{h}\left(v+\frac{1}{2}\right)\hat{\tilde{h}}(v) + \hat{g}\left(v+\frac{1}{2}\right)\hat{\tilde{g}}(v) = 0 \tag{1.26}$$

– *Exact restoration:*

$$\hat{h}(v)\hat{\tilde{h}}(v) + \hat{g}(v)\hat{\tilde{g}}(v) = 1 \tag{1.27}$$

In the decomposition, the function is successively convolved with the filters h (low frequencies) and g (high frequencies). Each resulting function is decimated by suppression of one sample out of two. The high frequency signal is left, and we iterate with the low frequency signal (upper part of Fig. 1.4). In the reconstruction, we restore the sampling by inserting a 0 between each sample, then we convolve with the conjugate filters $\tilde{h}$ and $\tilde{g}$, we add the resulting functions and we multiply the result by 2. We iterate up to the smallest scale (lower part of Fig. 1.4).

Figure 1.4 A filter bank associated with multiresolution analysis.

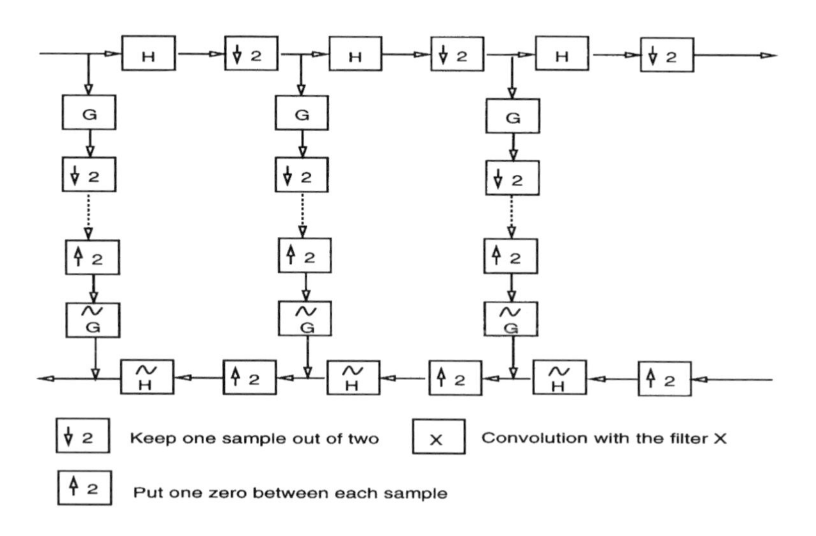

Orthogonal wavelets correspond to the restricted case where:

$$\hat{g}(\nu) = e^{-2\pi i\nu}\hat{h}^*\left(\nu + \frac{1}{2}\right) \tag{1.28}$$

$$\hat{\tilde{h}}(\nu) = \hat{h}^*(\nu) \tag{1.29}$$

$$\hat{\tilde{g}}(\nu) = \hat{g}^*(\nu) \tag{1.30}$$

and

$$|\ \hat{h}(\nu)\ |^2 + |\ \hat{h}\left(\nu + \frac{1}{2}\right)\ |^2 = 1 \tag{1.31}$$

It can be seen that this set satisfies the two basic relations (1.26) and (1.27). Daubechies wavelets are the only compact solutions. For biorthogonal wavelets (Cohen *et al.*, 1992; Meyer, 1993, p. 59) we have the relations:

$$\hat{g}(\nu) = e^{-2\pi i\nu}\hat{\tilde{h}}^*\left(\nu + \frac{1}{2}\right) \tag{1.32}$$

$$\hat{\tilde{g}}(\nu) = e^{2\pi i\nu}\hat{h}^*\left(\nu + \frac{1}{2}\right) \tag{1.33}$$

and

$$\hat{h}(\nu)\hat{\tilde{h}}(\nu) + \hat{h}^*\left(\nu + \frac{1}{2}\right)\hat{\tilde{h}}^*\left(\nu + \frac{1}{2}\right) = 1 \tag{1.34}$$

This satisfies also relations (1.26) and (1.27). A large class of compact wavelet functions can be derived. Many sets of filters have been proposed, especially for coding. It has been shown (Daubechies, 1988) that the choice of these filters must be guided by the regularity of the scaling and the wavelet functions. The computational complexity is proportional to N for an N-length input signal. This algorithm, involving decimation, provides a pyramid of the N elements.

1.4.2 Mallat's horizontal and vertical analyses

This two-dimensional algorithm is based on separate variables leading to prioritizing of x and y directions (see Fig. 1.5). The scaling function is defined by:

$$\phi(x, y) = \phi(x)\phi(y) \tag{1.35}$$

The passage from one resolution to the next is achieved by:

$$c_{j+1}(k_x, k_y) = \sum_{l_x=-\infty}^{+\infty}\sum_{l_y=-\infty}^{+\infty} h(l_x - 2k_x)h(l_y - 2k_y)f_j(l_x, l_y) \tag{1.36}$$

The detail signal is obtained from three wavelets:

– vertical wavelet :

$$\psi^1(x, y) = \phi(x)\psi(y)$$

– horizontal wavelet:

$$\psi^2(x, y) = \psi(x)\phi(y)$$

– diagonal wavelet:

$$\psi^3(x, y) = \psi(x)\psi(y)$$

which leads to three subimages:

$$w^1_{j+1}(k_x, k_y) = \sum_{l_x=-\infty}^{+\infty} \sum_{l_y=-\infty}^{+\infty} g(l_x - 2k_x)h(l_y - 2k_y)f_j(l_x, l_y)$$

$$w^2_{j+1}(k_x, k_y) = \sum_{l_x=-\infty}^{+\infty} \sum_{l_y=-\infty}^{+\infty} h(l_x - 2k_x)g(l_y - 2k_y)f_j(l_x, l_y)$$

$$w^3_{j+1}(k_x, k_y) = \sum_{l_x=-\infty}^{+\infty} \sum_{l_y=-\infty}^{+\infty} g(l_x - 2k_x)g(l_y - 2k_y)f_j(l_x, l_y)$$

The wavelet transform can be interpreted as frequency decomposition, with each set having a spatial orientation.

Figures 1.7 and 1.9 show the wavelet transform of a galaxy, Fig. 1.6 (NGC 2997), and a commonly-used test image (Lena), Fig. 1.8. For better visualization, we represent the normalized absolute values of

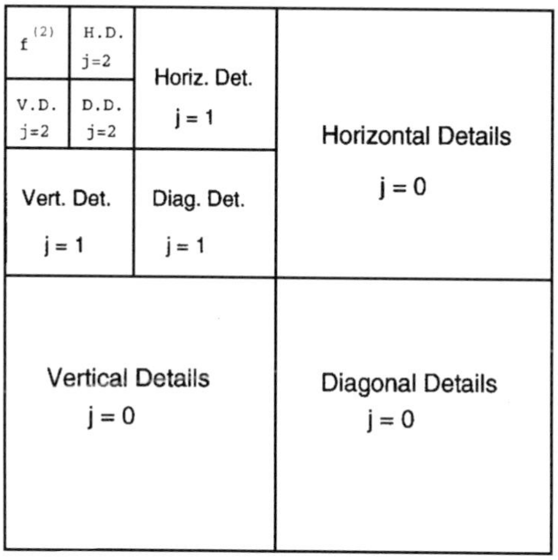

Figure 1.5 Mallat's wavelet transform representation of an image.

$f^{(2)}$ | H.D. j=2 | V.D. j=2 | D.D. j=2
Horiz. Det. j = 1
Vert. Det. j = 1
Diag. Det. j = 1
Horizontal Details j = 0
Vertical Details j = 0
Diagonal Details j = 0

the wavelet coefficients. We chose a look-up table (LUT) such that zero values were white, and the maximal value was black. We notice that this algorithm allows contours in the Lena image to be detected. However, with an astronomical image where we do not have contours, it is not easy to analyze the wavelet coefficients.

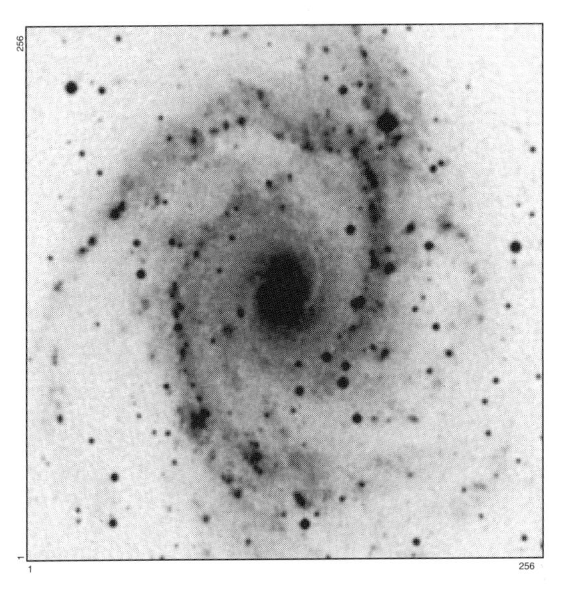

Figure 1.6 Galaxy NGC 2997.

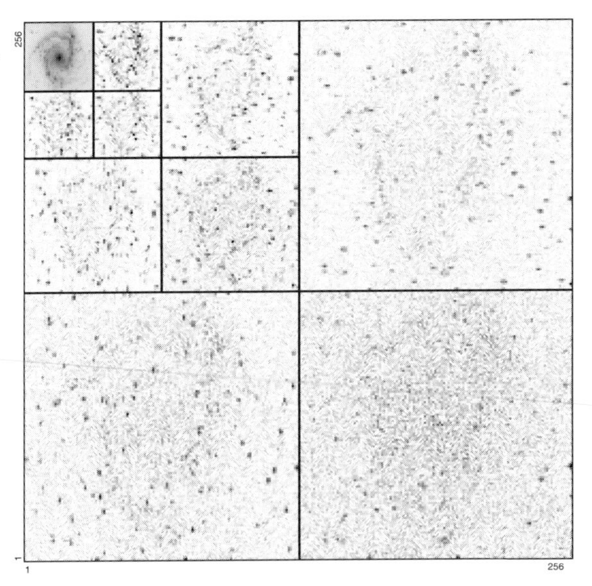

Figure 1.7 Wavelet transform of NGC 2997 by Mallat's algorithm.

1.4.3 Non-dyadic resolution factor

Feauveau (1990) introduced *quincunx* analysis, based on Adelson's work (Adelson, Simoncelli and Hingorani, 1987). This analysis is not dyadic and allows an image decomposition with a resolution factor equal to $\sqrt{2}$.

Figure 1.8 A widely-used test image, 'Lena'.

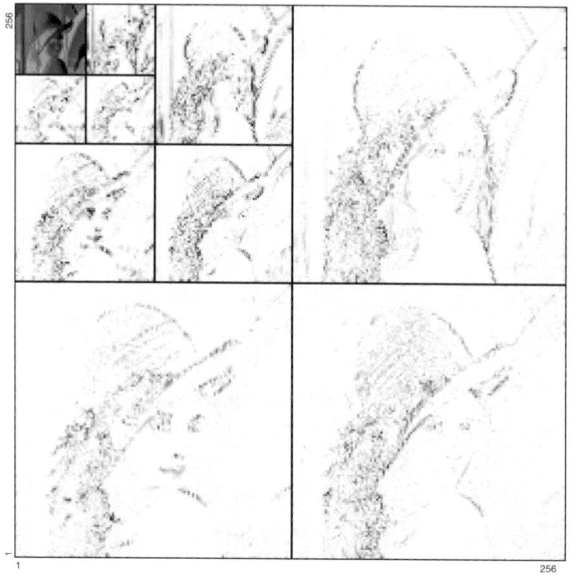

Figure 1.9 Wavelet transform of Lena by Mallat's algorithm.

The advantage is that only one wavelet is needed. At each step, the image is undersampled by two in one direction (x and y, alternatively). This undersampling is made by keeping one pixel out of two, alternatively even and odd. The following conditions must be satisfied by the filters:

$$\hat{h}\left(u+\frac{1}{2},v+\frac{1}{2}\right)\hat{\tilde{h}}(u,v)+\hat{g}\left(u+\frac{1}{2},v+\frac{1}{2}\right)\hat{\tilde{g}}(u,v)=0 \quad (1.37)$$

$$\hat{h}(u,v)\hat{\tilde{h}}(u,v)+\hat{g}(u,v)\hat{\tilde{g}}(u,v)=1 \quad (1.38)$$

Using this method, we have only one wavelet image at each scale, and not three like the previous method. Figure 1.10 shows the organization of the wavelet subimages and the smoothed image. For more effectively visualizing the entire transformation, coefficients are reorganized in a compact way. Figure 1.12 shows this reorganization. At points denoted by '**x**', we center the low-pass filter h which furnishes the image at the lower resolution, and at the points '**o**' we center the high-pass filter g which allows the wavelet coefficients to be obtained. The shift due to the filter g is made when we undersample.

Figure 1.11 shows the non-null coefficients of the two-dimensional

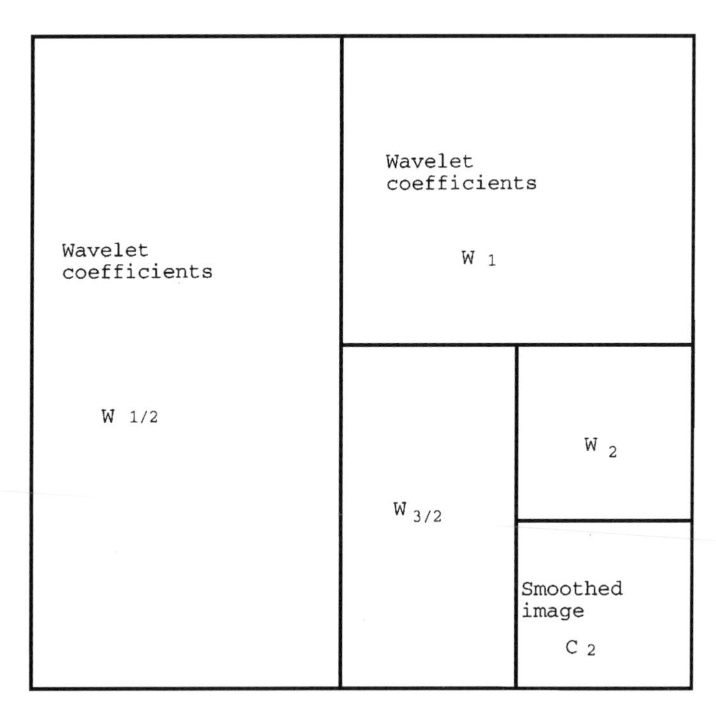

Figure 1.10 Feauveau's wavelet transform representation of an image.

filters. Figure 1.12 shows the overall schema of the multiresolution algorithm. Figure 1.13 shows the wavelet transform of the test image, Fig. 1.8.

1.4.4 The à trous algorithm

A wavelet transform for discrete data is provided by the particular version known as the *à trous* (with holes) algorithm (Holschneider *et al.*, 1989; Shensa, 1992). This is a 'stationary' or redundant transform, i.e. decimation is not carried out.

One assumes that the sampled data $\{c_0(k)\}$ are the scalar products, at pixels k of the function $f(x)$, with a scaling function $\phi(x)$ which corresponds to a low-pass filter.

If the wavelet function $\psi(x)$ obeys the dilation equation:

$$\frac{1}{2}\psi\left(\frac{x}{2}\right) = \sum_l g(l)\phi(x-l) \tag{1.39}$$

We compute the scalar products $\frac{1}{2^j}\langle f(x), \psi(\frac{x-k}{2^j})\rangle$, i.e. the discrete wavelet coefficients, with:

$$w_j(k) = \sum_l g(l)c_{j-1}(k + 2^{j-1}l) \tag{1.40}$$

Generally, the wavelet resulting from the difference between two

Figure 1.11 Coefficients of the two-dimensional filters.

```
            a
         f  b  f
      j  g  c  g  c
   f  g  i  d  i  g  f
a  b  c  d  e  d  c  b  a
   f  g  i  d  i  g  f
      j  g  c  g  c
         f  b  f
            a
```

	h	$\tilde{h}$
a	0.001671	–
b	−0.002108	−0.005704
c	−0.019555	−0.007192
d	0.139756	0.164931
e	0.687859	0.586315
f	0.006687	–
g	−0.006324	−0.017113
i	−0.052486	−0.014385
j	0.010030	–

successive approximations is applied:

$$w_j(k) = c_{j-1}(k) - c_j(k) \tag{1.41}$$

The first filtering is then performed by a twice-magnified scale leading to the $\{c_1(k)\}$ set. The signal difference $\{c_0(k)\} - \{c_1(k)\}$ contains the information between these two scales and is the discrete set associated with the wavelet transform corresponding to $\phi(x)$. The associated wavelet is $\psi(x)$.

$$\frac{1}{2}\psi\left(\frac{x}{2}\right) = \phi(x) - \frac{1}{2}\phi\left(\frac{x}{2}\right) \tag{1.42}$$

The distance between samples increasing by a factor 2 (see Fig. 1.14) from scale $(j-1)$ $(j > 0)$ to the next, $c_j(k)$, is given by:

$$c_j(k) = \sum_l h(l)c_{j-1}(k + 2^{j-1}l) \tag{1.43}$$

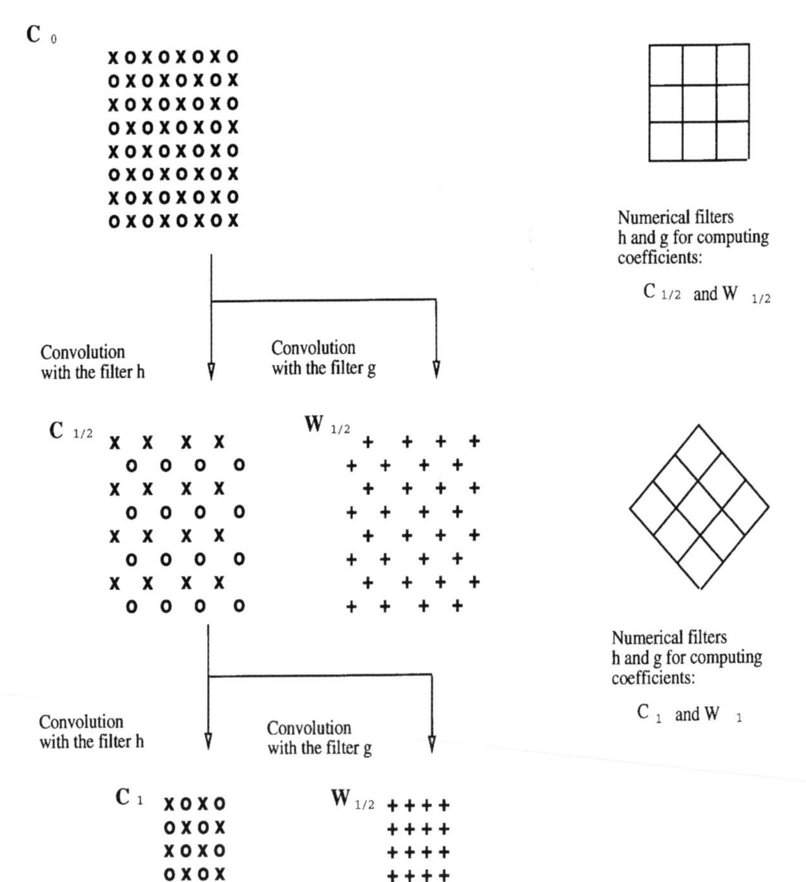

Figure 1.12 Feauveau multiresolution algorithm schema.

The coefficients $\{h(k)\}$ derive from the scaling function $\phi(x)$:

$$\frac{1}{2}\phi\left(\frac{x}{2}\right) = \sum_{l} h(l)\phi(x-l) \tag{1.44}$$

The algorithm allowing us to rebuild the data-frame is immediate:

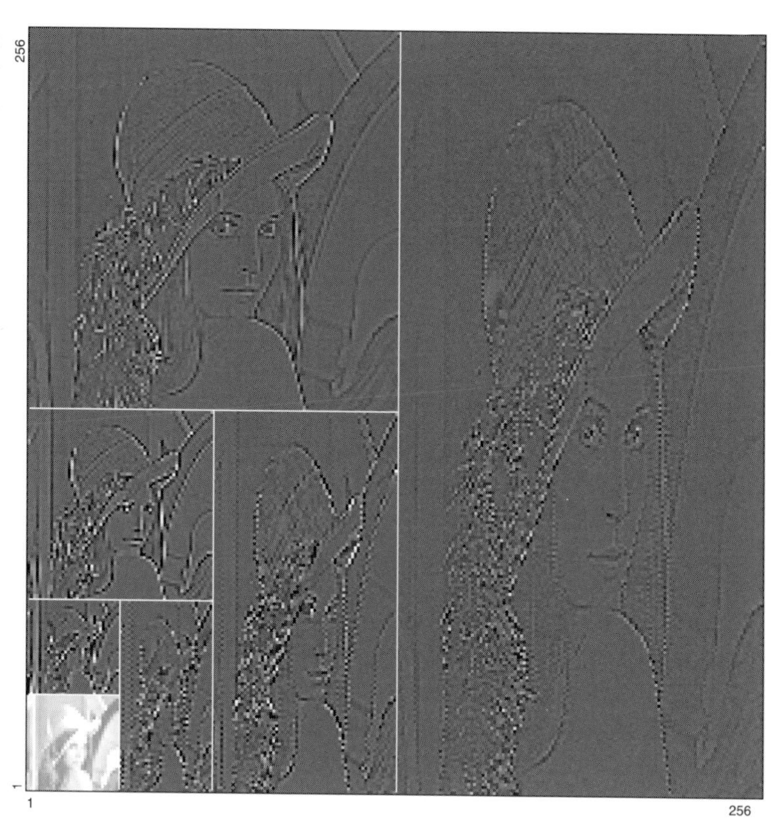

Figure 1.13 Wavelet transform of Lena by Feauveau's algorithm.

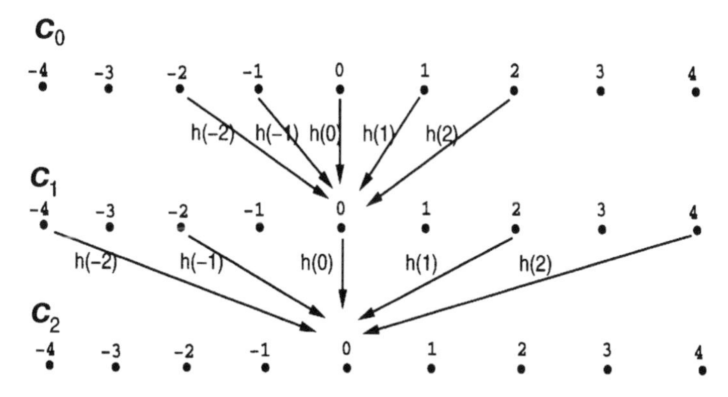

Figure 1.14 Passage from c_0 to c_1, and from c_1 to c_2.

the last smoothed array c_{n_p} is added to all the differences, w_j.

$$c_0(k) = c_{n_p}(k) + \sum_{j=1}^{n_p} w_j(k) \tag{1.45}$$

Choosing the triangle function as the scaling function ϕ (see Fig. 1.15) leads to piecewise linear interpolation:

$$\begin{array}{ll} \phi(x) = 1 - \mid x \mid & \text{if } x \in [-1, 1] \\ \phi(x) = 0 & \text{if } x \notin [-1, 1] \end{array}$$

We have:

$$\frac{1}{2}\phi\left(\frac{x}{2}\right) = \frac{1}{4}\phi(x+1) + \frac{1}{2}\phi(x) + \frac{1}{4}\phi(x-1) \tag{1.46}$$

c_1 is obtained from:

$$c_1(k) = \frac{1}{4}c_0(k-1) + \frac{1}{2}c_0(k) + \frac{1}{4}c_0(k+1) \tag{1.47}$$

Figure 1.15 Triangle function ϕ.

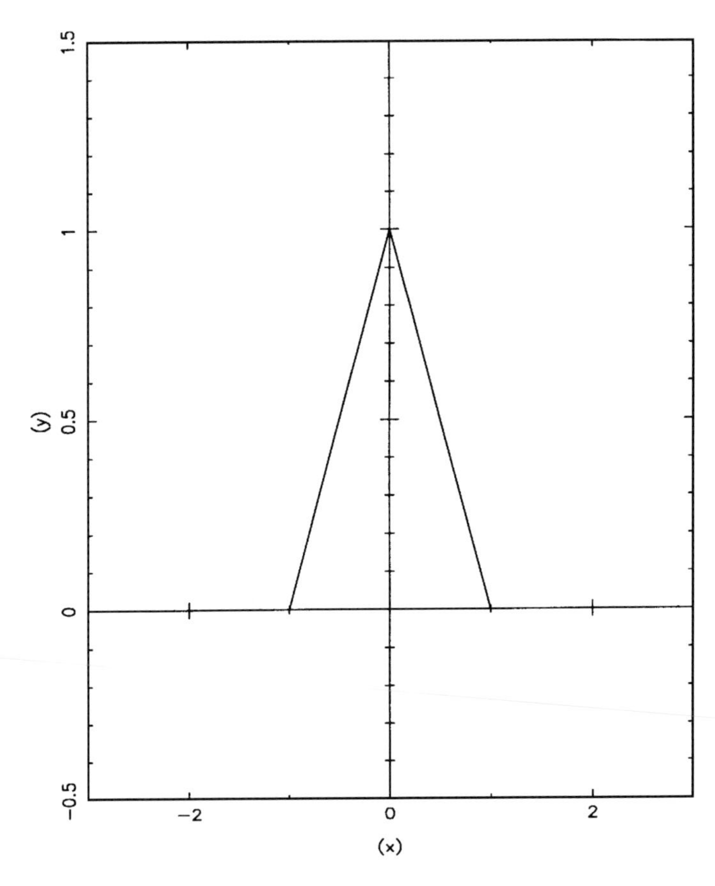

and c_{j+1} is obtained from c_j by:

$$c_{j+1}(k) = \frac{1}{4}c_j(k-2^j) + \frac{1}{2}c_j(k) + \frac{1}{4}c_j(k+2^j) \qquad (1.48)$$

Figure 1.16 shows the wavelet associated with the scaling function. The wavelet coefficients at scale j are:

$$w_{j+1}(k) = -\frac{1}{4}c_j(k-2^j) + \frac{1}{2}c_j(k) - \frac{1}{4}c_j(k+2^j) \qquad (1.49)$$

The above à trous algorithm is easily extended to two-dimensional space. This leads to a convolution with a mask of 3×3 pixels for the wavelet associated with linear interpolation. The coefficients of the mask are:

$$\begin{pmatrix} 1/4 & 1/2 & 1/4 \end{pmatrix} \otimes \begin{pmatrix} 1/4 \\ 1/2 \\ 1/4 \end{pmatrix}$$

where $\otimes$ is the Kronecker product.

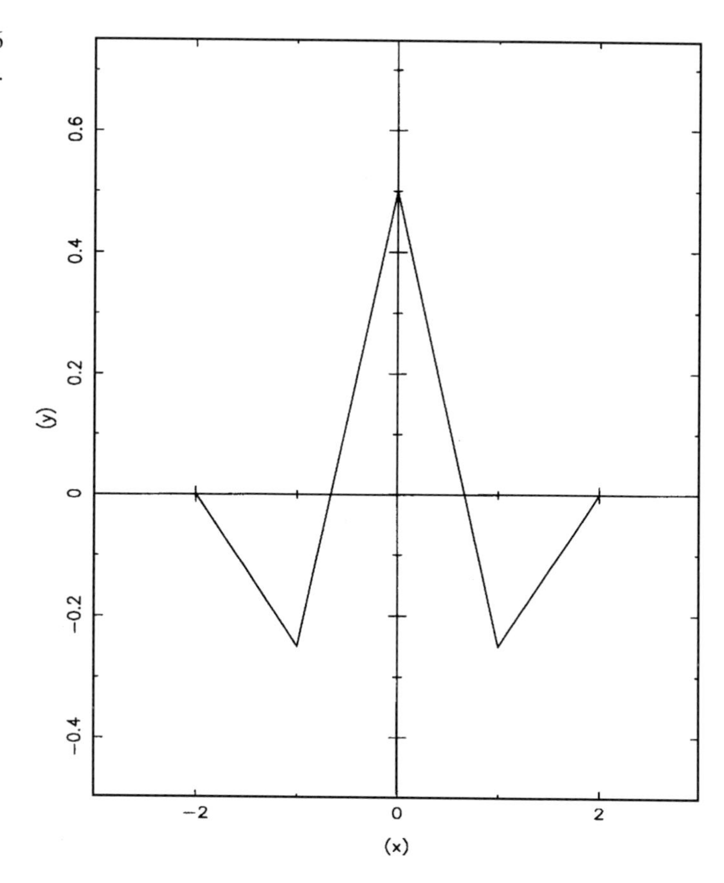

Figure 1.16 Wavelet ψ.

At each scale j, we obtain a set $\{w_j(k,l)\}$ which we will call a wavelet plane in the following. A wavelet plane has the same number of pixels as the image.

Spline functions, piecewise polynomials, have data approximation properties which are highly-regarded (Strang and Nguyen, 1996). If we choose a B_3-spline for the scaling function, the coefficients of the convolution mask in one dimension are $(\frac{1}{16}, \frac{1}{4}, \frac{3}{8}, \frac{1}{4}, \frac{1}{16})$, and in two dimensions:

$$\begin{pmatrix} 1/16 & 1/4 & 3/8 & 1/4 & 1/16 \end{pmatrix} \otimes \begin{pmatrix} 1/16 \\ 1/4 \\ 3/8 \\ 1/4 \\ 1/16 \end{pmatrix}$$

To facilitate computation, a simplification of this wavelet is to assume separability in the two-dimensional case. In the case of the B_3-spline, this leads to a row-by-row convolution with $(\frac{1}{16}, \frac{1}{4}, \frac{3}{8}, \frac{1}{4}, \frac{1}{16})$; followed by column-by-column convolution.

The most general way to handle the boundaries is to consider that $c(k+N) = c(N-k)$ (mirror). But other methods can be used such as periodicity ($c(k+N) = c(k)$), or continuity ($c(k+N) = c(N)$).

Figure 1.17 shows the à trous transform of the galaxy NGC 2997. Three wavelet scales are shown (upper left, upper right, lower left) and the final smoothed plane (lower right). The original image is given exactly by the sum of these four images.

Figure 1.18 shows the same sequence of images for the Lena image.

Figure 1.19 shows each scale as a perspective plot. Figure 1.20 is the same, with stacked plots. Figure 1.21 shows the first scale of the wavelet transform of NGC 2997 as a gray-level image; as an isophot plot; as a perspective plot; and the associated histogram of wavelet coefficients.

In Fig. 1.22, each scale is replaced with a contour level, and the contours of all scales are shown together. The net effect is to show an aspect of each resolution level at the same time. This representation is an example of the multiresolution support, a data structure which will be introduced and discussed in the next chapter.

Figures 1.23 and 1.24 show the nebula NGC 40 and its wavelet transform. This last figure shows each scale displayed as a contour plot, and the inter-scale connections clearly show the hierarchy of structure in the scales.

Figure 1.17 Wavelet transform of NGC 2997 by the à trous algorithm.

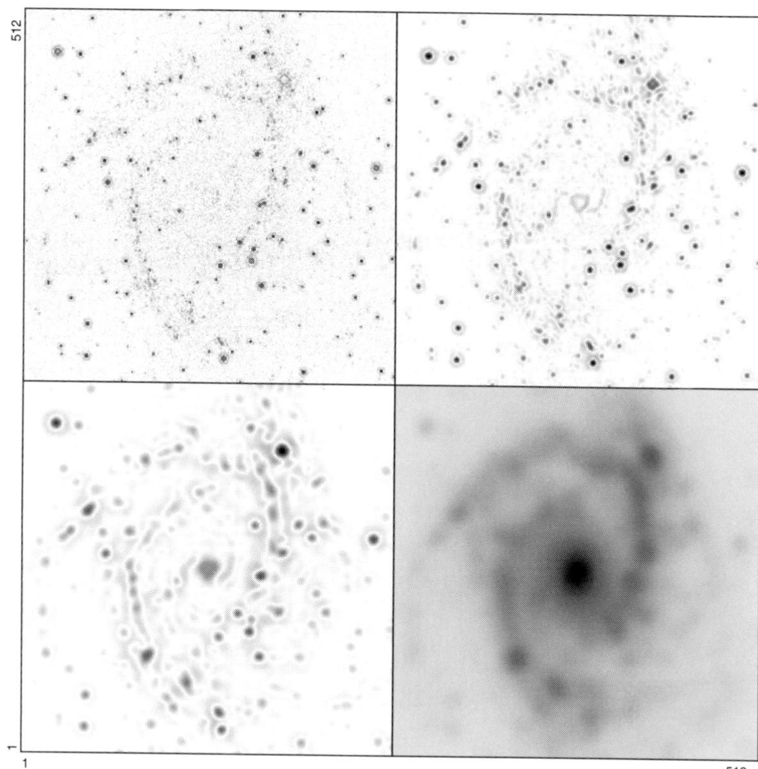

Figure 1.18 Wavelet transform of Lena by the à trous algorithm.

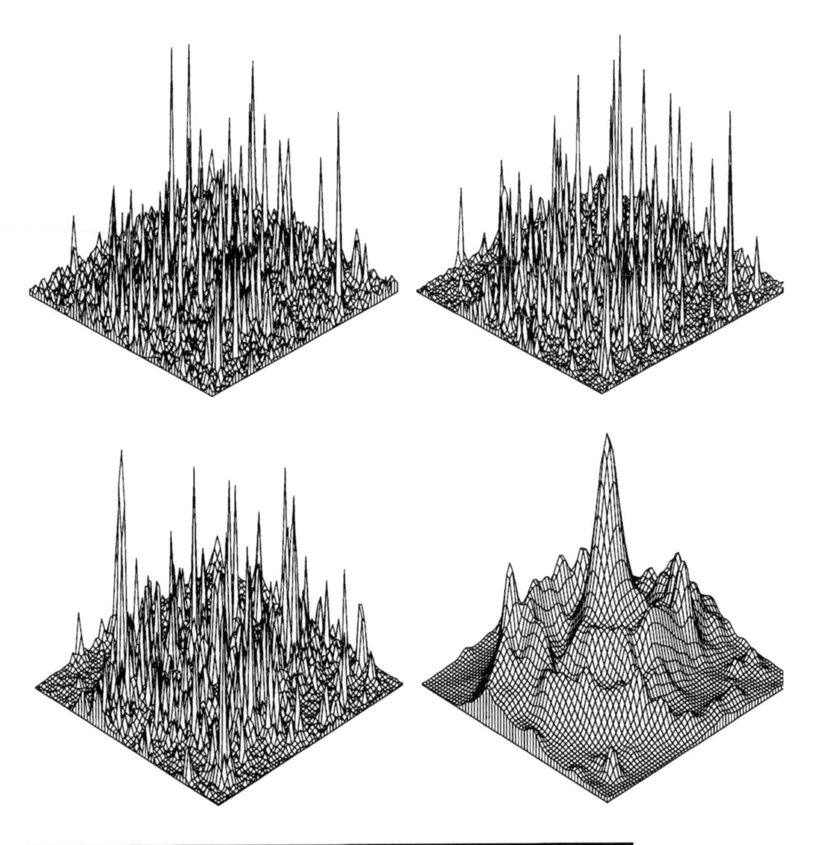

Figure 1.19
3D visualization of NGC 2997 wavelet transform (à trous algorithm).

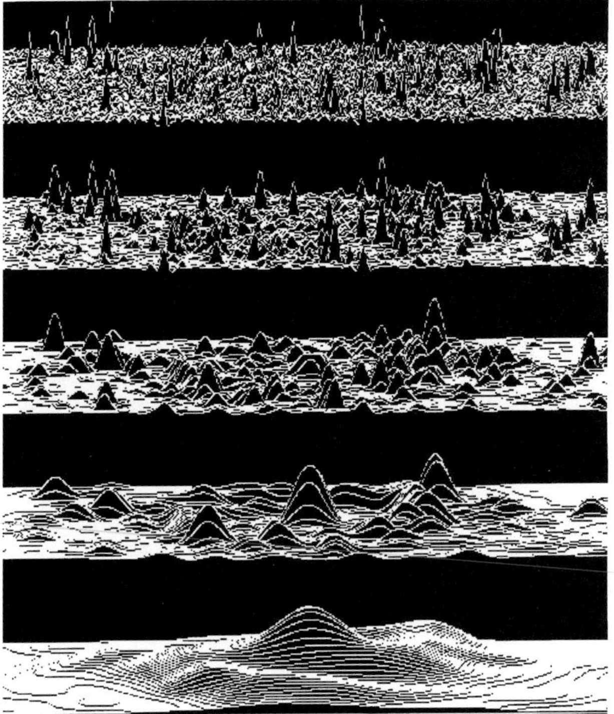

Figure 1.20
Superposition of NGC 2997 wavelet scales.

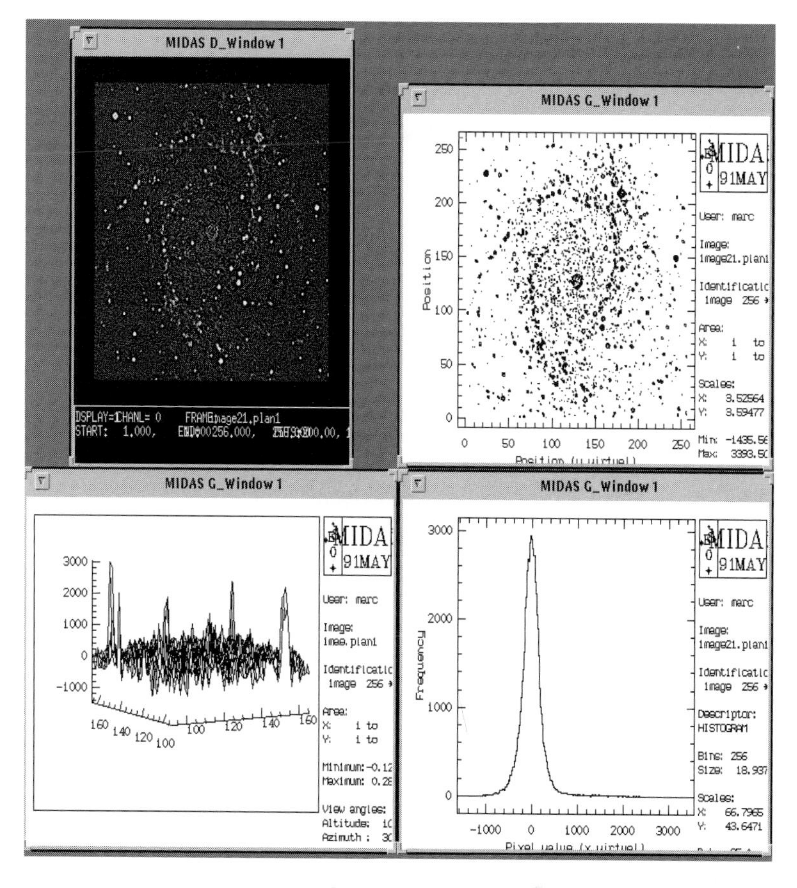

Figure 1.21 Visualization of one wavelet scale in gray level, isophot, perspective plot, and histogram.

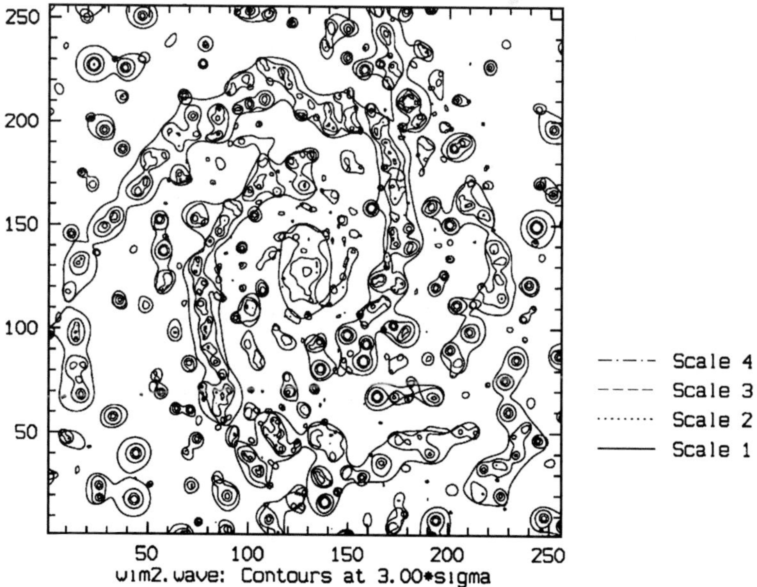

Figure 1.22 NGC 2997: one contour per scale is plotted.

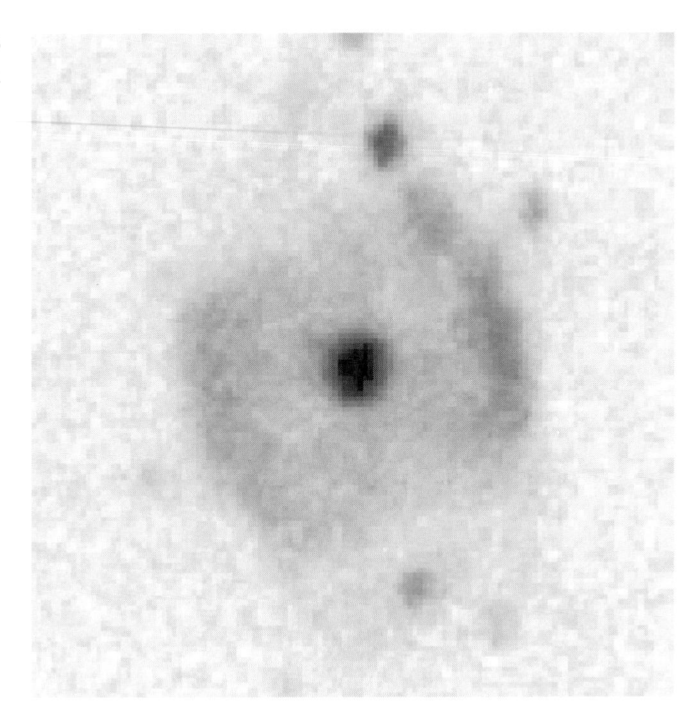

Figure 1.23 Nebula NGC 40.

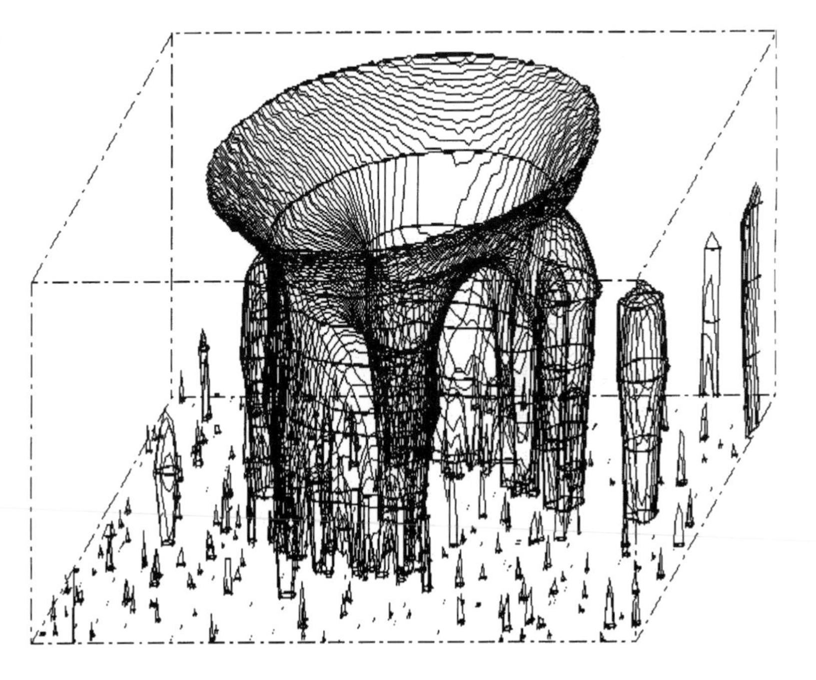

Figure 1.24 NGC 40 wavelet coefficients.

1.4.5 Pyramidal algorithm

The Laplacian pyramid. The Laplacian pyramid was developed by Burt and Adelson in 1981 (Burt and Adelson, 1983) in order to compress images. The term Laplacian was applied by Burt to the difference between two successive levels in a pyramid, defined itself in turn by repeatedly applying a low-pass (smoothing) filtering operation. After the filtering, only one sample out of two is kept. The number of pixels decreases by a factor 2 at each scale. The difference between images is obtained by expanding (or interpolating) one of the pair of images in the sequence associated with the pyramid.

The convolution is carried out with the filter h by keeping one sample out of two (see Fig. 1.25):

$$c_{j+1}(k) = \sum_l h(l - 2k) c_j(l) \tag{1.50}$$

To reconstruct c_j from c_{j+1}, we need to calculate the difference signal w_{j+1}:

$$w_{j+1}(k) = c_j(k) - \tilde{c}_j(k) \tag{1.51}$$

where $\tilde{c}_j$ is the signal reconstructed by the following operation (see Fig. 1.26):

$$\tilde{c}_j(k) = 2 \sum_l h(k - 2l) c_j(k) \tag{1.52}$$

In two dimensions, the method is similar. The convolution is carried out by keeping one sample out of two in the two directions. We have:

$$c_{j+1}(n, m) = \sum_{k,l} h(k - 2n, l - 2m) c_j(k, l) \tag{1.53}$$

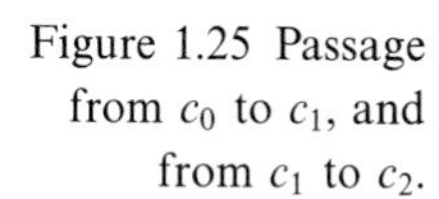

Figure 1.25 Passage from c_0 to c_1, and from c_1 to c_2.

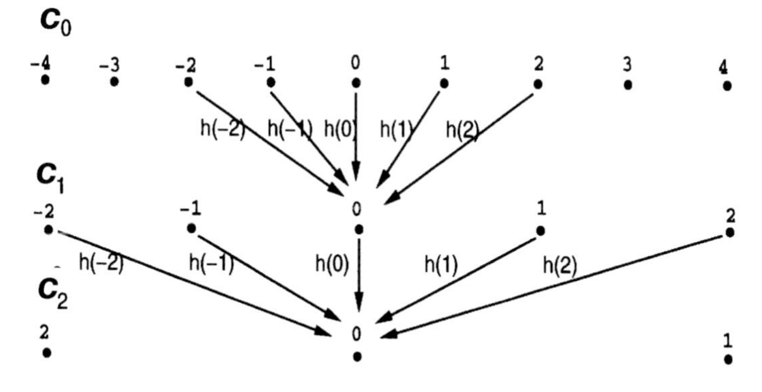

and $\tilde{c}_j$ is:

$$\tilde{c}_j(n,m) = 2\sum_{k,l} h(n-2l, m-2l)c_{j+1}(k,l) \tag{1.54}$$

The number of samples is divided by four. If the image size is $N \times N$, then the pyramid size is $\frac{4}{3}N^2$. We get a pyramidal structure (see Fig. 1.27).

The Laplacian pyramid leads to an analysis with four wavelets (Bijaoui, 1991b) and there is no invariance to translation.

Pyramidal algorithm with one wavelet. To modify the previous algorithm in order to have an isotropic wavelet transform, we compute the difference signal by:

$$w_{j+1}(k) = c_j(k) - \tilde{c}_j(k) \tag{1.55}$$

but $\tilde{c}_j$ is computed without reducing the number of samples:

$$\tilde{c}_j(k) = \sum_l h(k-l)c_j(k) \tag{1.56}$$

and c_{j+1} is obtained by:

$$c_{j+1}(k) = \sum_l h(l-2k)c_j(l) \tag{1.57}$$

The reconstruction method is the same as with the Laplacian pyramid, but the reconstruction is not exact. However, the exact reconstruction can be performed by an iterative algorithm. If P_0 represents the wavelet coefficient pyramid, we look for an image such that the wavelet transform of this image gives P_0. Van Cittert's iterative algorithm (1931) gives:

$$P_{n+1} = P_0 + P_n - R(P_n) \tag{1.58}$$

where

– P_0 is the pyramid to be reconstructed

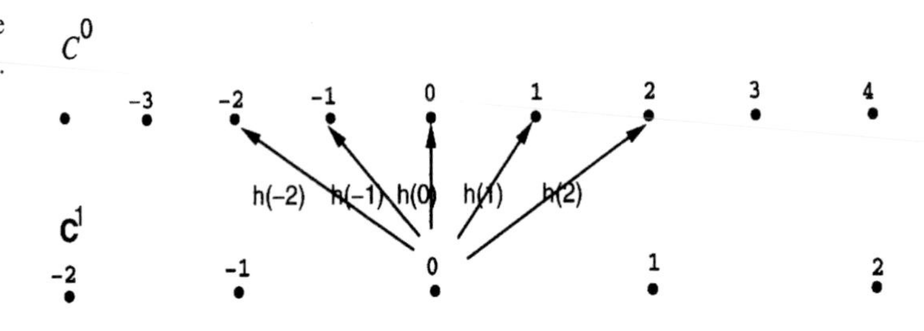

Figure 1.26 Passage from c_1 to c_0.

- P_n is the pyramid after n iterations
- R is an operator which consists of a reconstruction followed by a wavelet transform.

The solution is obtained by reconstructing the pyramid P_n.

Normally, we need no more than seven or eight iterations to converge. Another way to have a pyramidal wavelet transform with an isotropic wavelet is to use a scaling function with a cut-off frequency.

1.4.6 Scaling functions with a frequency cut-off

Wavelet transform using the Fourier transform. We start with the set of scalar products $c_0(k) = \langle f(x), \phi(x-k)\rangle$. If $\phi(x)$ has a cut-off frequency $\nu_c \leq \frac{1}{2}$ (Starck, 1992; Starck and Bijaoui, 1994a,b; Starck *et al.*, 1994), the data are correctly sampled. The data at resolution $j = 1$ are:

$$c_1(k) = \langle f(x), \frac{1}{2}\phi\left(\frac{x}{2} - k\right)\rangle \tag{1.59}$$

Figure 1.27
Pyramidal structure.

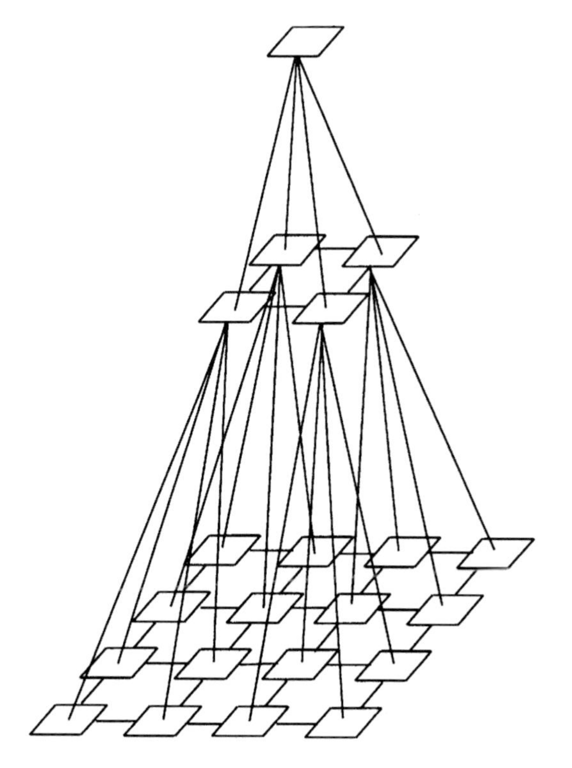

and we can compute the set $c_1(k)$ from $c_0(k)$ with a discrete filter $\hat{h}(\nu)$:

$$\hat{h}(\nu) = \begin{cases} \frac{\hat{\phi}(2\nu)}{\hat{\phi}(\nu)} & \text{if } \mid \nu \mid < \nu_c \\ 0 & \text{if } \nu_c \leq \mid \nu \mid < \frac{1}{2} \end{cases} \tag{1.60}$$

and

$$\forall \nu, \forall n \;\; \hat{h}(\nu + n) = \hat{h}(\nu) \tag{1.61}$$

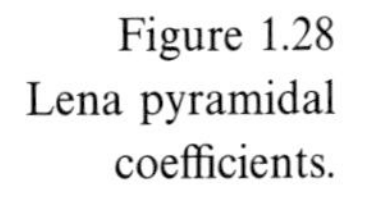
Figure 1.28 Lena pyramidal coefficients.

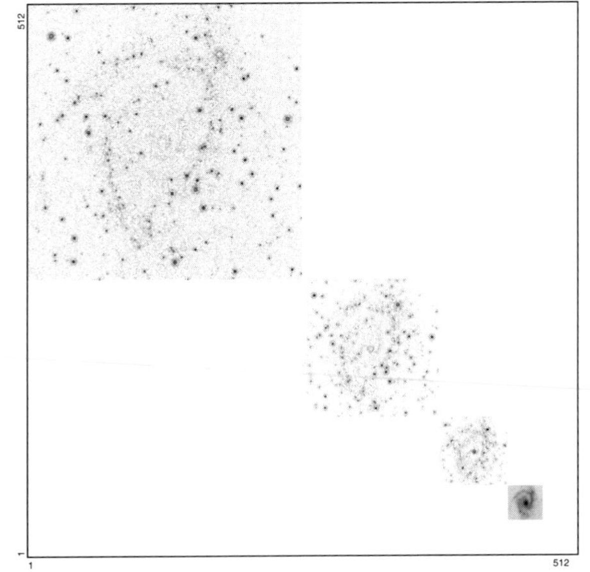

Figure 1.29 NGC 2997 pyramidal coefficients.

where n is an integer. So:

$$\hat{c}_{j+1}(\nu) = \hat{c}_j(\nu)\hat{h}(2^j\nu) \tag{1.62}$$

The cut-off frequency is reduced by a factor 2 at each step, allowing a reduction of the number of samples by this factor.

The wavelet coefficients at scale $j+1$ are:

$$w_{j+1}(k) = \langle f(x), 2^{-(j+1)}\psi(2^{-(j+1)}x - k)\rangle \tag{1.63}$$

and they can be computed directly from $c_j(k)$ by:

$$\hat{w}_{j+1}(\nu) = \hat{c}_j(\nu)\hat{g}(2^j\nu) \tag{1.64}$$

where g is the following discrete filter:

$$\hat{g}(\nu) = \begin{cases} \frac{\hat{\psi}(2\nu)}{\hat{\phi}(\nu)} & \text{if } \mid \nu \mid < \nu_c \\ 1 & \text{if } \nu_c \leq \mid \nu \mid < \frac{1}{2} \end{cases} \tag{1.65}$$

and

$$\forall \nu, \forall n \;\; \hat{g}(\nu + n) = \hat{g}(\nu) \tag{1.66}$$

The frequency band is also reduced by a factor 2 at each step. Applying the sampling theorem, we can build a pyramid of $N + \frac{N}{2} + \ldots + 1 = 2N$ elements. For an image analysis the number of elements is $\frac{4}{3}N^2$. The overdetermination is not very high.

The B-spline functions are compact in direct space. They correspond to the autoconvolution of a square function. In Fourier space we have:

$$\hat{B}_l(\nu) = \left(\frac{\sin \pi\nu}{\pi\nu}\right)^{l+1} \tag{1.67}$$

$B_3(x)$ is a set of four polynomials of degree 3. We choose the scaling function $\phi(\nu)$ which has a $B_3(x)$ profile in Fourier space:

$$\hat{\phi}(\nu) = \frac{3}{2}B_3(4\nu) \tag{1.68}$$

In direct space we get:

$$\phi(x) = \frac{3}{8}\left[\frac{\sin\frac{\pi x}{4}}{\frac{\pi x}{4}}\right]^4 \tag{1.69}$$

This function is quite similar to a Gaussian and converges rapidly to 0. For two dimensions the scaling function is defined by $\hat{\phi}(u, v) = \frac{3}{2}B_3(4r)$, with $r = \sqrt{u^2 + v^2}$. This is an isotropic function.

The wavelet transform algorithm with n_p scales is the following:

1 Start with a B_3-spline scaling function and derive ψ, h and g numerically.
2 Compute the corresponding FFT image. Name the resulting complex array T_0.
3 Set j to 0. Iterate:
4 Multiply T_j by $\hat{g}(2^j u, 2^j v)$. We get the complex array W_{j+1}. The inverse FFT gives the wavelet coefficients at scale 2^j;
5 Multiply T_j by $\hat{h}(2^j u, 2^j v)$. We get the array T_{j+1}. Its inverse FFT gives the image at scale 2^{j+1}. The frequency band is reduced by a factor 2.
6 Increment j.
7 If $j \leq n_p$, go back to step 4.
8 The set $\{w_1, w_2, \ldots, w_{n_p}, c_{n_p}\}$ describes the wavelet transform.

If the wavelet is the difference between two resolutions, i.e.

$$\hat{\psi}(2\nu) = \hat{\phi}(\nu) - \hat{\phi}(2\nu) \tag{1.70}$$

and:

$$\hat{g}(\nu) = 1 - \hat{h}(\nu) \tag{1.71}$$

then the wavelet coefficients $\hat{w}_j(\nu)$ can be computed by $\hat{c}_{j-1}(\nu) - \hat{c}_j(\nu)$.

Reconstruction. If the wavelet is the difference between two resolutions, an evident reconstruction for a wavelet transform $\mathcal{W} = \{w_1, \ldots, w_{n_p}, c_{n_p}\}$ is:

$$\hat{c}_0(\nu) = \hat{c}_{n_p}(\nu) + \sum_j \hat{w}_j(\nu) \tag{1.72}$$

But this is a particular case, and other alternative wavelet functions can be chosen. The reconstruction can be made step-by-step, starting from the lowest resolution. At each scale, we have the relations:

$$\hat{c}_{j+1} = \hat{h}(2^j \nu)\hat{c}_j(\nu) \tag{1.73}$$

$$\hat{w}_{j+1} = \hat{g}(2^j \nu)\hat{c}_j(\nu) \tag{1.74}$$

We look for c_j knowing c_{j+1}, w_{j+1}, h and g. We restore $\hat{c}_j(\nu)$ based on a least mean square estimator:

$$\hat{p}_h(2^j \nu) \mid \hat{c}_{j+1}(\nu) - \hat{h}(2^j \nu)\hat{c}_j(\nu) \mid^2 + \hat{p}_g(2^j \nu) \mid \hat{w}_{j+1}(\nu) - \hat{g}(2^j \nu)\hat{c}_j(\nu) \mid^2 \tag{1.75}$$

is to be minimum. The weight functions $\hat{p}_h(\nu)$ and $\hat{p}_g(\nu)$ permit a general solution to the restoration of $\hat{c}_j(\nu)$. From the derivation of

$\hat{c}_j(\nu)$ we get:

$$\hat{c}_j(\nu) = \hat{c}_{j+1}(\nu)\hat{\tilde{h}}(2^j\nu) + \hat{w}_{j+1}(\nu)\hat{\tilde{g}}(2^j\nu) \tag{1.76}$$

where the conjugate filters have the expression:

$$\hat{\tilde{h}}(\nu) = \left[\hat{p}_h(\nu)\hat{h}^*(\nu)\right] / \left[\hat{p}_h(\nu) \mid \hat{h}(\nu) \mid^2 + \hat{p}_g(\nu) \mid \hat{g}(\nu) \mid^2\right] \tag{1.77}$$

$$\hat{\tilde{g}}(\nu) = \left[\hat{p}_g(\nu)\hat{g}^*(\nu)\right] / \left[\hat{p}_h(\nu) \mid \hat{h}(\nu) \mid^2 + \hat{p}_g(\nu) \mid \hat{g}(\nu) \mid^2\right] \tag{1.78}$$

It is easy to see that these filters satisfy the exact reconstruction eqn. (1.27). In fact, eqns. (1.77) and (1.78) give the general solution to this equation. In this analysis, the Shannon sampling condition is always respected. No aliasing exists, so that the dealiasing condition (1.26) is not necessary.

The denominator is reduced if we choose:

$$\hat{g}(\nu) = \sqrt{1- \mid \hat{h}(\nu) \mid^2}$$

This corresponds to the case where the wavelet is the difference between the square of two resolutions:

$$\mid \hat{\psi}(2\nu) \mid^2 = \mid \hat{\phi}(\nu) \mid^2 - \mid \hat{\phi}(2\nu) \mid^2 \tag{1.79}$$

In Fig. 1.30 the chosen scaling function derived from a B-spline of degree 3, and its resulting wavelet function, are plotted in frequency space. Their conjugate functions are plotted in Fig. 1.31.

Figure 1.30 On the left, the interpolation function $\hat{\phi}$ and, on the right, the wavelet $\hat{\psi}$.

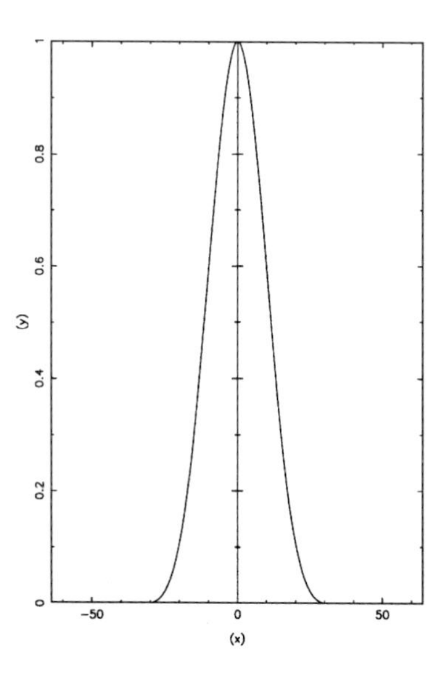

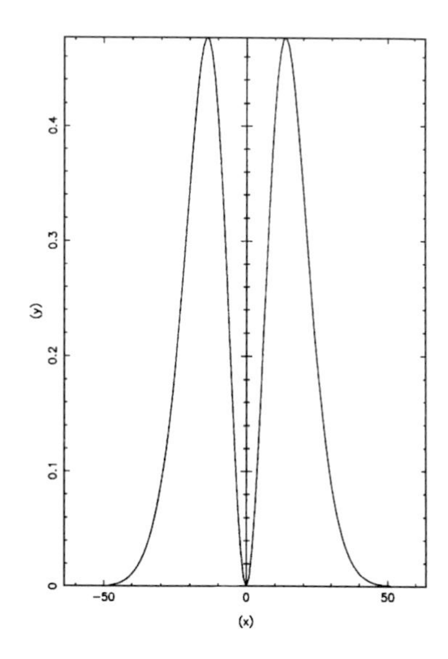

The reconstruction algorithm is:

1 Compute the FFT of the image at the low resolution.
2 Set j to n_p. Iterate:
3 Compute the FFT of the wavelet coefficients at scale j.
4 Multiply the wavelet coefficients $\hat{w}_j$ by $\hat{\hat{g}}$.
5 Multiply the image at the lower resolution $\hat{c}_j$ by $\hat{\hat{h}}$.
6 The inverse Fourier transform of the addition of $\hat{w}_j\hat{\hat{g}}$ and $\hat{c}_j\hat{\hat{h}}$ gives the image c_{j-1}.
7 Set $j = j - 1$ and return to step 3.

The use of a scaling function with a cut-off frequency allows a reduction of sampling at each scale, and limits the computing time and the memory size.

1.4.7 Discussion of the wavelet transform

We will look at limitations of the wavelet transform. This will justify the additional consideration in subsequent chapters of other similar multiresolution transforms.

Anisotropic wavelet. The two-dimensional extension of Mallat's algorithm leads to a wavelet transform with three wavelet functions (we have at each scale three wavelet coefficient subimages) which does not simplify the analysis and the interpretation of the wavelet coefficients.

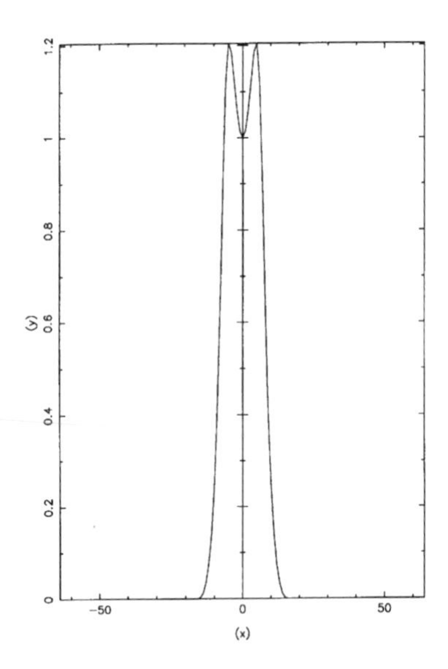

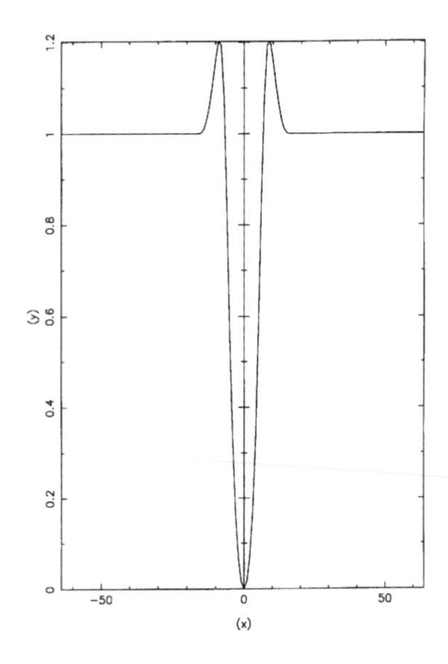

Figure 1.31 On the left, the filter $\hat{\hat{h}}$, and on the right the filter $\hat{\hat{g}}$.

An isotropic wavelet seems more appropriate in astronomical imaging and in other domains where objects are often isotropic (e.g. stars).

Invariance by translation. Mallat's and Feauveau's methods provide a remarkable framework to code a signal, and especially an image, with a pyramidal set of values. But contrary to the continuous wavelet transform, these analyses are not covariant under translation. At a given scale, we derive a decimated number of wavelet coefficients. We cannot restore the intermediary values without using the approximation at this scale and the wavelet coefficients at smaller scales. Since the multiresolution analysis is based on scaling functions without cut-off frequency, the application of the Shannon interpolation theorem is not possible. The interpolation of the wavelet coefficients can only be carried out after reconstruction and shift. This has no importance for a signal coding which does not modify the data, but the situation is not the same in a strategy in which we want to analyze or restore an image.

Scale separation. If the image I we want to analyze is the convolution product of an object O by a point spread function (PSF), $I = P * O$, we have:

$$\hat{w}^{(I)}(a,u,v) = \sqrt{a}\hat{\psi}^*(au,av)\hat{I}(u,v) \tag{1.80}$$

where $w^{(z)}$ are the wavelet coefficients of z, and a is the scale parameter. We deduce:

$$\hat{w}^{(I)}(a,u,v) = \sqrt{a}\hat{\psi}^*(au,av)\hat{P}(u,v)\hat{O}(u,v) \tag{1.81}$$

$$= \hat{O}(u,v)\hat{w}^{(P)}(a,u,v) \tag{1.82}$$

Then we can directly analyze the object from the wavelet coefficients of the image. But due to decimation effects in Mallat's, Feauveau's, and pyramidal methods, this equation becomes false. The wavelet transform using the FFT (which decimates using Shannon's theorem) and the à trous algorithm (which does not decimate) are the only ones that respect the scale separation property.

Negative values (bumps). By definition, the wavelet coefficient mean is null. Every time we have a positive structure at a scale, we have negative values surrounding it. These negative values often create artifacts during the restoration process, or complicate the analysis. For instance, if we threshold small values (noise, non-significant structures, etc.) in the wavelet transform, and if we reconstruct the image at full resolution, the structure's flux will be modified.

Furthermore, if an object is very high in intensity, the negative values will be large and will lead to detection of false structure.

Point objects. We often have bright point objects in astronomical imaging (stars, cosmic ray hits, etc.), and the convolution of a Dirac function by the wavelet function is equal to the wavelet function. Then, at each scale, and at each point source, we will have the wavelet. Therefore, cosmic rays for instance can pollute *all* the scales of the wavelet transform.

Multiresolution transforms. There is no ideal wavelet transform algorithm, and the selected one will depend on the application. Issues such as negative wavelet coefficient values, and the presence of spurious point objects, cannot be solved with the wavelet transform. These issues lead us to investigate and to develop other multiresolution tools which we now present.

1.5 Multiresolution based on the median transform

1.5.1 Multiresolution median transform

The search for new multiresolution tools has been motivated so far by problems related to the wavelet transform. It would be more desirable for a point structure (represented in one pixel in the image) to be present only at the first scale. It would also be desirable for a positive structure in the image to not create negative values in the multiresolution space. We will see how such an algorithm can be arrived at, using morphological filters such as the median filter.

The median transform is nonlinear, and offers advantages for robust smoothing (i.e. the effects of outlier pixel values are mitigated). Define the median transform of image f, with square kernel of dimensions $n \times n$, as $\text{med}(f, n)$. Let $n = 2s+1$; initially $s = 1$. The iteration counter will be denoted by j, and S is the user-specified number of resolution scales.

1 Let $c_j = f$ with $j = 1$.
2 Determine $c_{j+1} = \text{med}(f, 2s+1)$.
3 The multiresolution coefficients w_{j+1} are defined as: $w_{j+1} = c_j - c_{j+1}$.
4 Let $j \longleftarrow j+1$; $s \longleftarrow 2s$. Return to step 2 if $j < S$.

A straightforward expansion formula for the original image is given by:

$$f = c_p + \sum_j w_j \tag{1.83}$$

where c_p is the residual image.

In step 4, the set of resolution levels associated with s lead to a dyadic analysis. Other possibilities involving intermediate scales (e.g. $s \longleftarrow \sqrt{2}\, s$) can also be considered.

The multiresolution coefficient values, w_j, are evidently not necessarily of zero mean, and so the potential artifact-creation difficulties related to this aspect of wavelet transforms do not arise. Note of course that values of w can be negative.

For input integer image values, this transform can be carried out in integer arithmetic only, which may lead to computational savings.

Computational requirements are high, and these can be mitigated to some extent by decimation: one pixel out of two is retained at each scale. Here the transform kernel does not change from one iteration to the next, but the image on which this transform is applied does.

1.5.2 Pyramidal median transform

The Pyramidal Median Transform (PMT) is obtained by:

1 Let $c_j = f$ with $j = 1$.
2 Determine $c^*_{j+1} = \text{med}(c_j, 2s+1)$ with $s = 1$.
3 The pyramidal multiresolution coefficients w_{j+1} are defined as:

$$w_{j+1} = c_j - c^*_{j+1}$$

4 Let $c_{j+1} = \text{dec}(c^*_{j+1})$ (where the decimation operation, dec, entails 1 pixel replacing each 2×2 subimage).
5 Let $j \longleftarrow j+1$. Return to step 2 iff $j < S$.

Here the kernel or mask of dimensions $(2s+1) \times (2s+1)$ remains the same during the iterations. The image itself, to which this kernel is applied, becomes smaller.

While this algorithm aids computationally, the reconstruction formula (eqn. (1.83) above) is no longer valid. Instead we use the following algorithm based on B-spline interpolation:

1 Take the lowest scale image, c_p.
2 Interpolate c_p to determine the next resolution image (of twice the dimensionality in x and y). Call the interpolated image c'_p.
3 Calculate $c_{p-1} \longleftarrow c'_p + w_p$.
4 Set $p \longleftarrow p-1$. Go to step 2 if $p > 0$.

This reconstruction procedure takes account of the pyramidal sequence of images containing the multiresolution coefficients, w_j. It presupposes, though, that high-quality reconstruction is possible. We ensure that by use of the following refined version of the Pyramidal

Multi-Median Transform. Using iteration, the definition of the coefficients, $w_{j+1} = c_j - c_{j+1}$, is improved vis-à-vis their potential for reconstructing the input image.

1.5.3 Iterative pyramidal median transform

An iterative scheme can be proposed for reconstructing an image, based on pyramidal multi-median transform coefficients. Alternatively, the PMT algorithm, itself, can be enhanced to allow for better estimates of coefficient values. The following is an iterative algorithm for this objective:

1 $i \longleftarrow 0$. Initialize f^i with the given image, f. Initialize the multiresolution coefficients at each scale j, w_j^f, to 0.
2 Using the Pyramidal Multi-Median Transform, determine the set of transform coefficients, $w_j^{f_i}$.
3 $w_j^f \longleftarrow w_j^f + w_j^{f_i}$.
4 Reconstruct image f_{i+1} from w_j^f (using the interpolation algorithm described in the previous section).
5 Determine the image component which is still not reconstructible from the wavelet coefficients: $f_{i+1} \longleftarrow f - f_{i+1}$.
6 Set $i \longleftarrow i + 1$, and return to step 2.

The number of iterations is governed by when f_{i+1} in step 5 approaches a null image. Normally four or five iterations suffice. Note that the additivity of the multiresolution coefficients in step 3 is justified by the image decomposition in step 5 and the reconstruction formula used in step 4, both of which are based on additive operations.

1.5.4 Non-iterative pyramidal transform with exact reconstruction

A non-iterative version of the pyramidal median transform can be performed by decimating and interpolating the median images during the transform:

1 Let $c_j = f$ with $j = 1$.
2 Determine $c_{j+1} = \text{dec}[\text{med}(c_j, 2s + 1)]$.
3 Determine c^*_{j+1} = interpolation of c_{j+1} to size of c_j.
4 The pyramidal multiresolution coefficients w_{j+1} are defined as:

$$w_{j+1} = c_j - c^*_{j+1}$$

5 Let $j \longleftarrow j + 1$. Return to step 2 iff $j < S$.

This saves computation time in two ways. First, there is no need to iterate. Secondly, in step 2 one does not really calculate the median for all pixels and then decimate it; rather, one just calculates the median for the pixels to be left after decimation. Thus the medians are four times fewer. This algorithm is very close to the Laplacian pyramid developed by Burt and Adelson (1983). The reconstruction algorithm is the same as before, but the reconstructed image has no error. In the following, we will mean this version when referring to PMT.

1.5.5 Conclusion on multiscale median transforms

The multiscale median transform is well-suited to all applications where an image reconstruction from a subset of coefficients is needed (e.g. restoration, compression, partial reconstruction). The suppression of subsets of coefficients leads to fewer artifacts in the reconstructed image, because often the visual artifacts are due to the shape of the wavelet function (the negative ring, in particular, surrounding objects). For data analysis, the median transform is interesting because the shapes of structures in the scales are closer to those in the input image than would be the case with a wavelet transform. This is due to the non-linearity of the median filter. We will see later (Chapter 8) how such a transform can be used for object detection.

Other morphological tools can be used to perform a similar transform such as opening (N erosions followed by N dilations). However results were found to be better with the median filter. In the median-based transform, coefficients can be positive or negative. For some applications, it would be useful to have a decomposition into multiresolution coefficients which are all positive. This can be provided by mathematical morphology.

1.6 Multiresolution and mathematical morphology

Mathematical morphology has been surveyed e.g. in Serra (1982). The usual operators for gray-level images are (see Haralick, Sternberg and Xinhua Zhuang, 1987; Maragos and Shaffer, 1990):

- *erosion* which consists of replacing each pixel of an image by the minimum of its neighbors.
- *dilation* which consists of replacing each pixel of an image by the maximum of its neighbors.
- *opening* which consists of doing N erosions followed by N dilations.
- *closing* which consists of doing N dilations followed by N erosions.

Morphology processing was first introduced in astronomy by Lea (Lea and Keller, 1989) and by Huang (Huang and Bijaoui, 1991). A survey of multiresolution analysis using a scale-space approach is to be found in Jackway and Deriche (1996); and further multiresolution approaches in Liang and Wong (1993).

1.6.1 Multiresolution

Detail structure of a given size of objects can be obtained by taking the difference between the original image and its opening of order N, N being the parameter which characterizes detail sizes. A possible multiresolution algorithm is:

1 Define the neighborhood of a pixel (generally the 8 closest neighbors are chosen).
2 Initialize j to 0, and start from data $c_0 = f$.
3 open_N being the function which gives an opening of order N, we compute $c_{j+1} = \text{open}_N(c_j)$ and the coefficients at scale j by:

$$w_{j+1} = c_j - c_{j+1} \tag{1.84}$$

4 Double the opening order: $N = 2 * N$.
5 $j = j + 1$.
6 if j is less than the number of resolution levels we need, return to step 3.

Using this algorithm, we have a positive multiresolution transform. This was the strategy chosen by Appleton (Appleton, Siqueira and Basart, 1993) for cirrus (due to intergalactic dust phenomena) filtering in infrared images.

An undersampling or decimation can be introduced, just as for the median transform, which leads to a pyramidal algorithm.

1.6.2 Pyramidal morphological transform

The goal of this transform is to allow the decomposition of an image I into a sum of components with different sizes (similarly to the pyramidal median transform), each of these components being positive.

The algorithm performing this is:

1 Define the neighborhood of a pixel (generally the 8 closest neighbors).
2 Initialize j to 0, and start from data $c_0 = f$.

3 open_1 is the function which gives an opening of order 1, and $m_{j+1} = \text{open}_1(c_j)$ is computed, with the coefficients at scale j given by the following:

$$w_{j+1} = c_j - m_{j+1} \tag{1.85}$$

4 Undersample m_{j+1} in order to obtain c_{j+1}.
5 $j = j + 1$.
6 If j is less than the number of resolution levels needed, return to step 3.

The reconstruction is not exact. To make it so, we need to modify c_{j+1} in the following way:

1 Initialize the error image $E^{(0)}$ to f : $E^{(0)} = f$.
2 $n = 0$.
3 Compute the coefficients $c_j^{(n)}$ of $E^{(n)}$ by the previous algorithm.
4 Reconstruct $\tilde{f}^{(n)}$.
5 $E^{(n)} = f - \tilde{f}^{(n)}$.
6 If $\| E^{(n)} \| > \epsilon$, then return to step 3.

Thus we obtain a set of positive coefficients, and the reconstruction is exact.

1.6.3 Conclusions on non-wavelet multiresolution approaches

Nonlinear multiresolution transforms are complementary to the wavelet transform. According to the application, one or other may be used.

We note in passing one particularly interesting use of the PMT: the final scale often provides a good estimate of the image background. This is especially so for images containing small structures, e.g. astronomical wide-field images.

2 Multiresolution support and filtering

2.1 Noise modeling

2.1.1 Definition of significant coefficients

Images generally contain noise. Hence the wavelet coefficients are noisy too. In most applications, it is necessary to know if a coefficient is due to signal (i.e. it is significant) or to noise. Generally noise in astronomical images follows a Gaussian or a Poisson distribution, or a combination of both. So we consider these three possibilities, in some depth, in the following. We consider separately the Poisson distribution case where we have a small number of counts or photons (less than 20 per pixel). We define the multiresolution support of an image and how we can derive it from our noise modeling. Finally, we consider various other cases relating to non-stationary noise.

The wavelet transform yields a set of resolution-related views of the input image. A wavelet image scale at level j has coefficients given by $w_j(x, y)$. If we obtain the distribution of the coefficient $w_j(x, y)$ for each plane, based on the noise, we can introduce a statistical significance test for this coefficient. The procedure is the classical significance-testing one. Let $\mathscr{H}'$ be the hypothesis that the image is locally constant at scale j. Rejection of hypothesis $\mathscr{H}'$ depends (for a positive coefficient value) on:

$$P = \text{Prob}(W_N > w_j(x, y)) \tag{2.1}$$

and if the coefficient value is negative

$$P = \text{Prob}(W_N < w_j(x, y)) \tag{2.2}$$

Given a threshold, ϵ, if $P > \epsilon$ the null hypothesis is not excluded. Although non-null, the value of the coefficient could be due to noise. On the other hand, if $P < \epsilon$, the coefficient value cannot be due only to the noise alone, and so the null hypothesis is rejected. In this case, a significant coefficient has been detected.

2.1.2 Gaussian noise

Significant level. If the distribution of $w_j(x,y)$ is Gaussian, with zero mean and standard deviation σ_j, we have the probability density

$$p(w_j(x,y)) = \frac{1}{\sqrt{2\pi}\sigma_j} e^{-w_j(x,y)^2/2\sigma_j^2} \quad (2.3)$$

Rejection of hypothesis $\mathscr{H}'$ depends (for a positive coefficient value) on:

$$P = \text{Prob}(w_j(x,y) > W) = \frac{1}{\sqrt{2\pi}\sigma_j} \int_{w_j(x,y)}^{+\infty} e^{-W^2/2\sigma_j^2} dW \quad (2.4)$$

and if the coefficient value is negative, it depends on

$$P = \text{Prob}(w_j(x,y) < W) = \frac{1}{\sqrt{2\pi}\sigma_j} \int_{-\infty}^{w_j(x,y)} e^{-W^2/2\sigma_j^2} dW \quad (2.5)$$

Given stationary Gaussian noise, it suffices to compare $w_j(x,y)$ to $k\sigma_j$. Often k is chosen as 3, which corresponds approximately to ϵ (cf. the previous subsection) $= 0.002$. If $w_j(x,y)$ is small, it is not significant and could be due to noise. If $w_j(x,y)$ is large, it is significant:

$$\begin{array}{lll} \text{if } \mid w_j \mid \geq k\sigma_j & \text{then } w_j \text{ is significant} \\ \text{if } \mid w_j \mid < k\sigma_j & \text{then } w_j \text{ is not significant} \end{array} \quad (2.6)$$

So we need to estimate, in the case of Gaussian noise models, the noise standard deviation at each scale. These standard deviations can be determined analytically in the case of some transforms, including the à trous transform, but the calculations can become complicated.

Estimation of noise standard deviation at each scale. The appropriate value of σ_j in the succession of wavelet planes is assessed from the standard deviation of the noise σ_I in the original image I, and from study of the noise in the wavelet space. This study consists of simulating an image containing Gaussian noise with a standard deviation equal to 1, and taking the wavelet transform of this image. Then we compute the standard deviation σ_j^e at each scale. We get a curve σ_j^e as a function of j, giving the behavior of the noise in the wavelet space. (Note that if we had used an orthogonal wavelet transform, this curve would be linear.) Due to the properties of the wavelet transform, we have $\sigma_j = \sigma_I \sigma_j^e$. The standard deviation of the noise at a scale j of the image is equal to the standard deviation of the noise of the image multiplied by the standard deviation of the noise of scale j of the wavelet transform.

An alternative, here, would be to estimate the standard deviation of the noise σ_1 of the first plane from the histogram of w_1. The values of the wavelet image w_1 are due mainly to the noise. A histogram shows a Gaussian peak around 0. A 3-sigma clipping (i.e. robust estimation of the variance by defining it on the basis only of values within 3 standard deviations of the mean) is then used to reject pixels where the signal is significantly large. The standard deviation of the noise σ_j is estimated from σ_1. This is based on the study of noise variation between two scales, as described above.

A final alternative (Lee, 1983) to be mentioned here relates to multiple images of the same scene. In this case, a pixel-dependent specification of the noise threshold is used, rather than one that is just level-dependent. A wavelet transform of each of the N images is determined. Thus we have N wavelet coefficients at each position, (x, y), and at each scale, j. From the N values, $w_j(x, y)$, the standard deviation, $\sigma_j(x, y)$, is calculated. The significance threshold is then defined by:

$$k\sigma_j(x, y)/\sqrt{N} \qquad (2.7)$$

(the denominator is explained by the error of the mean of N Gaussian values varying as $1/\sqrt{N}$).

Automatic noise estimation in an image. There are different ways to estimate the standard deviation of Gaussian noise in an image (Bracho and Sanderson, 1985; Lee, 1981; Lee and Hoppel, 1989; Mastin, 1985; Meer, Jolian and Rosenfeld, 1990; Vorhees and Poggio, 1987). Olson (1993) made an evaluation of six methods and showed that the average method was best, and this is also the simplest method. This method consists of filtering the image I with the average filter and subtracting the filtered image from I. Then a measure of the noise at each pixel is computed.

To keep image edges from contributing to the estimate, the noise measure is disregarded if the magnitude of the intensity gradient is larger than some threshold T. The threshold value may be found from the accumulated histogram of the magnitude of the intensity gradient. An alternative is to apply a k-sigma clipping.

2.1.3 Poisson noise

If the noise in the data I is Poisson, the Anscombe transform

$$t(I(x, y)) = 2\sqrt{I(x, y) + \frac{3}{8}} \tag{2.8}$$

acts as if the data arose from a Gaussian white noise model (Anscombe, 1948), with $\sigma = 1$, under the assumption that the mean value of I is large.

It is useful to show just how large the mean value of I must be, for the above transformation to be valid. For each integer i between 1 and 100, we generated 100 Poisson values with parameter (i.e. mean) i. Figure 2.1a shows the variance of these Poisson sets. Figure 2.1b shows the variance of the corresponding Anscombe-transformed sets. Figure 2.1b shows no noticeable degradation as the Poisson parameter becomes small. Oscillation around a value of 1, over the entire range of Poisson parameter values (i.e. mean value of I ranging from 100 down to 1), is clear. The range of variation (in Fig. 2.1b, roughly from a high of 1.5, and a low of 0.6) is unnoticeable when plotted on the vertical scale used in Fig. 2.1a. Anscombe (1948, eqn. (2.9))

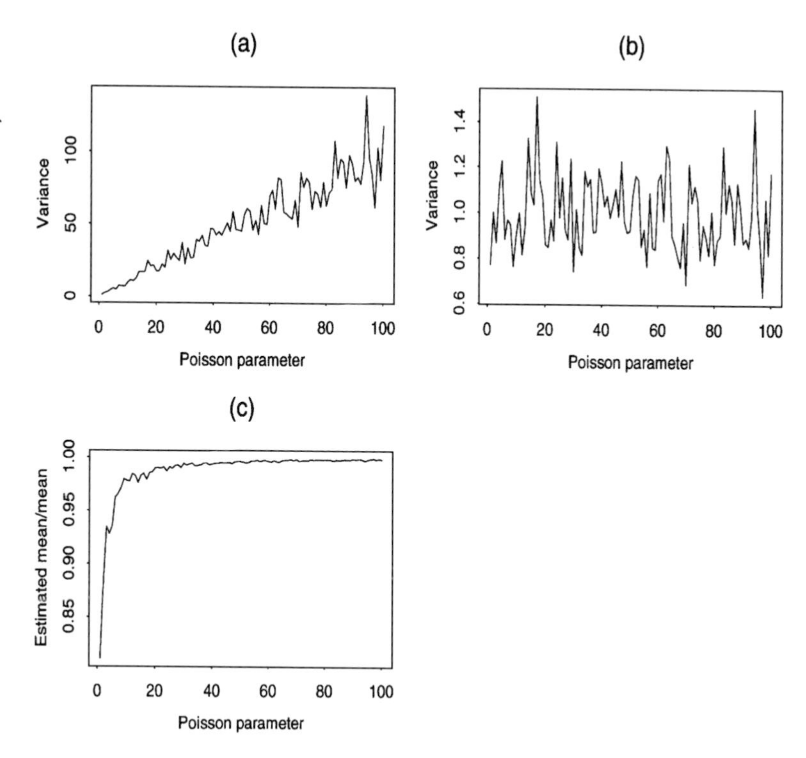

Figure 2.1 Parts (a) and (b) show, respectively, the variance of a series of Poisson-distributed data sets, and the variance of the Anscombe-transformed data sets. Part (c) shows the bias of this transformation.

gives an approximate formula for the variance where it is seen to be roughly constant even for very small values of the Poisson parameter. We conclude that the variance of the stabilized Poisson data is, from a practical point of view, equal to 1 irrespective of the mean value of I.

Figure 2.1c examines bias (see Anscombe, 1948, eqn. (2.10)). Figure 2.1c shows mean values estimated from the stabilized Poisson data, with each such value divided by the mean value of the original Poisson data. Thus, for a mean value m_i of the Poisson variates generated with parameter i, and for a mean value m_i^* of these transformed Poisson variates, the corresponding point on this graph is $((m_i^*/2)^2 - 3/8)/m_i$. We can see practical identity of both means from around a Poisson parameter value of 30 upwards.

For Poisson parameter values between about 10 and 30, the curve shown in Fig. 2.1c could be used to correct the image restoration results. We would simply correct, in a multiplicative manner, for the consistent underestimate of results in this range of intensity values, as evidenced by Fig. 2.1c. Such situations may be particularly relevant for the outlying parts of extended objects where intensity values are close to low background values. We expect the centers of extended astronomical objects, or point sources (respectively galaxies and stars, for instance), to have sufficient associated counts such that a bias correction is not needed.

For Poisson parameter values under about 10, the Anscombe transformation looses control over the bias. In this case, an alternative approach to variance stabilization is needed. An approach for very small numbers of counts, including frequent zero cases, has been described in Bijaoui, Bury and Slezak (1994b), Bury (1995) and Slezak, de Lapparent and Bijaoui (1993), and will be described in subsection 2.1.5 below. Small numbers of detector counts will most likely be associated with the image background. Note that errors related to small values carry the risk of removing real objects, but not of amplifying noise. This is seen in Fig. 2.1c, where at increasingly low values, the pixel value is increasingly underestimated.

2.1.4 Gaussian and Poisson noise

Variance stabilization. The arrival of photons, and their expression by electron counts, on CCD detectors may be modeled by a Poisson distribution. In addition, there is additive Gaussian read-out noise. The Anscombe transformation (eqn. (2.8)) has been extended (Bijaoui, 1994) to take this combined noise into account: see Appendix 1. As an approximation, consider the signal's value, $I(x, y)$, as a sum of a Gaussian variable, γ, of mean g and standard-deviation σ; and a Poisson variable, n, of mean m_0: we set $I(x, y) = \gamma + \alpha n$ where α is the gain.

The generalization of the variance stabilizing Anscombe formula, which also generalizes a transformation due to Bartlett (1936), is derived in Appendix 1 as:

$$t = \frac{2}{\alpha}\sqrt{\alpha I(x, y) + \frac{3}{8}\alpha^2 + \sigma^2 - \alpha g} \tag{2.9}$$

With appropriate values of α, σ and g, this reduces to Anscombe's transformation (eqn. (2.8)).

Figure 2.2 shows an update of Fig. 2.1 for the combined Poisson and Gaussian case. Values were generated as $I = \alpha n + \gamma$ where α is the gain, n the Poisson-distributed component, and γ the Gaussian-distributed component. The gain was taken as 7.5 e^-/DN. The mean and standard deviation of the Gaussian read-out noise were taken respectively as 0.0 and 13 e^- (or 1.733 DN). For each value of the Poisson component's parameter (shown on the horizontal axes in Fig. 2.2), 100 realizations were made of the signal, x. Figure 2.2a shows the variance of $I(x, y)$ as a function of the Poisson parameter. Figure 2.2b shows the variance,

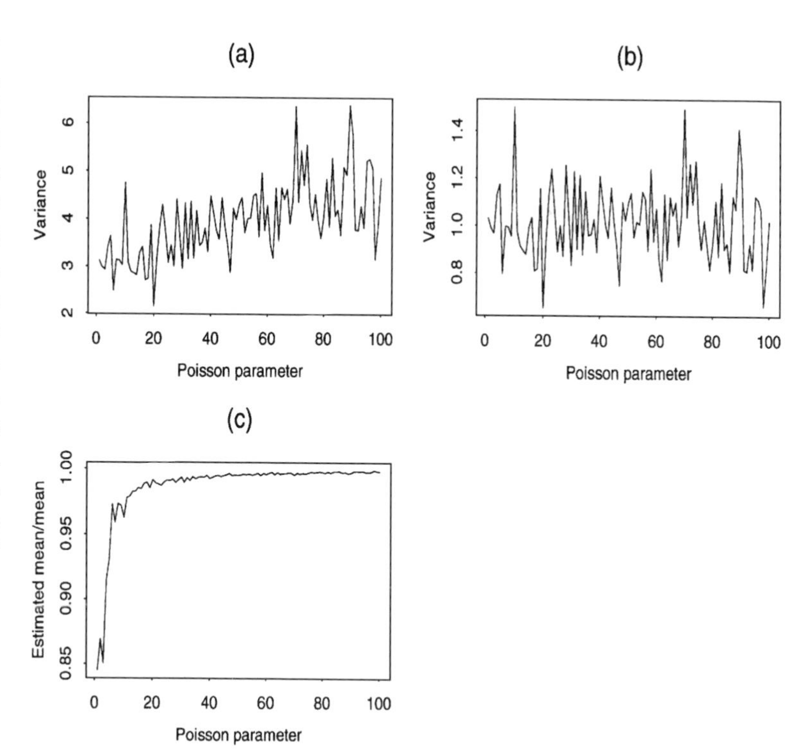

Figure 2.2 Similarly to previous figure, noise with combined Poisson and Gaussian components is analyzed. Parts (a) and (b) show, respectively, the variance of a series of data sets, and the variance of this data following use of the generalized Anscombe transformation. Part (c) shows the bias of this transformation.

following the generalized Anscombe transformation (eqn. (2.9)). We can see that this variance oscillates around 1. Figure 2.2c shows the bias which, as in the purely Poisson case, indicates that there is no underestimation of transformed values, from a signal value of around 20–30 upwards.

Automatic estimation of standard deviation of read-out noise. If the sigma (standard deviation) of the read-out noise is not known, we can use the variance stabilization transform to estimate it. This is achieved by finding the transformation t such that the standard deviation of $t(I)$ is equal to 1. The algorithm is:

1 Set n to 0, S_{min} to 0, and S_{max} to $\sigma(I)$.
2 Set r_n to $(S_{min} + S_{max})/2$.
3 Compute the transform of I with a standard deviation read-out noise equal to r_n.
4 Estimate the standard deviation of the noise σ_S in $t(I)$ by the average method described in subsection 2.1.2 above.
5 If $\sigma_S < 1$ then $S_{min} = r_n$ else $S_{max} = r_n$.
6 If $S_{max} - S_{min} > \epsilon$ then $n = n + 1$ and go to step 2.
7 $r_n = (S_{min} + S_{max})/2$ is a good estimation of the standard deviation of the read-out noise.

The same method can be applied if it is the gain which is not known, and the standard deviation of the read-out noise is known.

2.1.5 Poisson noise with few photons or counts

A wavelet coefficient at a given position and at a given scale j is

$$w_j(x, y) = \sum_{k \in K} n_k \psi\left(\frac{x_k - x}{2^j}, \frac{y_k - y}{2^j}\right) \tag{2.10}$$

where K is the support of the wavelet function ψ and n_k is the number of events which contribute to the calculation of $w_j(x, y)$ (i.e. the number of photons included in the support of the dilated wavelet centered at (x,y)).

If a wavelet coefficient $w_j(x, y)$ is due to the noise, it can be considered as a realization of the sum $\sum_{k \in K} n_k$ of independent random variables with the same distribution as that of the wavelet function (n_k being the number of photons or events used for the calculation of $w_j(x, y)$). Then we compare the wavelet coefficient of the data to the values which can be taken by the sum of n independent variables.

The distribution of one event in the wavelet space is directly given by the histogram H_1 of the wavelet ψ. Since independent events are considered, the distribution of the random variable W_n (to be associated with a wavelet coefficient) related to n events is given by n autoconvolutions of H_1

$$H_n = H_1 \otimes H_1 \otimes ... \otimes H_1 \tag{2.11}$$

Figure 2.3 shows the shape of a set of H_n. For a large number of events, H_n converges to a Gaussian.

In order to facilitate the comparisons, the variable W_n of distribution H_n is reduced by

$$c = \frac{W_n - E(W_n)}{\sigma(W_n)} \tag{2.12}$$

and the cumulative distribution function is

$$F_n(c) = \int_{-\infty}^{c} H_n(u)du \tag{2.13}$$

From F_n, we derive $c_{\min}$ and $c_{\max}$ such that $F(c_{\min}) = \epsilon$ and $F(c_{\max}) = 1 - \epsilon$.

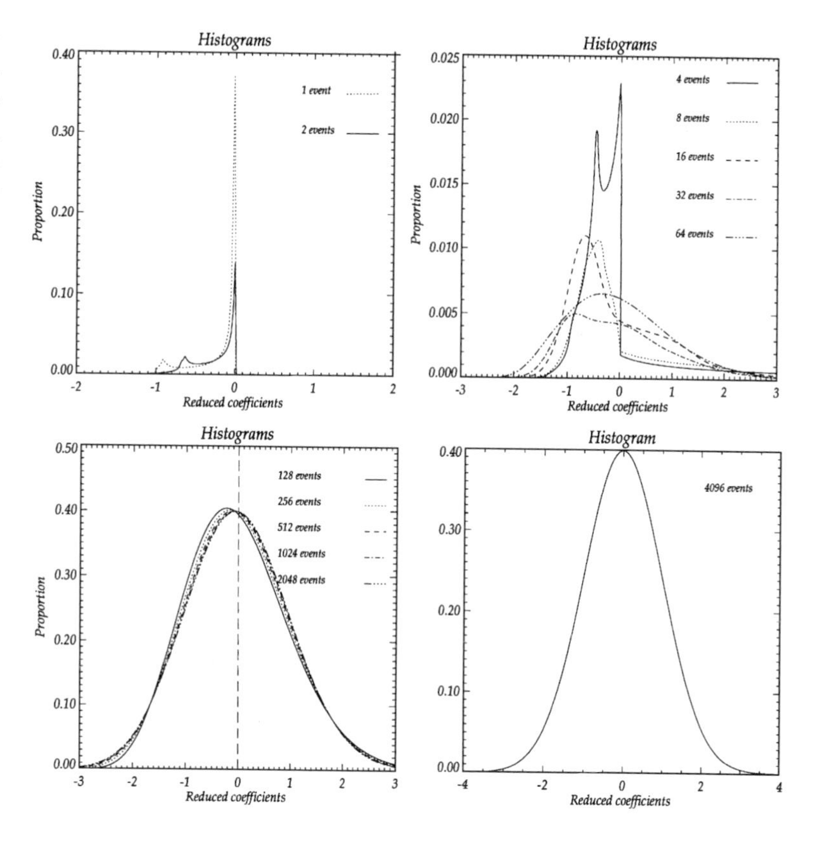

Figure 2.3 Autoconvolution histograms for the wavelet associated with a B_3 spline scaling function for one and 2 events (top left), 4 to 64 events (top right), 128 to 2048 (bottom left), and 4096 (bottom right).

Therefore a reduced wavelet coefficient $w_j^r(x, y)$, calculated from $w_j(x, y)$, and resulting from n photons or counts is significant if:

$$F(w^r) > c_{\max} \tag{2.14}$$

or

$$F(w^r) < c_{\min} \tag{2.15}$$

and $w_j^r(x, y)$ is obtained by

$$w_j^r(x, y) = \frac{w_j(x, y)}{\sqrt{n}\sigma_{\psi_j}} \tag{2.16}$$

$$= \frac{w_j(x, y)}{\sqrt{n}\sigma_{\psi}} 4^j \tag{2.17}$$

where σ_ψ is the standard deviation of the wavelet function, and σ_{ψ_j} is the standard deviation of the dilated wavelet function ($\sigma_{\psi_j} = \sigma_\psi / 4^j$).

2.1.6 Other types of noise

For any type of noise, an analogous study can be carried out in order to find the detection level at each scale and at each position. The types of noise considered so far in this chapter correspond to the general cases in astronomical imagery. We now describe briefly methods which can be used for non-uniform and multiplicative noise.

Additive non-uniform noise. If the noise is additive, but non-uniform, we cannot estimate a standard deviation for the whole image. However, we can often assume that the noise is locally Gaussian, and we can compute a local standard deviation of the noise for each pixel. In this way, we obtain a standard deviation map of the noise, $I_\sigma(x, y)$. A given wavelet coefficient $w_j(x, y)$ is calculated from the pixels of the input image I in the range $I(x-l \dots x+l, y-l \dots y+l)$ where l is dependent on the wavelet transform algorithm, the wavelet function, and the scale j. An upper limit $u_j(x, y)$ for the noise associated with $w_j(x, y)$ is found by just considering the maximum value in $I_\sigma(x-l \dots x+l, y-l \dots y+l)$ and by multiplying this value by the constant σ_j^e which was defined in subsection 2.1.2 ('Estimation of noise standard deviation at each scale').

$$u_j(x, y) = \max(I_\sigma(x - l \dots x + l, y - l \dots y + l))\sigma_j^e \tag{2.18}$$

The detection level is not constant over each scale.

Multiplicative noise. If the noise is multiplicative, the image can be transformed by taking its logarithm. In the resulting image, the noise is additive, and a hypothesis of Gaussian noise can be used in order to find the detection level at each scale.

Multiplicative non-uniform noise. In this case, we take the logarithm of the image, and the resulting image is treated as for additive non-uniform noise above.

Unknown noise. If the noise does not follow any known distribution, we can consider as significant only wavelet coefficients which are greater than their local standard deviation multiplied by a constant: $w_j(x, y)$ is significant if

$$\mid w_j(x, y) \mid > k\sigma(w_j(x - l \dots x + l, y - l \dots y + l)) \tag{2.19}$$

2.2 Multiresolution support

2.2.1 Definition

The multiresolution support (Starck, Murtagh and Bijaoui, 1995) of an image describes in a logical or boolean way whether an image I contains information at a given scale j and at a given position (x, y). If $M^{(I)}(j, x, y) = 1$ (or *true*), then I contains information at scale j and at the position (x, y). M depends on several parameters:

- The input image.
- The algorithm used for the multiresolution decomposition.
- The noise.
- All constraints we want the support additionally to satisfy.

Such a support results from the data, the treatment (noise estimation, etc.), and from knowledge on our part of the objects contained in the data (size of objects, alignment, etc.). In the most general case, a priori information is not available to us.

2.2.2 Multiresolution support from the wavelet transform

The wavelet transform of an image by an algorithm such as the à trous one produces a set $\{w_j\}$ at each scale j. This has the same number of pixels as the image. The original image c_0 can be expressed as the sum of all the wavelet planes and the smoothed array c_p

$$c_0 = c_p + \sum_{j=1}^{p} w_j \tag{2.20}$$

and a pixel at position x, y can be expressed also as the sum over all the wavelet coefficients at this position, plus the smoothed array:

$$c_0(x, y) = c_p(x, y) + \sum_{j=1}^{p} w_j(x, y) \tag{2.21}$$

The multiresolution support will be obtained by detecting at each scale the significant coefficients. The multiresolution support is defined by:

$$M(j, x, y) = \begin{cases} 1 & \text{if } w_j(x, y) \text{ is significant} \\ 0 & \text{if } w_j(x, y) \text{ is not significant} \end{cases} \tag{2.22}$$

2.2.3 Algorithm

The algorithm to create the multiresolution support is as follows:

1 We compute the wavelet transform of the image.
2 We estimate the noise standard deviation at each scale. We deduce the statistically significant level at each scale.
3 The binarization of each scale leads to the multiresolution support.
4 Modification using a priori knowledge (if desired).

Step 4 is optional. A typical use of a priori knowledge is the suppression of isolated pixels in the multiresolution support in the case where the image is obtained with a point spread function (PSF) of more than one pixel. Then we can be sure that isolated pixels are residual noise which has been detected as significant coefficients. If we use a pyramidal algorithm or a nonlinear multiresolution transform, the same method can be used.

In order to visualize the support, we can create an image S defined by:

$$S(x, y) = \sum_{j=1}^{p} 2^j M(j, x, y) \tag{2.23}$$

Figure 2.4 shows such a multiresolution support visualization of an image of galaxy NGC 2997.

2.2.4 Gaussian noise estimation from the multiresolution support

In subsection 2.1.2 ('Automatic noise estimation in an image') it was shown how a Gaussian noise σ_I can be estimated automatically in an image I. This estimation is very important because all the noise standard deviations σ_j at scales j are derived from σ_I. Thus an error in σ_I will introduce an error in all of σ_j. This measure of σ_I can be refined by the use of the multiresolution support. Indeed, if we consider the set of pixels $\mathcal{S}$ in the image which are due only to the noise, and if we take the standard deviation of them, we should find the same value σ_I. This set is easily obtained from the multiresolution support. We say that a pixel (x, y) belongs to the noise if $M(j, x, y) = 0$ for all j (i.e. there is no significant coefficient at any scale). The new estimation of σ_I is then computed by the following iterative algorithm:

1 Estimate the standard deviation of the noise in I: we have $\sigma_I^{(0)}$.

2 Compute the wavelet transform (à trous algorithm) of the image I with p scales: we have

$$I(x, y) = c_p(x, y) + \sum_{j=1}^{p} w_j(x, y)$$

where w_j are the wavelet scales, and c_p is the low frequency part of I. The noise in c_p is negligible.

3 Set n to 0.

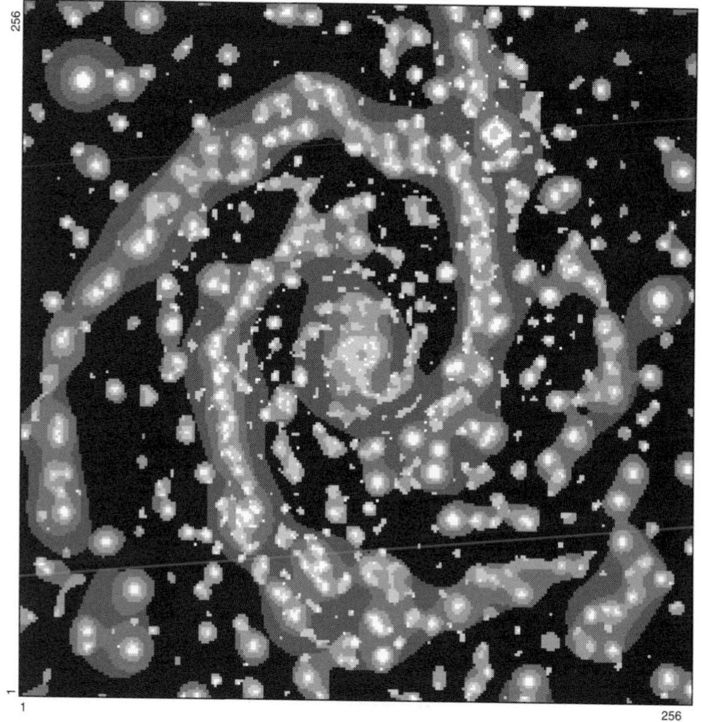

Figure 2.4 Multiresolution support representation of a spiral galaxy.

4 Compute the multiresolution support M which is derived from the wavelet coefficient and from $\sigma_I^{(n)}$.

5 Select the pixels which belong to the set $\mathcal{S}$: if $M(j,x,y) = 0$ for all j in $1 \ldots p$, then the pixel $(x,y) \in \mathcal{S}$.

6 For all the selected pixels (x,y), compute the values $I(x,y) - c_p(x,y)$ and compute the standard deviation $\sigma_I^{(n+1)}$ of these values (the difference between I and c_p is computed in order not to take the background into account in the noise estimation).

7 $n = n + 1$.

8 If $(\mid \sigma_I^{(n)} - \sigma_I^{(n-1)} \mid)/\sigma_I^{(n)} > \epsilon$ then go to step 4.

This method converges in a few iterations, and allows the noise estimate to be improved.

Figure 2.5 shows a simulated image mostly of galaxies, as used by Caulet and Freudling (1993; Freudling and Caulet, 1993). Gaussian noise is added, leading to varying signal-to-noise ratios (SNR, defined as $10 \times \log_{10}(\sigma(\text{input image})/\sigma(\text{added noise}))$. Figure 2.6 shows an image with SNR −0.002. This image was produced by adding noise of standard deviation arbitrarily chosen to be equal to the overall standard deviation of the original image. For varying SNRs, the following three methods were assessed. Firstly, a median-smoothed version of

Figure 2.5 Simulated image of galaxies.

the image (providing a crude estimate of the image background) is subtracted, and the noise is estimated from this difference image. This noise estimate is made more robust by 3-sigma clipping. Secondly, the multiresolution support approach described above was used, where the specification of the multiresolution support was iteratively refined. Thirdly, the noise statistics in a running block or square window are used, and the averages of these values are returned.

Table 2.1 gives results obtained. The excellent estimation capability of the iterative multiresolution support method can be seen.

Table 2.1 *Noise added (standard deviation) followed by automatic noise estimates. The SNR associated with the added noise is also given.*

Noise	Med3	Multres	Block	SNR
24.1356	23.9787	24.4685	21.6591	13.8246
291.2640	282.3600	293.9950	244.7540	3.0083
582.5280	561.9350	585.3710	490.8740	−0.0020
1165.0600	1126.9600	1171.5200	979.6810	−3.0123
2330.1100	2250.4300	2339.1500	1949.1300	−6.0226

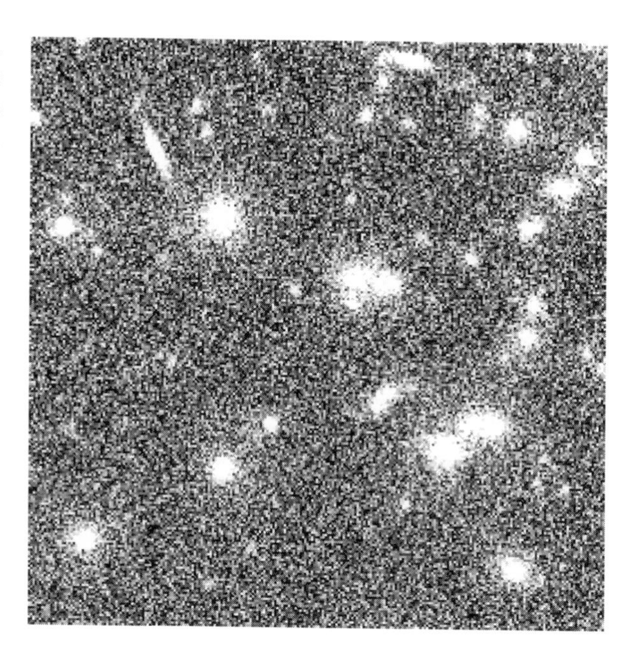

Figure 2.6 Simulated image of galaxies with Gaussian noise (sigma 582.5) added.

2.2.5 Concluding remarks on the multiresolution support and noise

The multiresolution support allows us to integrate, in a visualizable manner, and in a way which is very suitable for ancillary image alteration, information coming from data, knowledge, and processing. We will see in the next section how we can use this in image filtering and, in a later chapter, in image restoration. The multiresolution support depends completely on the noise.

Figure 2.7 summarizes how the multiresolution support is derived from the data and our noise-modeling. The first step consists of

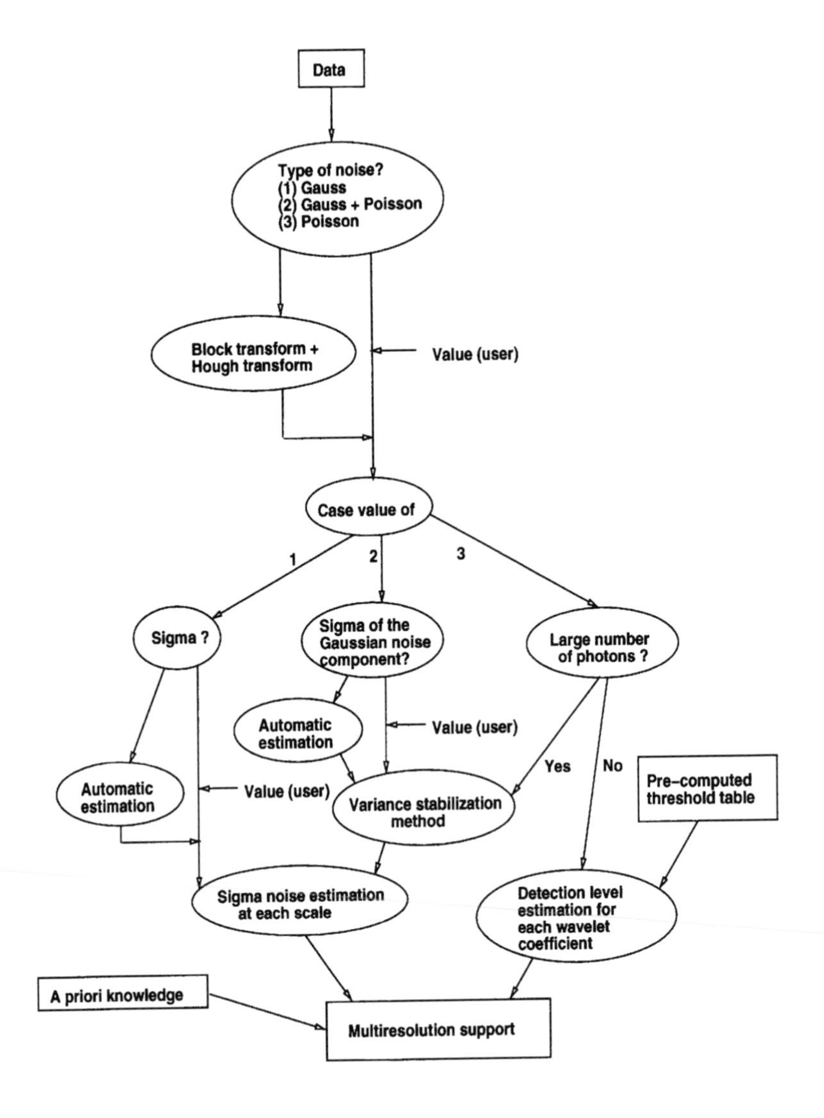

Figure 2.7 Determination of multiresolution support from noise modeling.

estimating if the noise follows a Gaussian distribution, a Poisson distribution, a combination of both, or some other distribution. This step can be carried out automatically based on a Hough transform of a local mean-squared versus standard deviation diagram (Lee and Hoppel, 1989). Astronomers generally know which kind of noise they have in the data and this point is not critical. If the noise follows a pure Poisson distribution, and if the number of photons per pixel is low, then the method described in subsection 2.1.5 above, has to be employed. In other cases (Poisson, or Gaussian plus Poisson, with sufficient counts per pixel), the variance stabilization method is convenient. If we have the Gaussian plus Poisson case, and if we do not know exactly the standard deviation of the Gaussian component, this can be automatically computed. In the case of pure Gaussian noise, the standard deviation can be automatically estimated too by a large number of methods. The wavelet transform and multiresolution support help to refine the estimation.

2.3 Filtering

The suppression of noise is one of the most important tasks in astronomical image processing. Frequently, linear techniques are used because linear filters are easy to implement and design (Sucher, 1995). But these methods modify the morphology of the structures in the images. To overcome these shortcomings, a large number of nonlinear methods have been presented in the literature. The most popular is the median filter, which is computationally efficient and has proved extremely successful for removing noise of an impulsive (shot) nature. Clearly, it suffers from the fact that with increasing window size the signal detail becomes blurred. Therefore, many generalizations have been developed, e.g. order statistic filters and filters based on threshold decomposition (see Pitas and Venetsanopoulos, 1990, for a list of references). We describe in this section how the modeling of the noise, developed previously, can be used for filtering purposes.

2.3.1 Convolution using the continuous wavelet transform

We will examine here the computation of a convolution by using the continuous wavelet transform in order to get a framework for linear smoothing. Let us consider the convolution product of two functions:

$$h(x) = \int_{-\infty}^{+\infty} f(u)g(x-u)dx \tag{2.24}$$

We introduce two real wavelet functions $\psi(x)$ and $\chi(x)$ such that:

$$C = \int_0^{+\infty} \frac{\hat{\psi}^*(\nu)\hat{\chi}(\nu)}{\nu} d\nu \tag{2.25}$$

is defined. $W_g(a,b)$ denotes the wavelet transform of g with the wavelet function $\psi(x)$:

$$W_g(a,b) = \frac{1}{\sqrt{a}} \int_{-\infty}^{+\infty} g(x)\psi^*\left(\frac{x-b}{a}\right) dx \tag{2.26}$$

We restore $g(x)$ with the wavelet function $\chi(x)$:

$$g(x) = \frac{1}{C} \int_0^{+\infty} \int_{-\infty}^{+\infty} \frac{1}{\sqrt{a}} W_g(a,b)\chi\left(\frac{x-b}{a}\right) \frac{da\, db}{a^2} \tag{2.27}$$

The convolution product can be written as:

$$h(x) = \frac{1}{C} \int_0^{+\infty} \frac{da}{a^{\frac{5}{2}}} \int_{-\infty}^{+\infty} W_g(a,b)db \int_{-\infty}^{+\infty} f(u)\chi\left(\frac{x-u-b}{a}\right) du \tag{2.28}$$

Let us denote $\tilde{\chi}(x) = \chi(-x)$. The wavelet transform $W_f(a,b)$ of $f(x)$ with the wavelet $\tilde{\chi}(x)$ is:

$$\tilde{W}_f(a,b) = \frac{1}{\sqrt{a}} \int_{-\infty}^{+\infty} f(x)\tilde{\chi}\left(\frac{x-b}{a}\right) dx \tag{2.29}$$

This leads to:

$$h(x) = \frac{1}{C} \int_0^{+\infty} \frac{da}{a^2} \int_{-\infty}^{+\infty} \tilde{W}_f(a, x-b) W_g(a,b) db \tag{2.30}$$

We get the final result:

$$h(x) = \frac{1}{C} \int_0^{+\infty} \tilde{W}_f(a,x) \otimes W_g(a,x) \frac{da}{a^2} \tag{2.31}$$

In order to compute a convolution with the continuous wavelet transform:

- We compute the wavelet transform $\tilde{W}_f(a,b)$ of the function $f(x)$ with the wavelet function $\tilde{\chi}(x)$.
- We compute the wavelet transform $W_g(a,b)$ of the function $g(x)$ with the wavelet function $\psi(x)$.
- We sum the convolution product of the wavelet transforms, scale by scale.

The wavelet transform permits us to perform any linear filtering. Its efficiency depends on the number of terms in the wavelet transform associated with $g(x)$ for a given signal $f(x)$. If we have a filter where the number of significant coefficients is small for each scale, the complexity of the algorithm is proportional to N. For classical convolution, the complexity is also proportional to N, but the number of operations is proportional to the length of the convolution mask. The main advantage of the present technique lies in the possibility of having a filter with long scale terms without computing the convolution in a large window. If we carry out the convolution with the FFT algorithm, the complexity is of order $N \log_2 N$. The computing time is greater than that obtained with the wavelet transform if we concentrate the energy in a very small number of coefficients.

2.3.2 Wiener-like filtering in wavelet space

Let us consider a wavelet coefficient w_j at the scale j. We assume that its value, at a given scale and a given position, results from a noisy process, based on a Gaussian distribution with mathematical expectation W_j, and a standard deviation B_j:

$$P(w_j \mid W_j) = \frac{1}{\sqrt{2\pi} B_j} \exp\left(-\frac{(w_j - W_j)^2}{2B_j^2} \right) \tag{2.32}$$

Now, we assume that the set of coefficients W_j for a given scale also follow a Gaussian distribution, with zero mean and standard deviation S_j:

$$P(W_j) = \frac{1}{\sqrt{2\pi} S_j} \exp\left(-\frac{W_j^2}{2S_j^2} \right) \tag{2.33}$$

The null mean value results from the wavelet property:

$$\int_{-\infty}^{+\infty} \psi^*(x) dx = 0 \tag{2.34}$$

We want to get an estimate of W_j knowing w_j. Bayes' theorem gives:

$$P(W_j \mid w_j) = \frac{P(W_j) P(w_j \mid W_j)}{P(w_j)} \tag{2.35}$$

We get:

$$P(W_j \mid w_j) = \frac{1}{\sqrt{2\pi} \beta_j} \exp\left(-\frac{(W_j - \alpha_j w_j)^2}{2\beta_j^2} \right) \tag{2.36}$$

where:

$$\alpha_j = \frac{S_j^2}{S_j^2 + B_j^2} \tag{2.37}$$

The probability $P(W_j \mid w_j)$ follows a Gaussian distribution with mean:

$$m = \alpha_j w_j \tag{2.38}$$

and variance:

$$\beta_j^2 = \frac{S_j^2 B_j^2}{S_j^2 + B_j^2} \tag{2.39}$$

The mathematical expectation of W_j is $\alpha_j w_j$.

With a simple multiplication of the coefficients by the constant α_j, we get a linear filter. The algorithm is:

1 Compute the wavelet transform of the data. We get w_j.

2 Estimate the standard deviation of the noise B_0 of the first plane from the histogram of w_0. As we process oversampled images, the values of the wavelet image corresponding to the first scale (w_0) are due mainly to the noise. The histogram shows a Gaussian peak around 0. We compute the standard deviation of this Gaussian function, with a 3-sigma clipping, rejecting pixels where the signal could be significant.

3 Set i to 0.

4 Estimate the standard deviation of the noise B_j from B_0. This is achieved from study of the variation of the noise between two scales, with the hypothesis of a white Gaussian noise.

5 $S_j^2 = s_j^2 - B_j^2$ where s_j^2 is the variance of w_j.

6 $\alpha_j = \frac{S_j^2}{S_j^2+B_j^2}$.

7 $W_j = \alpha_j w_j$.

8 Assign $i+1$ to i and go to step 4.

9 Reconstruct the image from W_j.

2.3.3 Hierarchical Wiener filtering

In the above process, we do not use the information between the wavelet coefficients at different scales. We modify the previous algorithm by introducing a prediction w_h of the wavelet coefficient from the upper scale. This prediction could be determined from the regression (Antonini, 1991) between the two scales but better results are obtained when we only set w_h to W_{j+1}. Between the expectation coefficient W_j and the prediction, a dispersion exists which we assume to follow a Gaussian distribution:

$$P(W_j \mid w_h) = \frac{1}{\sqrt{2\pi}T_j} e^{-\frac{(W_j - w_h)^2}{2T_j^2}} \tag{2.40}$$

The relation which gives the coefficient W_j knowing w_j and w_h is:

$$P(W_j \mid w_j \text{ and } w_h) = \frac{1}{\sqrt{2\pi}\beta_j} e^{-\frac{(W_j - \alpha_j w_j)^2}{2\beta_j^2}} \frac{1}{\sqrt{2\pi}T_j} e^{-\frac{(W_j - w_h)^2}{2T_j^2}} \tag{2.41}$$

with:

$$\beta_j^2 = \frac{S_j^2 B_j^2}{S^2 + B_j^2} \tag{2.42}$$

and:

$$\alpha_j = \frac{S_j^2}{S_j^2 + B_j^2} \tag{2.43}$$

This follows a Gaussian distribution with a mathematical expectation:

$$W_j = \frac{T_j^2}{B_j^2 + T_j^2 + Q_j^2} w_j + \frac{B_j^2}{B_j^2 + T_j^2 + Q_j^2} w_h \tag{2.44}$$

with:

$$Q_j^2 = \frac{T_j^2 B_j^2}{S_j^2} \tag{2.45}$$

W_j is the barycenter (center of gravity) of the three values w_j, w_h, 0 with the weights T_j^2, B_j^2, Q_j^2. The particular cases are:

- If the noise is large ($S_j \ll B_j$) and even if the correlation between the two scales is good (T_j is low), we get $W_j \rightarrow 0$.
- If $B_j \ll S_j \ll T$ then $W_j \rightarrow w_j$.
- If $B_j \ll T_j \ll S$ then $W_j \rightarrow w_j$.
- If $T_j \ll B_j \ll S$ then $W_j \rightarrow w_h$.

At each scale, by changing all the wavelet coefficients w_j of the plane by the estimate value W_j, we get a Hierarchical Wiener Filter. The algorithm is:

1 Compute the wavelet transform of the data. We get w_j.
2 Estimate the standard deviation of the noise B_0 of the first plane from the histogram of w_0.
3 Set j to the index associated with the last plane: $j = n$.
4 Estimate the standard deviation of the noise B_j from B_0.
5 $S_j^2 = s_j^2 - B_j^2$ where s_j^2 is the variance of w_j.
6 Set w_h to W_{j+1} and compute the standard deviation T_j of $w_j - w_h$.
7 $W_j = \frac{T_j^2}{B_j^2+T_j^2+Q_j^2} w_j + \frac{B_j^2}{B_j^2+T_j^2+Q_j^2} w_h$
8 $j = j - 1$. If $i > 0$ go to step 4.
9 Reconstruct the image.

2.3.4 Adaptive filtering

Filtering from significant coefficients. In the preceding algorithm we have assumed the properties of the signal and the noise to be stationary. The wavelet transform was first used to obtain an algorithm which is faster than classical Wiener filtering. Subsequently we took into account the correlation between two different scales. In this way we got a filtering with stationary properties. In fact, these hypotheses were too simple, because in general the signal may not arise from a Gaussian stochastic process. Knowing the noise distribution, we can determine the statistically significant level at each scale of the measured wavelet coefficients. If $w_j(x)$ is very weak, this level is not significant and could be due to noise. Then the hypothesis that the value $W_j(x)$ is null is not ruled out. In the opposite case where $w_j(x)$ is significant, we keep its value.

It has been seen in section 2.1 how significant wavelet coefficients are detected in an image. Reconstruction, after setting non-significant coefficients to zero, at full resolution leads to adaptive filtering (Starck and Bijaoui, 1994a). The restored image is

$$\tilde{I}(x,y) = c_p(x,y) + \sum_{j=1}^{p} M(j,x,y)w_j(x,y) \tag{2.46}$$

where M is as defined in eqn. (2.22).

This noise-related thresholding may be compared with Donoho's work (Donoho and Johnstone, 1993). A difference between this approach and our's is the wavelet transform used; the use of the à trous algorithm allows artifacts to be suppressed, which would arise due to decimation related to orthogonal wavelet transforms.

Iterative filtering from significant coefficients. In the method just described, we obtain an image $\tilde{I}$ by reconstructing the thresholded coefficients. A satisfactory filtering implies that the error image $E = I - \tilde{I}$, obtained as the difference between the original image and the filtered image, contains only noise and no 'structure'. Such is not the case in practice with the approach described. However, we can easily arrive at this objective by iterating a few times:

1 $n \leftarrow 0$.
2 Initialize the solution, $I^{(0)}$, to zero.
3 Estimate the significance level (e.g. 3-sigma) at each scale.
4 Determine the error, $E^{(n)} = I - I^{(n)}$ (where I is the input image, to be filtered).
5 Determine the wavelet transform of $E^{(n)}$.

6 Threshold: only retain the significant coefficients.
7 Reconstruct the thresholded error image. This yields the image $\tilde{E}^{(n)}$ containing the significant residuals of the error image.
8 Add this residual to the solution: $I^{(n)} \leftarrow I^{(n)} + \tilde{E}^{(n)}$.
9 If $\mid (\sigma_{E^{(n-1)}} - \sigma_{E^{(n)}})/\sigma_{E^{(n)}} \mid > \epsilon$ then $n \leftarrow n+1$ and go to step 4.
10 $I^{(n)}$ contains the filtered image, and $I - I^{(n)}$ is our estimation of the noise.

At each iteration, we extract the residual image of significant structures and we introduce them into the solution. We generally used between 6 and 10 iterations. On termination, we are certain that there are no further significant structures in the residual images.

If the noise justifies it, variance stabilization or an analogous transformation may be used as appropriate.

Iterative filtering from a multiresolution support. From the iterative algorithm described in the preceding section, we reconstruct a filtered image $\tilde{I}$ such that, for all pixels, we have

$$\mid I(x, y) - \tilde{I}(x, y) \mid < k\sigma_I \tag{2.47}$$

where σ_I is the standard deviation of the noise contained in the image. This filtering is effective, but does not always correspond to what is wanted. In astronomy, for example, we would prefer not to touch a pixel if it generates a significant coefficient at any one scale. In general, we say that if a multiresolution coefficient of the original image is significant (i.e. $\mid w_j^{(I)}(x, y) \mid > K$, where K is the significance threshold), then the multiresolution coefficient of the error image (i.e. $w_j^{(E^{(n)})}$) must satisfy the following exactly:

$$w_j^{(E^{(n)})}(x, y) = 0 \quad \text{if} \quad \mid w_j^{(I)}(x, y) \mid > K \tag{2.48}$$

To arrive at this objective, we use the multiresolution support of the image, and the algorithm becomes:

1 $n \leftarrow 0$.
2 Initialize the solution, $I^{(0)}$, to zero.
3 Determine the multiresolution support of the image.
4 Estimate the significance level (e.g. 3-sigma) at each scale.
5 Determine the error, $E^{(n)} = I - I^{(n)}$ (where I is the input image, to be filtered).
6 Determine the multiresolution transform of $E^{(n)}$.
7 Threshold: only retain the coefficients which belong to the support.
8 Reconstruct the thresholded error image. This yields the image $\tilde{E}^{(n)}$ containing the significant residuals of the error image.

9 Add this residual to the solution: $I^{(n)} \leftarrow I^{(n)} + \tilde{E}^{(n)}$.
10 If $\mid (\sigma_{E^{(n-1)}} - \sigma_{E^{(n)}})/\sigma_{E^{(n)}} \mid > \epsilon$ then $n \leftarrow n + 1$ and go to step 4.

Thus the regions of the image which contain significant structures at any level are not modified by the filtering. The residual will contain the value zero over all of these regions. The support can also be enriched by any available a priori knowledge. For example, if artifacts exist around objects, a simple morphological opening of the support can be used to eliminate them.

When the noise associated with image I is Poisson, we can apply Anscombe's transformation and this is discussed in the next subsection. It typifies what we need to do to handle other types of image noise properties.

Filtering based on Poisson noise. If the noise associated with image I is Poisson, the following transformation acts as if the data came from a Gaussian process with a noise of standard deviation 1, subject to a sufficiently large mean value of image I:

$$T(I(x, y)) = 2\sqrt{I(x, y) + 3/8} \tag{2.49}$$

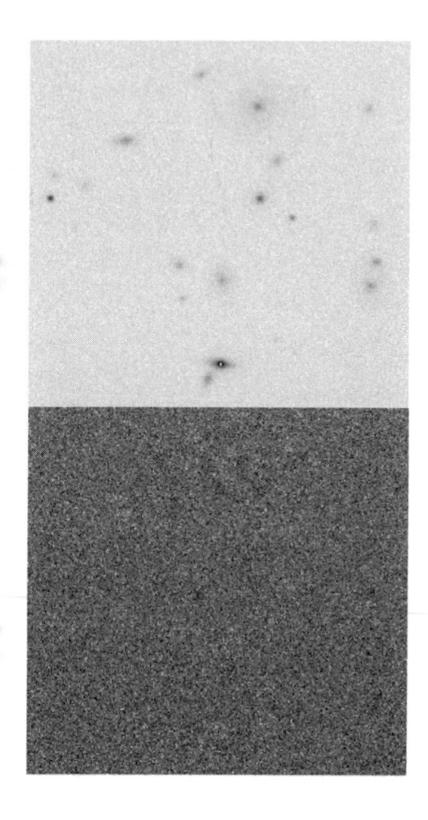

Figure 2.8 Simulated image (top left), simulated image and Gaussian noise (top right), filtered image (bottom left), and residual image (bottom right).

Therefore the noise contained in $e^{(n)} = T(I) - T(I^{(n)})$ can be suppressed using the same principle as the suppression of noise in $E^{(n)} = I - I^{(n)}$. Image $e^{(n)}$ is decomposed into multiresolution coefficients (in the case of the multiresolution strategy), and only the significant coefficients, or the coefficients associated with the multiresolution support, are retained. The support is, of course, determined from $T(I)$ and not from I. Reconstruction then gives $\tilde{e}^{(n)}$. We have the following relations:

$$e^{(n)}(x,y) = T(I(x,y)) - T(I^{(n)}(x,y))$$

$$E^{(n)}(x,y) = I(x,y) - I^{(n)}(x,y)$$

Hence we have

$$[T(I(x,y))]^2 = [e^{(n)}(x,y) + T(I^{(n)}(x,y))]^2$$

$$= (e^{(n)}(x,y))^2 + 4\left(I^{(n)}(x,y) + \frac{3}{8}\right) + 4e^{(n)}(x,y)\sqrt{I^{(n)}(x,y) + \frac{3}{8}}$$

Figure 2.9 Simulated image.

and

$$[T(I(x,y))]^2 = 4\left(I(x,y) + \frac{3}{8}\right)$$

From these two equations, we deduce that $I(x, y)$ can be expressed by:

$$I(x,y) =$$

$$\frac{1}{4}\left[(e^{(n)}(x,y))^2 + 4\left(I^{(n)}(x,y) + \frac{3}{8}\right) + 4e^{(n)}(x,y)\sqrt{I^{(n)}(x,y) + \frac{3}{8}}\right] - \frac{3}{8}$$

Now, replacing I by its expression in $E^{(n)} = I(x,y) - I^{(n)}(x,y)$, we have:

$$E^{(n)}(x,y) = e^{(n)}(x,y)\left[\frac{e^{(n)}(x,y)}{4} + \sqrt{I^{(n)}(x,y) + \frac{3}{8}}\right]$$

Filtering $e^{(n)}$ by thresholding non-significant coefficients, or coefficients which are not contained in the support, we obtain $\tilde{e}^{(n)}$, and we

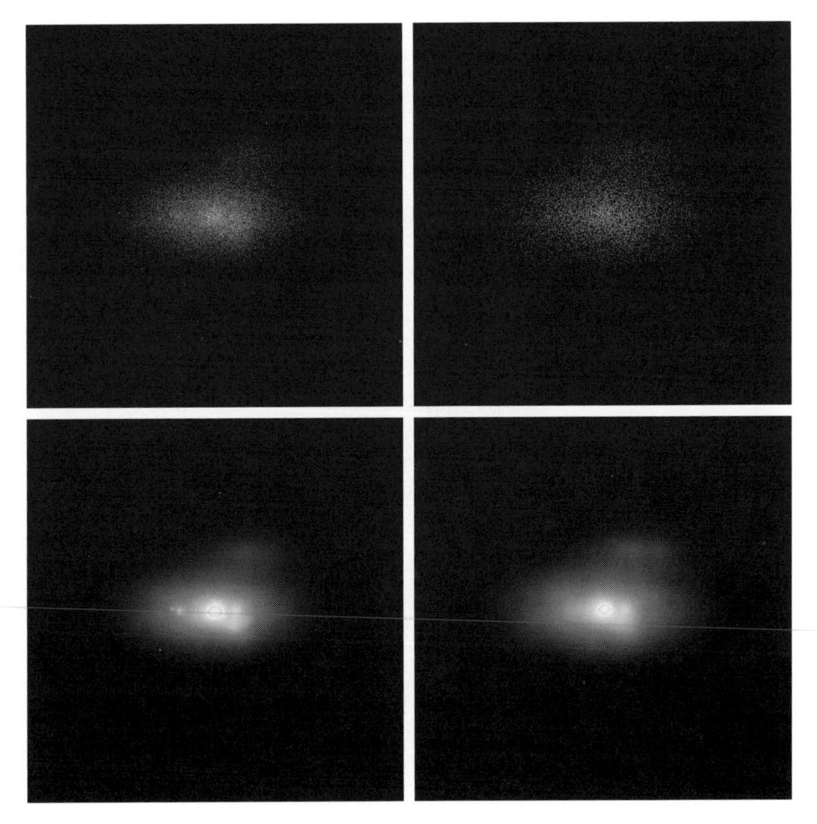

Figure 2.10 Simulated images (top left and right) with two different signal-to-noise ratios, and filtered images (bottom left and right).

then have

$$\tilde{E}^{(n)}(x,y) = \tilde{e}^{(n)}(x,y)\left[\frac{\tilde{e}^{(n)}(x,y)}{4} + \sqrt{I^{(n)}(x,y) + \frac{3}{8}}\right]$$

While this section deals with Poisson noise, the case of combined Poisson and Gaussian noise is handled in a similar way and we have

$$\tilde{E}^{(n)}(x,y) = \tilde{e}^{(n)}(x,y)\left[\frac{\alpha\tilde{e}^{(n)}(x,y)}{4} + \sqrt{\alpha I^{(n)}(x,y) + \frac{3}{8}\alpha^2 + \sigma^2 - \alpha g}\right]$$

where α is the gain, and g and σ are respectively the mean and the standard-deviation of the Gaussian component of the noise.

2.3.5 Examples

Simulation 1: image with Gaussian noise. A simulated image containing stars and galaxies is shown in Fig. 2.8 (top left). The simulated noisy image, the filtered image and the residual image are respectively

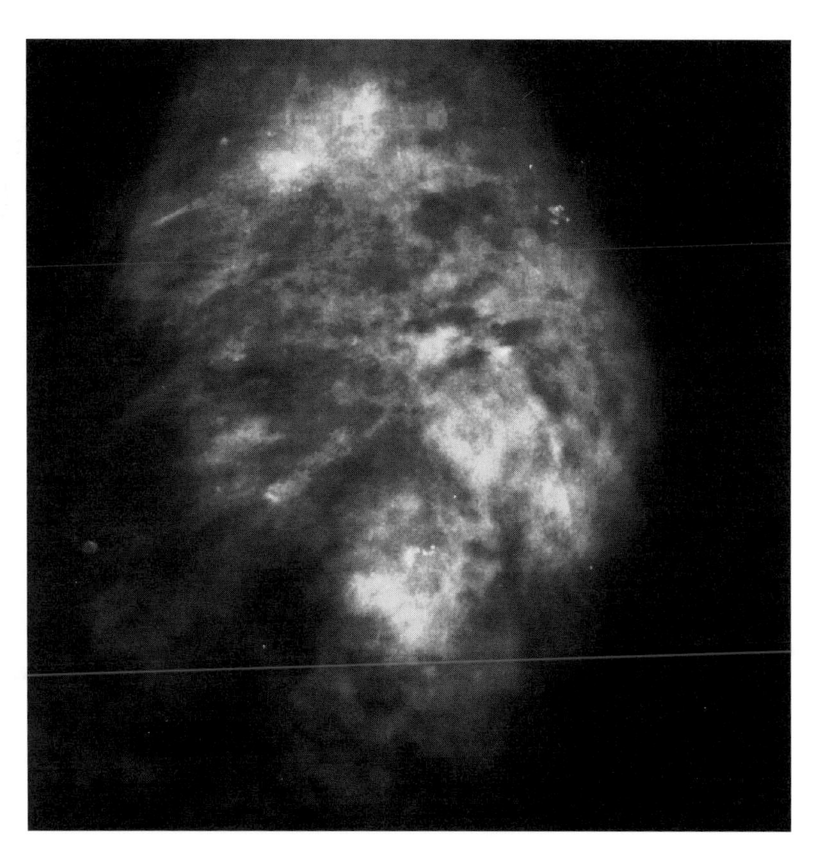

Figure 2.11 Mammographic image.

shown in Fig. 2.8 top right, bottom left, and bottom right. We can see that there is no structure in the residual image. The filtering was carried out using the multiresolution support.

Simulation 2: image with Poisson noise. Figure 2.9 shows a simulated image of a galaxy cluster. Two point sources are superimposed (on the left of the cluster), a cooling flow is at the center, a substructure on its left, and a group of galaxies at the top. From this image, two 'noisy' images are created, which correspond to two different signal-to-noise ratios (see Fig. 2.10 top). In both cases, the background level corresponds to 0.1 events per pixel. This corresponds typically to X-ray cluster observations. In the first noisy image, the maximum is equal to 23 events, and in the second, the maximum is 7. The background is not very relevant. The problem in this kind of image is the weak number of photons per object. It is really difficult to extract any information from them. Figure 2.10, bottom left and right, shows the filtering of images shown, top left and right respectively. Even if the

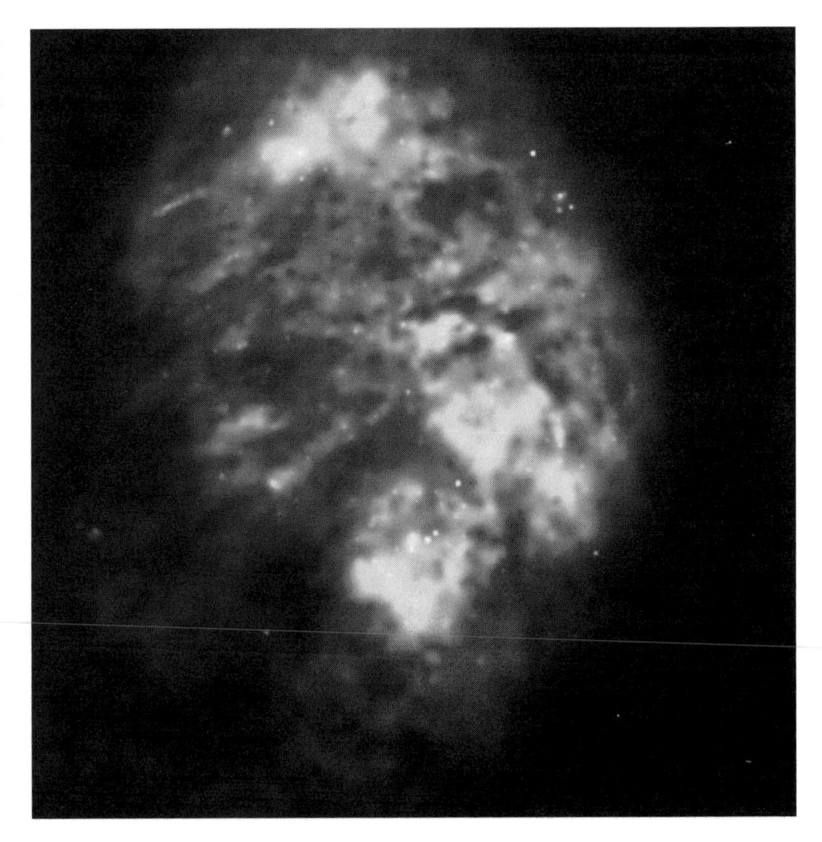

Figure 2.12 Noise-filtered mammographic image.

two point sources could not have been distinguished by eye in the noisy image, they were detected in the first case and correctly restored. In the second, the signal-to-noise was too faint and they have not been restored. But all other important structures are quite well restored.

Mammography image filtering. Figure 2.11 shows a radiological image used in testing for microcalcifications as symptoms of breast cancer. Information was not available on the image's noise characteristics, and in addition the wispy structures associated with faint tissue provided clutter which made more difficult the process of finding the small local (possibly grouped) points of greater intensity. For these reasons we assumed additive non-homogeneous noise. The pyramidal median transform was used in this case. This was motivated by its robust properties, designed to counteract rather than to faithfully reproduce the faint image features associated with multiple levels of tissue. The iterative multiresolution thresholding approach described in this chapter was used, to refine the initial estimate of the noise. The noise-filtered output image is shown in Fig. 2.12. Smoother structures are evident, as are also the greatly improved candidate set of microcalcifications.

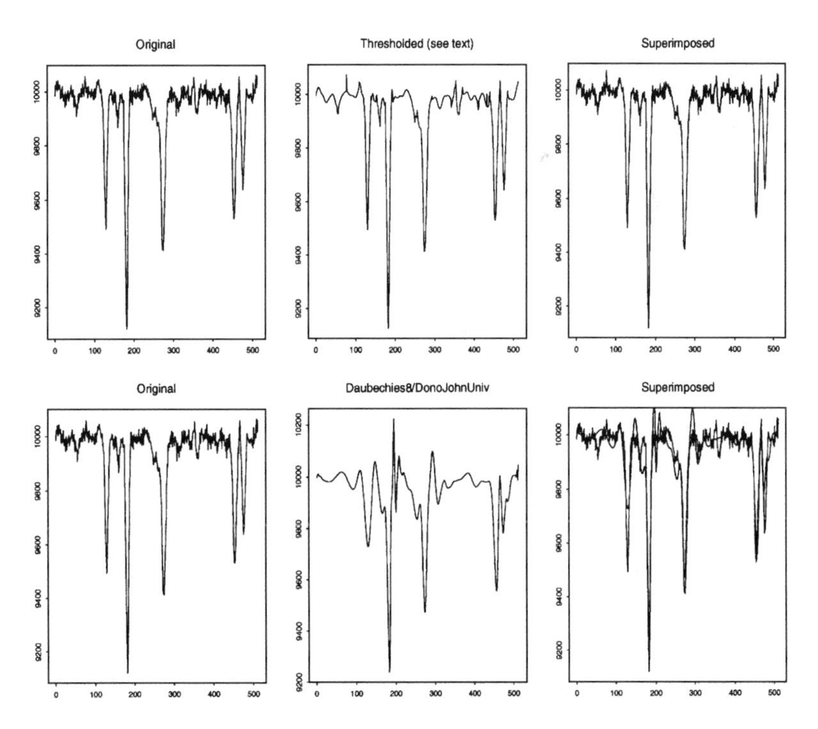

Figure 2.13 Top row: original noisy spectrum; filtered spectrum; both superimposed. Bottom row: original; filtered (using Daubechies coefficient 8, and Donoho and Johnstone 'universal' thresholding); both superimposed.

Spectrum filtering. Fig. 2.13 shows a noisy spectrum (upper left, repeated lower right). For the astronomer, the spectral lines – here mainly absorption lines extending downwards – are of interest. The continuum may also be of interest, i.e. the overall spectral tendency. The spectral lines are unchanged in the filtered version (upper center, and upper right). To illustrate the damage that can result from another wavelet transform, and another noise suppression policy, the lower center (and lower right) version shows the result of applying Daubechies' (19988) coefficient 8, a compactly-supported orthonormal wavelet. This was followed by thresholding based on estimated variance of the coefficients (Donoho and Johnstone, 1993), but not taking into account the image's noise properties as we have done (see Nason and Silverman, 1994). One sees immediately that a problem- (or image-) driven choice of wavelet and filtering strategy is indispensable.

2.4 Image comparison using a multiresolution quality criterion

In this section, we will briefly explore one application of multiresolution. It is sometimes useful, as in image restoration (or deblurring: see Chapter 3) where we want to evaluate the quality of the restoration, to compare images with an objective comparison criterion.

The correlation between the original image $I(i, j)$ and the restored

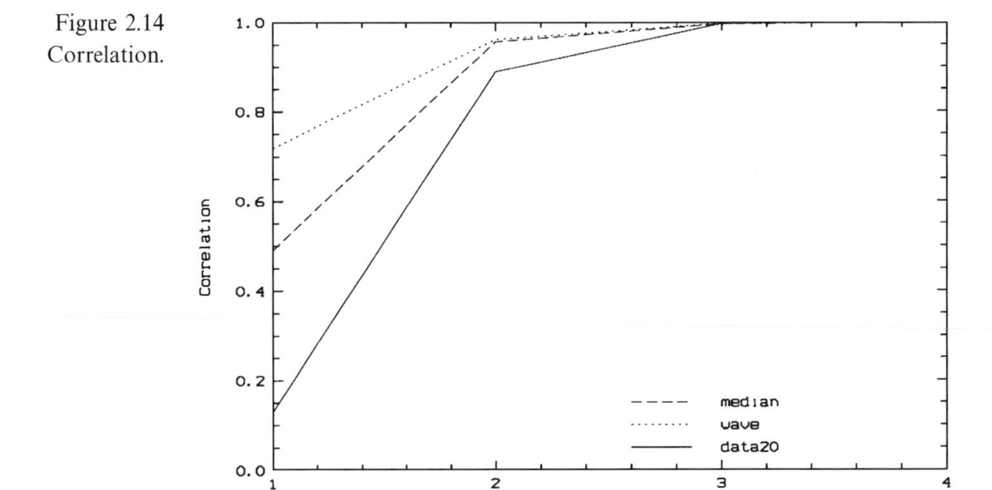

Figure 2.14 Correlation.

one $\tilde{I}(i,j)$ provides a classical criterion. The correlation coefficient is:

$$\rho = \frac{\sum_{i=1}^{N}\sum_{j=1}^{N} I(i,j)\tilde{I}(i,j)}{\sqrt{\sum_{i=1}^{N}\sum_{j=1}^{N} I^2(i,j)\sum_{i=1}^{N}\sum_{j=1}^{N} \tilde{I}^2(i,j)}} \tag{2.50}$$

The correlation is 1 if the images are identical, and less if some differences exist. Another way to compare two pictures is to determine the mean square error:

$$E_{\mathrm{ms}}^2 = \frac{1}{N^2}\sum_{i=1}^{N}\sum_{j=1}^{N}(I(i,j) - \tilde{I}(i,j))^2 \tag{2.51}$$

E_{ms}^2 can be normalized by:

$$E_{\mathrm{nms}}^2 = \frac{\sum_{i=1}^{N}\sum_{j=1}^{N}(I(i,j) - \tilde{I}(i,j))^2}{\sum_{i=1}^{N}\sum_{j=1}^{N} I^2(i,j)} \tag{2.52}$$

The signal-to-noise ratio (SNR) corresponding to the above error is:

$$\mathrm{SNR}_{\mathrm{dB}} = 10\log_{10}\frac{1}{E_{\mathrm{nms}}^2} \tag{2.53}$$

in units of decibels (dB).

These criteria are of limited use since they give no information on the resulting resolution. A more comprehensive criterion must take resolution into account. We can compute for each dyadic scale the correlation coefficient and the quadratic error between the wavelet transforms of the original and the restored images. Hence we can compare, for each resolution, the quality of the restoration.

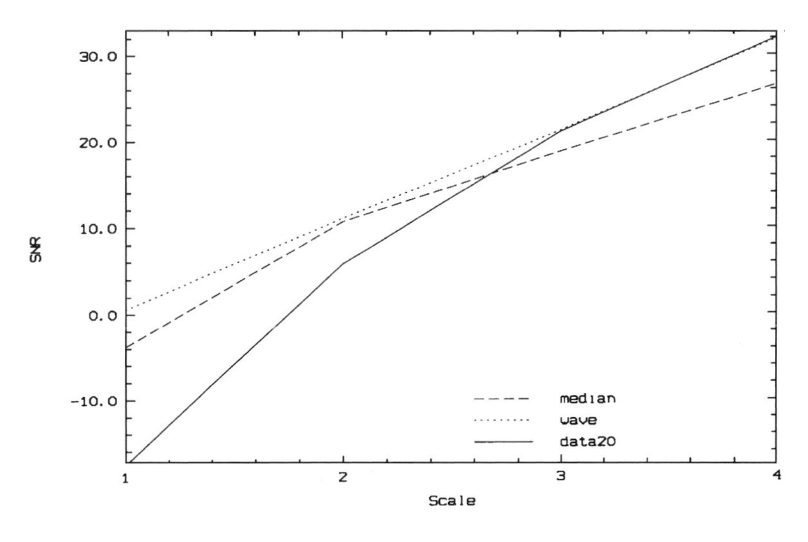

Figure 2.15
Signal-to-noise ratio.

Figures 2.14 and 2.15 show the comparison of three images with a reference image. *Data20* is a simulated noisy image, *median* and *wave* are the output images after respectively applying a median filter, and a thresholding in the wavelet space. These curves show that the thresholding in the wavelet space is better than the median at all scales.

3 Deconvolution

3.1 Introduction to deconvolution

Consider an image characterized by its intensity distribution (the 'data') $I(x, y)$, corresponding to the observation of a 'real image' $O(x, y)$ through an optical system. If the imaging system is linear and shift-invariant, the relation between the data and the image in the same coordinate frame is a convolution:

$$I(x, y) = (O * P)(x, y) + N(x, y) \tag{3.1}$$

where $P(x, y)$ is the point spread function (PSF) of the imaging system, and $N(x, y)$ is additive noise. In practice $O * P$ is subject to non-stationary noise which one can tackle by simultaneous object estimation and restoration (Katsaggelos, 1991). The issue of more extensive statistical modeling will not be further addressed here (see Llacer and Núñez, 1990; Lorenz and Richter, 1993; Molina, 1994), beyond noting that multiresolution frequently represents a useful framework, allowing the user to introduce a priori knowledge of objects of interest.

In Fourier space we have:

$$\hat{I}(u, v) = \hat{O}(u, v)\hat{P}(u, v) + \hat{N}(u, v) \tag{3.2}$$

We want to determine $O(x, y)$ knowing $I(x, y)$ and $P(x, y)$. This inverse problem has led to a large amount of work, the main difficulties being the existence of: (i) a cut-off frequency of the point spread function, and (ii) the additive noise (see for example Cornwell, 1989).

Equation (3.1) is usually in practice an ill-posed problem. This means that there is not a unique solution. If the noise is modeled as a Gaussian or Poisson process, then an iterative approach for computing maximum likelihood estimates may be used. Van Cittert

(1931) restoration involves:

$$O^{(n+1)}(x,y) = O^{(n)}(x,y) + \alpha(I(x,y) - (P * O^{(n)})(x,y)) \quad (3.3)$$

where α is a convergence parameter generally taken as 1. In this equation, the object distribution is modified by adding a term proportional to the residual. The algorithm generally diverges in the presence of noise.

The one-step gradient method is provided by the minimization of the norm $\| I(x,y) - (P * O)(x,y) \|$ (Landweber, 1951) and leads to:

$$O^{(n+1)}(x,y) = O^{(n)}(x,y) + \alpha P^*(x,y) * [I(x,y) - (P * O^{(n)})(x,y)] \quad (3.4)$$

where $P^*(x,y) = P(-x,-y)$. P^* is the transpose of the point spread function, and $O^{(n)}$ is the current estimate of the desired 'real image'.

This method is more robust than Van Cittert's. The conjugate gradient method provides a faster way to minimize this norm with a somewhat more complex algorithm.

The Richardson-Lucy method (Lucy, 1974; Richardson, 1972; see also Adorf, 1992; Katsaggelos, 1991) can be derived from Bayes' theorem on conditional probabilities. Given additive Poisson noise, Shepp and Vardi (1982) showed that a maximum likelihood solution was obtained, by use of an expectation-maximization algorithm. Richardson-Lucy image restoration leads to:

$$\begin{aligned} O^{(n+1)} &\longleftarrow O^{(n)}[(I/I^{(n)}) * P^*] \\ I^{(n)} &\longleftarrow P * O^{(n)} \end{aligned} \quad (3.5)$$

This method is commonly used in astronomy. Flux is preserved and the solution is always positive. The positivity of the solution can be obtained too with Van Cittert's and the one-step gradient methods by thresholding negative values in $O^{(n)}$ at each iteration. However all these methods have a severe drawback: noise amplification, which prevents the detection of weak objects, and leads to false detections. To resolve these problems, some constraints must be added to the solution (positivity is already one such constraint, but it is not enough). The addition of such constraints is called regularization. Several regularization methods exist: Tikhonov's regularization (Tikhonov *et al.*, 1987) consists of minimizing the term:

$$\| I(x,y) - (P * O)(x,y) \| + \lambda \| H * O \| \quad (3.6)$$

where H corresponds to a high-pass filter. This criterion contains two terms. The first, $\| I(x,y) - P(x,y) * O(x,y) \|$, expresses fidelity to the data $I(x,y)$, and the second, $\lambda \| H * O \|$, expresses smoothness of the restored image. λ is the regularization parameter and represents the trade-off between fidelity to the data and the smoothness of the restored image. Finding the optimal value λ necessitates use of numeric techniques such as cross-validation (Galatsanos and Katsaggelos, 1992; Golub, Heath and Wahba, 1979). This method works well, but computationally it is relatively lengthy and produces smoothed images. This second point can be a real problem when we seek compact structures such as is the case in astronomical imaging. Other methods have been proposed. The most popular is certainly the maximum entropy method. It generally produces images which are overly smooth, but some recent work seems to show that this drawback can be avoided.

In the first part of this chapter we present a way to regularize the classic iterative methods such as Van Cittert, one-step gradient, or Richardson-Lucy, by applying a multiresolution support constraint to the solution. We show that this constraint is strong enough to regularize the deconvolution of images dominated by the noise, even if the deconvolution method normally diverges (i.e. Van Cittert). The method consists of searching for information only at positions and at scales where signal has been detected in the data, or in the residual. A range of examples illustrates the powerfulness of the method. In the second part of the chapter, we show that the wavelet transform can enter into the framework of entropic methods, and permits estimation of the different parameters (α, model, etc.). Furthermore by introducing the multiresolution support in the definition of the entropy, significant features can be preserved from the regularization. In the last part of the chapter, the special case of deconvolution in aperture synthesis is analyzed, and in particular the CLEAN method and its use in wavelet space.

3.2 Regularization using the multiresolution support

3.2.1 Noise suppression based on the wavelet transform

We have noted how, in using an iterative deconvolution algorithm such as Van Cittert or Richardson-Lucy, we define $R^{(n)}(x, y)$, the residual at iteration n:

$$R^{(n)}(x, y) = I(x, y) - (P * O^{(n)})(x, y) \tag{3.7}$$

By using the à trous wavelet transform algorithm (Bijaoui, Starck and Murtagh, 1994b; Starck, 1992, 1993), $R^{(n)}$ can be defined as the sum of its p wavelet scales and the last smooth array:

$$R^{(n)}(x,y) = c_p(x,y) + \sum_{j=1}^{p} w_j(x,y) \tag{3.8}$$

where the first term on the right is the last smoothed array, and w denotes a wavelet scale.

The wavelet coefficients provide a mechanism to extract only the significant structures from the residuals at each iteration. Normally, a large part of these residuals is statistically non-significant. The significant residual (Murtagh and Starck, 1994; Starck and Murtagh, 1994) is then:

$$\bar{R}^{(n)}(x,y) = c_p(x,y) + \sum_{j=1}^{p} T(w_j(x,y))w_j(x,y) \tag{3.9}$$

T is a function which is defined by:

$$T(w) = \begin{cases} 1 & \text{if } w \text{ is significant} \\ 0 & \text{if } w \text{ is non-significant} \end{cases} \tag{3.10}$$

Assuming that the noise follows a given distribution, methods discussed in Chapter 2 allow us to define if w is significant.

3.2.2 Noise suppression based on the multiresolution support

In the approach presented in the preceding section, a wavelet coefficient is significant if it is above a threshold. Therefore a coefficient which is less than this threshold is not considered, even if a significant coefficient had been found at the same scale as this coefficient during previous iterations; and consequently we were justified in thinking that we had found signal at this scale, and at this position. Arising out of this approach, it follows that the wavelet coefficients of the residual image could contain signal, above the set threshold, which is ignored.

In order to conserve such signal, we use the notion of multiresolution support. Whenever we find signal at a scale j and at a position (x,y), we will consider that this position in the wavelet space belongs to the multiresolution support of the image.

Equation (3.9) becomes:

$$\bar{R}^{(n)}(x,y) = c_p(x,y) + \sum_{j=1}^{p} M(j,x,y)\, w_j(x,y) \tag{3.11}$$

An alternative approach was outlined in Murtagh, Starck and Bijaoui (1995) and Starck *et al.* (1995): the support was initialized to zero, and built up at each iteration of the restoration algorithm. Thus in eqn. (3.11) above, $M(j,x,y)$ was additionally indexed by n, the iteration number. In this case, the support was specified in terms of significant pixels at each scale, j; and in addition pixels could become significant as the iterations proceeded, but could not be made non-significant. In practice, we have found both of these strategies to be equally acceptable.

3.2.3 Regularization of Van Cittert's algorithm

Van Cittert's iteration (1931) is:

$$O^{(n+1)}(x,y) = O^{(n)}(x,y) + \alpha R^{(n)}(x,y) \tag{3.12}$$

with $R^{(n)}(x,y) = I^{(n)}(x,y) - (P * O^{(n)})(x,y)$. Regularization using significant structures leads to:

$$O^{(n+1)}(x,y) = O^{(n)}(x,y) + \alpha \bar{R}^{(n)}(x,y) \tag{3.13}$$

The basic idea of our method consists of detecting, at each scale, structures of a given size in the residual $R^{(n)}(x,y)$ and putting them in the restored image $O^{(n)}(x,y)$. The process finishes when no more structures are detected. Then, we have separated the image $I(x,y)$ into two images $\tilde{O}(x,y)$ and $R(x,y)$. $\tilde{O}$ is the restored image, which ought not to contain any noise, and $R(x,y)$ is the final residual which ought not to contain any structure. R is our estimate of the noise $N(x,y)$.

3.2.4 Regularization of the one-step gradient method

The one-step gradient iteration is:

$$O^{(n+1)}(x,y) = O^{(n)}(x,y) + P(-x,-y) * R^{(n)}(x,y) \tag{3.14}$$

with $R^{(n)}(x,y) = I(x,y) - (P * O^{(n)})(x,y)$. Regularization by significant structures leads to:

$$O^{(n+1)}(x,y) = O^{(n)}(x,y) + P(-x,-y) * \bar{R}^{(n)}(x,y) \tag{3.15}$$

3.2.5 Regularization of the Richardson-Lucy algorithm

From eqn. (3.1), we have $I^{(n)}(x,y) = (P * O^{(n)})(x,y)$. Then $R^{(n)}(x,y) = I(x,y) - I^{(n)}(x,y)$, and hence $I(x,y) = I^{(n)}(x,y) + R^{(n)}(x,y)$.
The Richardson-Lucy equation is:

$$O^{(n+1)}(x,y) = O^{(n)}(x,y) \left[\frac{I^{(n)}(x,y) + R^{(n)}(x,y)}{I^{(n)}(x,y)} * P(-x,-y) \right] \tag{3.16}$$

and regularization leads to:

$$O^{(n+1)}(x,y) = O^{(n)}(x,y)\left[\frac{I^{(n)}(x,y)+\bar{R}^{(n)}(x,y)}{I^{(n)}(x,y)} * P(-x,-y)\right] \quad (3.17)$$

3.2.6 Convergence

The standard deviation of the residual decreases until no more significant structures are found. Convergence can be estimated from the residual. The algorithm stops when a user-specified threshold is reached:

$$(\sigma_{R^{(n-1)}} - \sigma_{R^{(n)}})/(\sigma_{R^{(n)}}) < \epsilon \quad (3.18)$$

3.2.7 Examples from astronomy

Application 1: wide-field digitized photograph. Figure 3.1 shows a digitized photographic wide-field image. The upper left is the original; the upper right shows a restored image following 5 accelerated iterations of the Richardson-Lucy method; the lower left shows a restored image following 10 accelerated Richardson-Lucy iterations; and the lower right shows the result of applying the method described in this article. One notices a much more noise-free outcome, which can be helpful for subsequent object detection. It may be noted that low-valued 'moats'

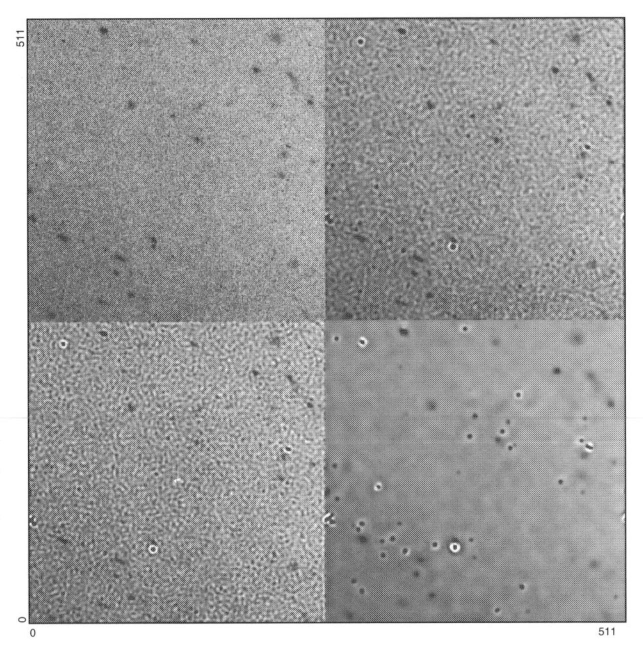

Figure 3.1 Digitized photographic wide-field image. Upper left: original. Upper right, and lower left: 5 and 10 accelerated iterations of the Richardson-Lucy method, respectively. Lower right: cosmic ray hits removed by median smoothing, followed by Richardson-Lucy restoration with noise suppression.

are created around bright objects: this could be prevented by dilating the multiresolution support.

Application 2: simulated Hubble Space Telescope image. A simulated Hubble Space Telescope Wide Field Camera image of a distant cluster of galaxies was used to assess how well the suppression of noise, inherent in the wavelet-based method, aids object detection (Fig. 3.2). The image used was one of a number described in Caulet and Freudling (1993) and Freudling and Caulet (1993). A spatially invariant point spread function was used. This is an approximation to the known spatially varying point spread function, but is not of great importance given the limited image dimensions, 256 × 256. The simulated image allowed us to bypass certain problems, such as cosmic ray hits and charge-coupled device (CCD) detector faults, and to concentrate on the general benefits of regularization.

The procedure followed was to detect objects in the simulated image, and also in the images restored by the wavelet-based (or regularized) Richardson-Lucy method, and the basic Richardson-Lucy method. The INVENTORY package in MIDAS (*Munich Image Data Analysis System*, a large image processing system, developed at the European Southern Observatory) was used for this. INVENTORY

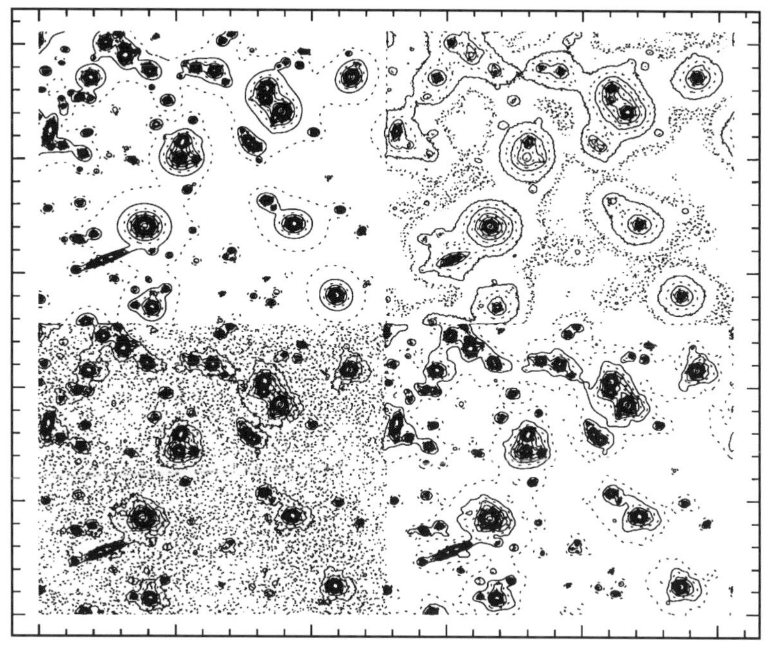

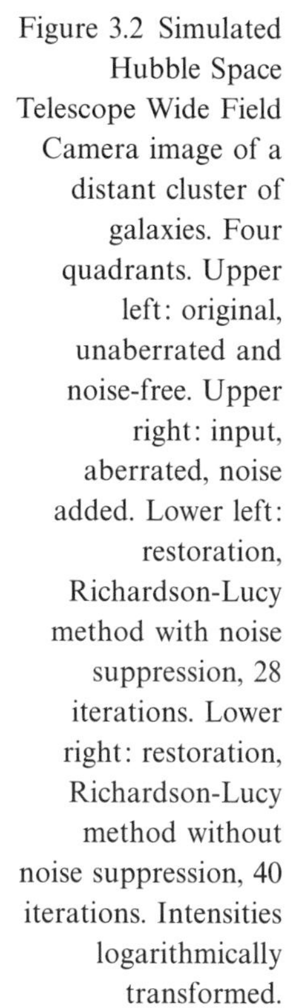

Figure 3.2 Simulated Hubble Space Telescope Wide Field Camera image of a distant cluster of galaxies. Four quadrants. Upper left: original, unaberrated and noise-free. Upper right: input, aberrated, noise added. Lower left: restoration, Richardson-Lucy method with noise suppression, 28 iterations. Lower right: restoration, Richardson-Lucy method without noise suppression, 40 iterations. Intensities logarithmically transformed.

detects objects by means of a local background threshold, which was varied.

A set of 122 objects was found, using INVENTORY, in the original, unaberrated, noise-free image (upper left, Fig. 3.2). This agrees well with the fact that 124 objects were used in the simulation (121 galaxies, 3 stars). With a somewhat different threshold in the case of the wavelet-based Richardson-Lucy method, 165 objects were obtained. With a very much raised threshold (to exclude noise peaks) in the case of the basic Richardson-Lucy method, 159 objects were obtained.

Detections of spurious objects were made in the case of both restorations. Given that we have 'ground truth' in this case, we simply selected the real objects among them. This was done by seeking good matches (less than 1 pixel separation) between objects found in the restored images, and the objects found in the original, unaberrated noise-free image. This led to 69 close matches, in the case of the wavelet-based Richardson-Lucy method; and to 53 close matches, in the case of the basic Richardson-Lucy method.

There was thus a greater number of object detections obtained with the wavelet-based Richardson-Lucy method. These were also more accurate: the mean square error was 0.349 pixel units as against 0.379 for the smaller number of detections obtained from the basic Richardson-Lucy method. For bright objects, photometric plots using aperture magnitudes were relatively similar in both cases; and for fainter objects neither was good. While the wavelet-based Richardson-Lucy method acquited itself well in these respects, its regularization property is clearly advantageous for object detection.

Application 3: Hubble Space Telescope distant galaxy image. Figure 3.3 (left) shows the distant galaxy 4C41.17, scarcely visible with this gray scale look-up table, left of the center, as imaged by the Hubble Space Telescope Wide Field Camera. A 256×256 image was taken, and roughly cleaned of cosmic ray hits. The image suffers from a significant background due to a nearby very bright star. The wavelet-based Richardson-Lucy method does not need to cater for a varying background, since it decides on object significance using *differences* of transformed images. It thus escapes an error-prone background estimation stage in the analysis. A restoration (40 unaccelerated iterations of the wavelet-based Richardson-Lucy method) is shown in the right panel of Fig. 3.3. The faint structures around the galaxy, with some variability, appear in the restored versions of a number of other, different images of the same object. This would seem to substantiate the fact that these faint structures are not noise-related (since noise-related artifacts appear differently in the different restorations). The faint structures around the galaxy appear to be consistent with radio images of this object.

Application 4: point sources. We examine a simulated globular cluster (see short description in Hanisch (1993)), corresponding to the pre-refurbished (before end 1993) Hubble Space Telescope Wide Field/Planetary Camera (WF/PC-1) image. The number of point sources used in the simulation was 467. Their centers are located at sub-pixel locations, which potentially complicates the analysis.

Figure 3.4 (upper left) shows the image following addition of noise and blurring due to the point spread function. Figure 3.4 (upper right) shows the result using multiresolution-based regularization and the generalized Anscombe (Poisson and Gaussian noise) formula. The regularized result following the assumption of Gaussian noise was visually much worse than the regularized result following the assumption of Poisson noise, and the latter is shown in Fig. 3.4 (lower left). Figure 3.4 (lower right) shows the Richardson-Lucy result with noise ignored. In these figures, the contours used were 4, 12, 20 and 28 (and note that the central regions of stars have been purposely omitted). The number of iterations used in all cases was 100. For regularization, a 3σ noise threshold was applied, and 4 levels were used in the à trous wavelet transform. Initialization of the support image also took the known object central positions into account (the practical usefulness of taking central positions, only, was thwarted by the spread of the objects).

The known 467 'truth' locations were used to carry out aperture photometry on these results. A concentric circular aperture of 3 pixels in diameter was used. The INVENTORY package in the MIDAS

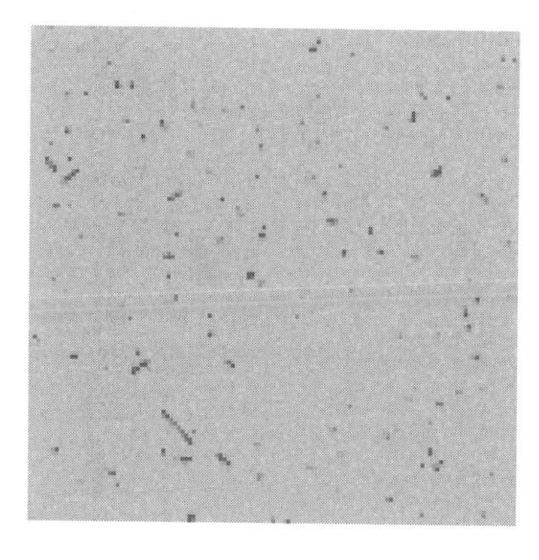

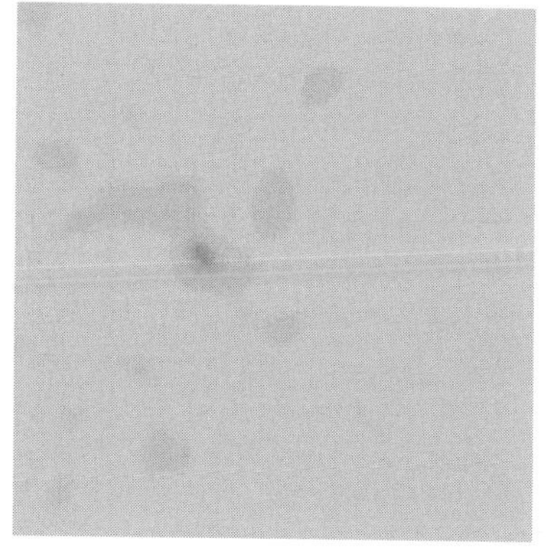

Figure 3.3 Left: Hubble Space Telescope Wide Field Camera image of galaxy 4C41.17; right: subimage, restoration, Richardson-Lucy method with noise suppression, 40 iterations.

image processing system was used for this and the results are shown in Fig. 3.5. The 'Noise ignored' case corresponds to the unregularized Richardson-Lucy restoration. In Fig. 3.5, an arbitrary zero magnitude correction was applied to the differences between derived aperture magnitudes, and the input magnitudes (vertical axes). Also, the horizontal line was drawn by eye in all cases.

The known point source positions were used to carry out aperture photometry. From the 467 objects, there was some loss due to unrecoverable faint objects. A further check on these objects' positions was subsequently made so that they corresponded to the real input information. The numbers of objects found, on which the photometry had been carried out, was respectively for the Poisson/Gauss, Noise ignored, Gaussian, and Poisson cases: 283, 288, 296, 285.

Clearly in Fig. 3.5, removing certain outlying objects would lead to a better linear fit in many instances; and there is no certain way to stop peeling away more and more outlying points. This implies that only a robust regression is appropriate for judging the best fit. We carried

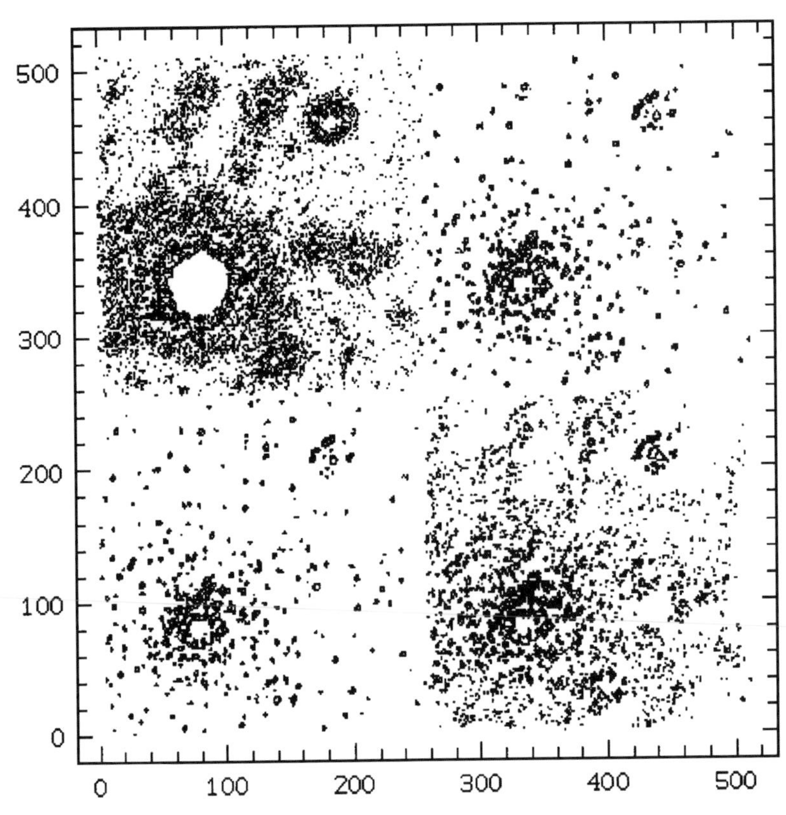

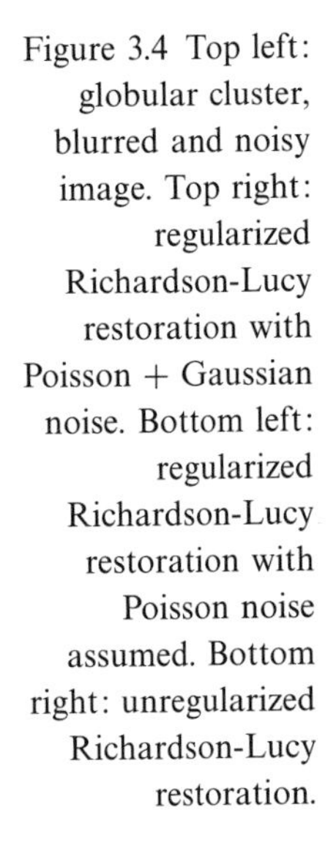
Figure 3.4 Top left: globular cluster, blurred and noisy image. Top right: regularized Richardson-Lucy restoration with Poisson + Gaussian noise. Bottom left: regularized Richardson-Lucy restoration with Poisson noise assumed. Bottom right: unregularized Richardson-Lucy restoration.

out a least median of squares linear regression (Rousseeuw and Leroy, 1987) on the set of input magnitudes versus obtained magnitudes, in all four cases. The usual least squares regression determines a linear fit which minimizes the sum of the squared residuals; the robust method used here minimizes the median of the squared residuals. It has a high breakpoint of almost 50%, i.e. almost half the data can be corrupted in an arbitrary fashion and the regression still follows the majority of the data. The robust version of the correlation coefficient between fitted line of best fit, and the data, was respectively for the Poisson/Gaussian, Noise ignored, Gaussian and Poisson cases: 0.986, 0.959, 0.966 and 0.982. Therefore the Poisson/Gaussian linearity performed best.

An investigation of object detectability now follows. Figure 3.4 gives a visual impression of how difficult it is to determine point source objects in either the unregularized case, or the case which assumes Gaussian noise. In both of these case, high intensity objects can be found, but as the intensity detection threshold is lowered, noise spikes are increasingly obtained.

Table 3.1 shows results obtained by varying the detection threshold. Other software package parameters (e.g. limits on closeness of detections, limits on proximity to image boundary) were not varied.

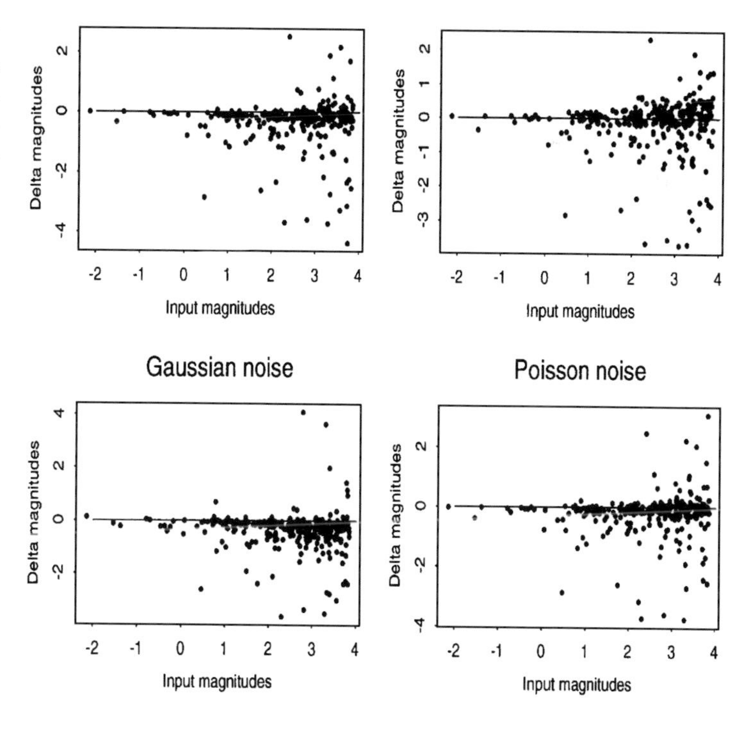

Figure 3.5 Linearity tests, using aperture photometry, for three different noise assumptions and for the unregularized ('noise ignored') Richardson-Lucy restoration. Zero lines are arbitrarily drawn.

The object detection package, INVENTORY, in the MIDAS image analysis system was used. To determine correct objects, the 'ground truth' coordinates of 467 point sources were matched against the object center positions found. For this, the translation and small rotation between every pair of ground-truth-coordinates and coordinates-found was determined, and the modal value of the translation vector and rotation matrix was determined. The bad results of the unregularized method (cf. last two columns of Table 3.1 – in particular, note the counter-intuitive decreases in the number of objects found to be correct) are indicative of noise confusion in finding ground truth coordinates among the obtained coordinates.

Figure 3.6 presents a particular summarization of Table 3.1 (models used: 'P+G': Poisson plus Gaussian; 'G': Gaussian; 'P': Poisson; 'Unreg': unregularized). The clear distinction between the 'P+G' and 'P' cases, on the one hand, and the 'G' and 'Unreg' cases, on the other, can be seen. The results of 'P+G' and 'P' are relatively very similar: 'P+G' is found to have a higher hit rate, at low detection thresholds, leading to both a higher number of correct objects, and a higher false alarm rate.

Application 5: extended object. We used the simulated elliptical galaxy available in the test image suite at anonymous ftp address stsci.edu: /software/stsdas/testdata/restore. This image is referred to there as 'Galaxy Number 2'. It has a simple elliptical shape. The brightness profile includes both bulge and exponential disk components. It has additional distortions introduced in isophote center, ellipticity and position angle. This image was convolved with a Hubble Space Telescope

Table 3.1 *Numbers of objects detected (hits), and numbers of these which were correct, for different noise models.*

	Poiss.+Gauss.		Poisson		Gaussian		Unreg.	
Threshold	Hits	Corr.	Hits	Corr.	Hits	Corr.	Hits	Corr.
30	62	56	52	42	88	80	52	41
20	77	63	58	44	131	87	71	49
10	103	81	90	66	182	76	110	60
5	174	145	147	118	272	191	183	83
2	264	192	223	176	402	231	366	64
1	351	219	304	197	498	219	449	65
0.5	416	226	356	221	559	231	482	17

Wide Field Camera point spread function, and Poisson and readout noise were added.

Under the assumption that the readout noise was small, we used a Poisson model for all noise in the image. We set negative values in the blurred, noisy input image to zero. This was the case in the background only, and was necessitated by the variance stabilizing algorithm used.

Figure 3.7 (left) shows contours formed in the truth image, overplotted with contours yielded by the regularized Richardson-Lucy method. Note that the truth image was not the one used as input for restoration; rather, it was the image on the basis of which the blurred, noisy

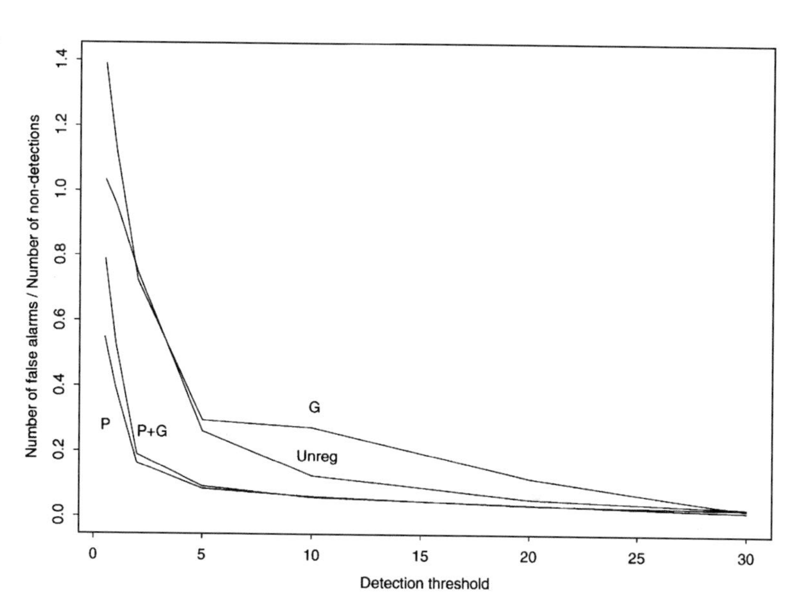

Figure 3.6 Plots of false alarms/non-detections for different noise models. See text for details.

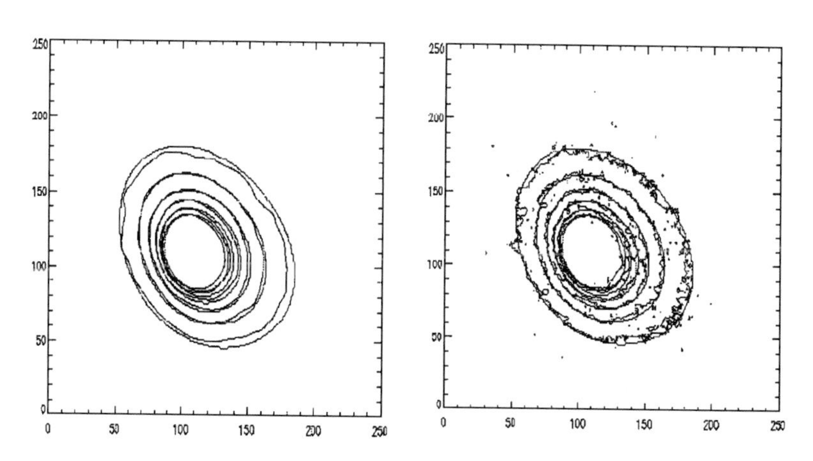

Figure 3.7 Isophotal contours corresponding to (left) 'truth image', and regularized Richardson-Lucy restoration; and (right) 'truth image', and unregularized Richardson-Lucy restoration.

input image was created. For the regularized restoration, a Poisson model was used for clipping wavelet coefficients. A 3σ threshold was chosen, above which (in wavelet space) a value was taken as significant. The multiresolution support algorithm was used, in order to prevent any untoward alteration to the galaxy. The plot shown in Fig. 3.7 (left) corresponds to just 5 iterations (unaccelerated) of the regularized Richardson-Lucy restoration method. Figure 3.7 (right) shows the same isophotes for the truth image, and those obtained by restoration following 5 iterations of the unregularized Richardson-Lucy method. Allowing further iterations (to convergence in the case of the regularized Richardson-Lucy method) yielded quite similar results in the case of the regularized restoration; but in the case of the unregularized restoration, the fitting of a point spread function to every noise spike made for a very unsmooth image.

3.2.8 Conclusion on regularization using the multiresolution support

The use of a multiresolution support has led to a versatile and powerful tool for such image processing tasks as restoration. As a byproduct, it also helps in object detection. The multiresolution support data structure is an important image processing tool.

The wavelet transform used could be replaced with some other multiresolution algorithm. However the à trous algorithm has acquitted itself well. The experimental results demonstrate the usefulness of this broad framework.

The regularized Richardson-Lucy method performed well on the restoration of point sources, and extended objects. The requirement that the user specify input parameters – e.g. a regularizing parameter – is avoided. The coherent framework, based on a stochastic multiresolution image model, is one which the astronomer is familiar with. He/she is asked only to decide on the appropriateness of Gaussian or Poisson assumptions for the analysis of a given image; and the level of the (smoothing-related) threshold, in terms of the standard deviation.

3.3 Multiscale entropy and image restoration

In the field of image deconvolution, one of the most popular techniques is the maximum entropy method (MEM) introduced initially by Burg (1967) and Ables (1974) for spectral analysis. Subsequently it was reformalized and improved to provide an efficient technique for deconvolution by Skilling and Gull (1984), and Skilling and Bryan

(1984). This algorithm is based on the idea that a given realization of a random variable (a non-deterministic signal for instance) carries a certain amount of information quantifiable by the entropy (Jaynes, 1957; Shannon, 1948). Thus, when trying to invert an ill-posed problem like the deconvolution of a signal, with the ill-posedness inherent in the presence of additive noise which is characteristic of an experimental signal, the entropy is used as a regularizing functional to constrain the solution, and give the simplest (in the sense of the amount of information contained) possible solution, which is compatible with the data.

However classical maximum entropy deconvolution gives rise to technical problems such as finding an optimal value of α, the relative weight between the goodness-of-fit and the entropy. It has been observed also that a 'low' value of α favors high frequency reconstructions, but gives a poorly regularized result, while a 'high' α leads to a restored image with good regularization, but in which the high frequency structures are under-reconstructed. Therefore Bontekoe, Koper and Kester (1994) introduced the concept of Pyramid Maximum Entropy reconstruction which is a special application of multi-channel maximum entropy image reconstruction (Gull and Skilling, 1991). They consider an image f as a weighted sum of a visible space pyramid of resolution $f = \sum_i f_i$, $i = 1 \ldots K$, which corresponds via a set of intrinsic correlation functions (ICFs) to a hidden-space pyramid h_i, $i = 1 \ldots K$ on which the constraint of maximum entropy is applied. A major difficulty arises when making the last operation. The channels must be summed, using an arbitrary weight for each. Another difficulty encountered lies in the choice of the default constant (model) in each channel. In order to circumvent these difficulties, we reformulated this idea, using the appropriate mathematical tool to decompose a signal into channels of spectral bands, the wavelet transform. We show that the default value (model) at each wavelet scale is linked physically to the standard deviation of the noise present at this scale. Introducing the concept of multiscale entropy, we show that we minimize a functional, dependent on the desired solution, regularized by minimizing the total amount of information contained at each resolution. We also use the concept of multiresolution support (Starck *et al.*, 1995) which leads to a fixed α for all types of images, removing the problem of its determination. Finally, we show that this method is very simple to use since there is no parameter to be determined by the user. We give examples of deconvolution of blurred astronomical images showing the power of the method, especially for reconstructing weak and strong structures at the same time. We point out that one can derive a very efficient filtering method from this approach.

3.3.1 Image restoration using the maximum entropy method

The data-image relation is

$$I(x, y) = (O * P)(x, y) + N(x, y) \tag{3.19}$$

where $P(x, y)$ is the point spread function of the imaging system, and $N(x, y)$ is additive noise. We want to determine $O(x, y)$ knowing $I(x, y)$ and $P(x, y)$. For this chapter, we consider Gaussian noise but in the case of Poisson, or Poisson plus Gaussian, we can reduce the problem to the Gaussian case using the Anscombe transform and its generalization (see Chapter 2). This inverse problem is ill-posed because of the noise, making the previous system of equations lack a unique solution. The maximum entropy method (MEM) which is a stochastic approach to the problem was initially proposed by Jaynes (1957). Among an infinite number of solutions to eqn. (3.19), it helps to choose the one which maximizes its entropy i.e. minimizes the amount of information contained in the solution. In other words, it is often said that this method gives the simplest solution compatible with the data, I.

Using Bayes' theorem to evaluate the probability of the realization of the original image O, knowing the data I, we have

$$\text{Prob}(O|I) = \frac{\text{Prob}(I|O).\text{Prob}(O)}{\text{Prob}(I)} \tag{3.20}$$

$\text{Prob}(I|O)$ is the conditional probability of getting the data I given an original image O, i.e. it represents the distribution of the noise. It is given, in the case of uncorrelated Gaussian noise with variance σ_I^2, by:

$$\text{Prob}(I|O) = \exp\left(-\sum_{\text{pixels}} \frac{(I - P * O)^2}{2{\sigma_I}^2}\right) \tag{3.21}$$

The eqn. (3.21) denominator is independent of O and is considered as a constant. $\text{Prob}(O)$ is the a priori distribution of the solution O. In the absence of any information on the solution O except its positivity, the maximum entropy principle suggests we take

$$\text{Prob}(O) = \exp(\alpha S(O)) \tag{3.22}$$

where $S(O)$ denotes the entropy of the image O.

Given the data, the most probable image is obtained by maximizing $\text{Prob}(O|I)$, or equivalently by maximizing the product of the two previous equations. Taking the logarithm, we thus need to maximize

$$\ln(\text{Prob}(O|I)) = \alpha S(O) - \sum_{\text{pixels}} \frac{(I - P * O)^2}{2\sigma_I^{\,2}} \tag{3.23}$$

which is a linear combination of two terms: the entropy of the image, and a quantity corresponding to χ^2 in statistics measuring the discrepancy between the data and the predictions of the model.

The solution is found by minimizing

$$J(O) = \sum_{\text{pixels}} \frac{(I - P * O)^2}{2\sigma_I^{\,2}} - \alpha S(O) = \frac{\chi^2}{2} - \alpha S(O) \tag{3.24}$$

where α is a parameter that can be seen alternatively as a Lagrangian parameter or a value fixing the relative weight between the goodness-of-fit and the entropy S. Several entropy definitions have been proposed:

- Burg (1978):

$$S_b = -\sum_{\text{pixels}} \ln(O) \tag{3.25}$$

- Frieden (1975):

$$S_f = -\sum_{\text{pixels}} O \ln(O) \tag{3.26}$$

- Gull and Skilling (1991):

$$S_g = \sum_{\text{pixels}} O - m - O \ln(O/m) \tag{3.27}$$

The last definition of the entropy has the advantage of having a zero maximum when O equals m, but requires the concept of a model, m, which is in practice the value of the background. The determination of the α parameter is not an easy task and in fact it is a very serious problem facing the maximum entropy method. In the historic MAXENT algorithm of Skilling and Gull, the choice of α is such that it must satisfy the ad hoc constraint $\chi^2 = N$ when the deconvolution is achieved, N being the number of degrees of freedom of the system, i.e. the number of pixels in image deconvolution problems. But this choice systematically leads to an under-fitting of the data (Titterington, 1985) which is clearly apparent for imaging problems with little blurring. In reality, the χ^2 statistic is expected to vary in the range $N \pm \sqrt{2N}$ from one data realization to another. In the Quantified Maximum Entropy point of view (Skilling, 1989), the optimum value of α is determined by including its probability $P(\alpha)$ in Bayes' equation and then by maximizing the marginal probability of having α, knowing the data and the model m. In practice, a value of α that is too large gives a resulting image which is too regularized with a large loss of resolution. A value that is too small leads to a poorly regularized solution showing unacceptable artifacts. Taking a flat model of the prior image softens the discontinuities, which may appear unacceptable for astronomical images, as these often contain stars and other point-like objects. Therefore the basic maximum entropy method appears to be not very appropriate for this kind of image which contains high and low spatial frequencies at the same time. Another point to be noted is a ringing effect of the maximum entropy method algorithm, producing artifacts around bright sources.

To solve these problems while still using the maximum entropy concept, some enhancements of the maximum entropy method have been proposed. Noticing that neighboring pixels of reconstructed images with MAXENT could have values differing a lot in expected flat regions (Charter, 1990), Gull and Skilling introduced the concepts of hidden image h and intrinsic correlation function, ICF, C (Gaussian or cubic spline-like) in the Preblur MAXENT algorithm.

The ICF describes a minimum scale length of correlation in the desired image O which is achieved by assuming that

$$O = C * h \tag{3.28}$$

This corresponds to imposing a minimum resolution on the solution O. Since the hidden space image h is not spatially correlated, this can be regularized by the entropy $S_g(h) = \sum h - m - h\ln(\frac{h}{m})$.

Since in astronomical images many scale lengths are present, the *Multi-channel Maximum Entropy Method*, developed by Weir (1991, 1992), uses a set of ICFs having different scale lengths, each defining a channel. The visible-space image is now formed by a weighted sum of the visible-space image channels O_j:

$$O = \sum_{j=1}^{K} p_j O_j \tag{3.29}$$

where K is the number of channels. As in Preblur MAXENT, each solution O_j is supposed to be the result of the convolution between a hidden image h_j with a low-pass filter (ICF) C_j:

$$O_j = C_j * h_j \tag{3.30}$$

But such a method has several drawbacks:

1 The solution depends on the width of the ICFs (Bontekoe *et al.*, 1994).
2 There is no rigorous way to fix the weights p_j (Bontekoe *et al.*, 1994).
3 The computation time increases linearly with the number of pixels.
4 The solution obtained depends on the choice of the models m_j ($j = 1 \ldots K$) which were chosen independently of the channel.

In 1993, Bontekoe *et al.* (1994) used a special application of this method which they called Pyramid Maximum Entropy on infrared image data. The pyramidal approach allows the user to have constant ICF width, and the computation time is reduced. They demonstrated that all weights can be fixed ($p_j = 1$ for each channel).

This method eliminates the first three drawbacks, and gives better reconstruction of the sharp and smooth structures. But in addition to the two last drawbacks, a new one is added: as the images O_j have different sizes (due to the pyramidal approach), the solution O is built by duplicating the pixels of the subimages O_j of each channel. This procedure is known to produce artifacts due to the appearance of high frequencies which are incompatible with the real spectrum of the true image $\hat{O}$.

However this problem can be easily overcome by duplicating the pixels before convolving with the ICF, or expanding the channels using linear interpolation. Thus the introduction of the 'pyramid of resolution' has solved some problems and brought lots of improvements to the classic maximum entropy method, but has also raised other questions. In the following, we propose another way to use the information at different scales of resolution using the mathematical tool, the wavelet transform. We show that the problems encountered by Bontekoe *et al.* are overcome with this approach, especially the reconstruction of the object O which becomes natural. Furthermore, the wavelet transform gives a good framework for noise modeling. This modeling allows significant wavelet coefficients (i.e. not due to the noise) to be preserved during regularization. Regularization becomes adaptive, dependent on both position in the image and scale.

3.3.2 Formalism of maximum entropy multiresolution

Multiscale entropy. The concept of entropy following Shannon's or Skilling and Gull's definition is a global quantity calculated on the whole image O. It is not appropriate for quantifying the distribution of the information at different scales of resolution. Therefore we propose the concept of multiscale entropy of a set of wavelet coefficients $\{w_j\}$:

$$S_{\mathrm{m}}(O) = \frac{1}{\sigma_I^2} \sum_{\text{scales } j} \sum_{\text{pixels}} \sigma_j \left(w_j(x,y) - m_j - |w_j(x,y)| \ln \frac{|w_j(x,y)|}{m_j} \right) \tag{3.31}$$

The multiscale entropy is the sum of the entropy at each scale.

The coefficients w_j are wavelet coefficients, and we take the absolute value of w_j in this definition because the values of w_j can be positive or negative, and a negative signal contains also some information in the wavelet transform. The advantage of such a definition of entropy is the fact we can use previous work concerning the wavelet transform and image restoration (Murtagh *et al.*, 1995; Starck and Bijaoui, 1994a; Starck *et al.*, 1995). The noise behavior has already been studied in the wavelet transform (see Chapter 2) and we can estimate the standard deviation of the noise σ_j at scale j. These estimates can be naturally introduced in our models m_j

$$m_j = k_m \sigma_j \tag{3.32}$$

The model m_j at scale j represents the value taken by a wavelet coefficient in the absence of any relevant signal and, in practice, it must be a small value compared to any significant signal value. Following the Gull and Skilling procedure, we take m_j as a fraction of the noise because the value of σ_j can be considered as a sort of physical limit under which a signal cannot be distinguished from the noise ($k_m = \frac{1}{100}$).

The term σ_j can be considered as a scale-dependent renormalization coefficient.

Multiscale entropy and multiresolution support. If the definition (3.31) is used for the multiscale entropy, the regularization acts on the whole image. We want to fully reconstruct significant structures, without imposing strong regularization, while efficiently eliminating the noise. Thus the introduction of the multiresolution support in another definition of the multiscale entropy leads to a functional that answers these requirements:

$$S_{\mathrm{ms}}(O) = \frac{1}{\sigma_I^2} \sum_{\text{scales } j} \sum_{\text{pixels}} A(j,x,y)\sigma_j \times \left(w_j(x,y) - m_j - |w_j(x,y)| \ln \frac{|w_j(x,y)|}{m_j} \right) \tag{3.33}$$

The A function of the scale j and the pixels (x,y) is $A(j,x,y) = 1 - M(j,x,y)$, i.e. the reciprocal of the multiresolution support M. In order to avoid discontinuities in the A function created by such a coarse threshold of $3\sigma_j$, one may possibly impose some smoothness by convolving it with a B-spline function with a full-width at half maximum (FWHM) varying with the scale j.

The degree of regularization will be determined at each scale j, and at each location (x, y), by the value of the function $A(j, x, y)$. If $A(j, x, y)$ has a value near 1 then we have strong regularization; and it is weak when A is around 0.

The entropy S_{ms} measures the amount of information only at scales and in areas where we have a low signal-to-noise ratio. We will show in the next section how these notions can be tied together to yield efficient methods for filtering and image deconvolution.

3.3.3 Deconvolution using multiscale entropy

Method. We assume that the blurring process of an image is linear. In our tests, the point spread function was space invariant but the method can be extended to space-variant point spread functions.

As in the maximum entropy method, we will minimize a functional of O, but considering an image as a pyramid of different resolution scales in which we try to maximize its contribution to the multiscale entropy. The functional to minimize is

$$J(O) = \sum_{\text{pixels}} \frac{(I - P * O)^2}{2\sigma_I^2} - \alpha S_{\mathrm{ms}}(O) \tag{3.34}$$

Then the final difficulty lies in finding an algorithm to minimize the functional $J(O)$. We use the iterative 'one-step gradient' method due to its simplicity.

The solution is found by computing the gradient

$$\begin{aligned} \nabla(J(O)) = & -P^* * \frac{(I - P * O)}{\sigma_I^2} \\ & + \alpha \frac{1}{\sigma_I^2} \sum_{\text{scale } j} \left[A(j)\sigma_j \operatorname{sgn}(w_j^{(O)}) \ln\left(\frac{\mid w_j^{(O)} \mid}{m_j} \right) \right] * \psi_j^* \end{aligned} \tag{3.35}$$

and performing the following iterative schema

$$O^{n+1} = O^n - \gamma \nabla(J(O^n)) \tag{3.36}$$

Note that the second part of eqn. (3.36) has always a zero mean value due to the fact that each scale is convolved with ψ_j^* (and ψ_j^* has a zero mean value due to the admissibility condition of the wavelet function). Thus the flux is not modified in the object O when applying the iterative scheme (eqn. (3.36)).

The positivity of the restored image can be assured during the iterative process of functional minimization by applying a simple positivity constraint (threshold) to the intermediate solution O^n. The iterative process is stopped when the standard deviation of the residuals shows no significant change (relative variation $\leq 10^{-3}$), and we check that the χ^2 value is in the range $N \pm \sqrt{2N}$ (this was always the case in experiments carried out; otherwise it would mean that the σ_I value is wrong).

Choice of the α parameter. In the classic maximum entropy method, the α parameter quantifies the relative weight between the goodness-of-fit, or χ^2 value, and the degree of smoothness introduced by the entropy. This parameter is generally constant over the whole image and therefore depends on the data (signal+noise). In our case, the degree of regularization applied to the non-significant structures at each wavelet scale j is controlled by the term $\frac{\alpha}{\sigma_I} A(j, x, y)$, and depends therefore on both the scale j and the location (x, y). Regularization is performed only at scales and positions where no signal is detected $(A(j, x, y) = 1)$. Note that the α parameter does not have the same importance as in the classical maximum entropy method: α has only to be high enough to regularize the solution at positions and scales where no signal has been detected. We found experimentally that

$$\alpha = 0.5 \max(\text{PSF}) \tag{3.37}$$

(PSF is the point spread function) produces good results, and for any kind of image.

3.3.4 Experiments

We tested our algorithm with simulated data. The simulated image contains an extended object, and several smaller sources (Fig. 3.8, upper left). It was convolved with a Gaussian point spread function ($\sigma = 2$), and Gaussian noise was added (Fig. 3.8, upper right). The ratio between the maximum in the convolved image and the standard deviation of the noise is 50. The results of the deconvolution by the maximum entropy method, and multiresolution maximum entropy, are shown in Fig. 3.8, lower left and right. The multiresolution maximum entropy method leads to better regularization, and the final resolution in the deconvolved image is better.

In order to give quantitative results on the possibilities of the multiscale maximum entropy method, we carried out tests on a simulated image of the ISOCAM mid-infrared camera on the infrared satellite ISO. The simulation consists of a faint extended object (galaxy, integrated flux = 1 Jy) near a bright star (point-like: 10 Jy): Fig. 3.9, upper left. This was blurred using the field of view point spread function, and noise was added (Fig. 3.9, upper right): the two objects are overlapping and the galaxy is barely detectable. After deconvolution using the multiscale maximum entropy method (see Fig. 3.9, bottom right), the two objects are separated. The restored star has a flux of 10.04 Jy, and the SNR of the reconstructed image is 22.4 dB (SNR $= 10\log_{10}(\sigma^2_{\text{Signal}}/\sigma^2_{\text{Residuals}})$). These results show that multiscale maximum entropy is very effective when we compare it to other methods (see Table 3.2), and prove the reliability of the photometry after deconvolution. It is clear in this example that photometric measurements cannot always be made directly on the data, and a deconvolution is often necessary, especially when objets are overlapping.

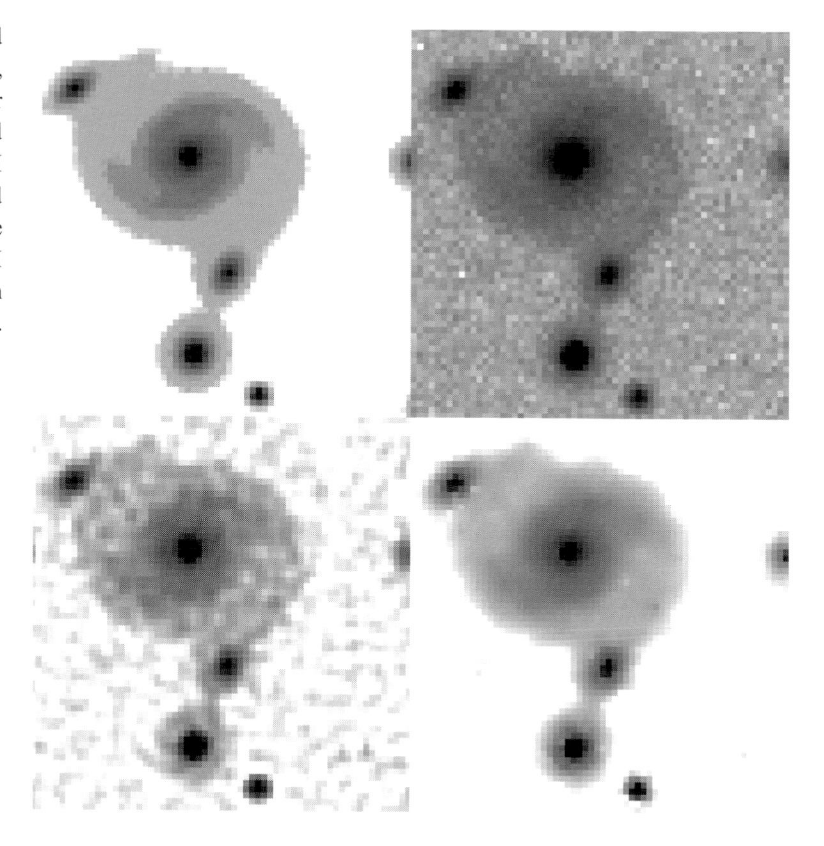

Figure 3.8 Simulated object (upper left), blurred image (upper right), deconvolved image by MEM (lower left), and deconvolved image by MEM multiresolution (lower right).

We tested our deconvolution method on astronomical 64 × 64 pixel images obtained with an mid-infrared camera, TIMMI, placed on the 3.6 ESO telescope (La Silla, Chile). The object studied is the β Pictoris dust disk. The image was obtained by integrating 5 hours on-source. The raw image has a peak signal-to-noise ratio of 80. It is strongly blurred by a combination of seeing, diffraction (0.7 arcsec on a 3m class telescope) and additive Gaussian noise. The initial

Table 3.2 *Quantitative results extracted from three images deconvolved by Lucy's method, maximum entropy, and multiscale maximum entropy. The first row gives the flux of the star. The second, the integrated flux in the extended object, and the last row, the signal-to-noise ratio of the deconvolved images. Since the two objects are mixed in the degraded image, it is impossible to attribute a photometric measure to each.*

	Orig. image	Data	Lucy	MEM	M. MEM
star flux (Jy)	10		4.27	14.1	10.04
ext. obj. flux (Jy)	1		0.33	1.33	0.94
SNR (dB)	∞	1.03	4.51	4.45	22.4

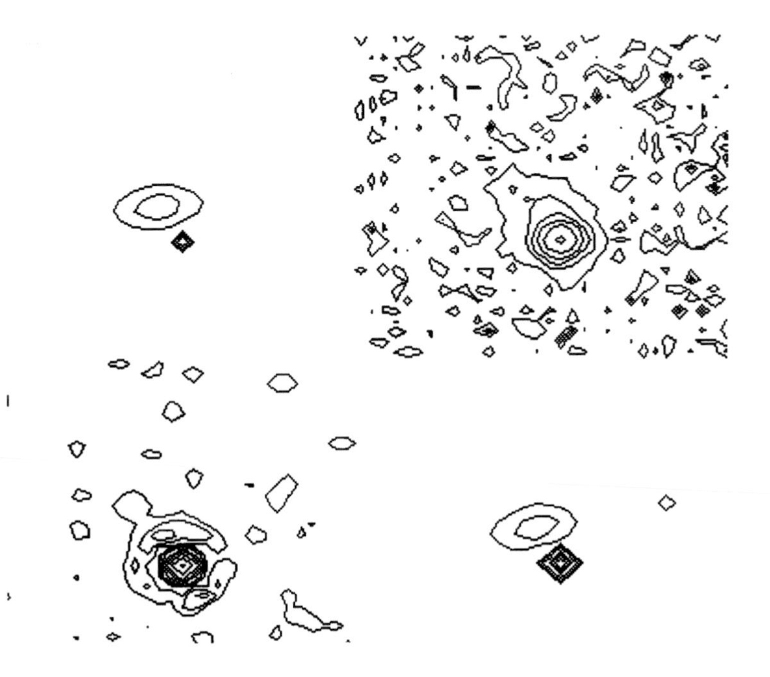

Figure 3.9 Simulated ISOCAM image (upper left), blurred image using the 1.5 arcsec pixel field of view point spread function (upper right), the deconvolved image by MEM (lower left), and deconvolved image by multiscale MEM (lower right).

disk shape in the original image has been lost after the convolution with the point spread function. Thus we need to deconvolve such an image to get the best information on this object, i.e. the exact profile and thickness of the disk, and subsequently to compare the results to models of thermal dust emission (Lagage and Pantin, 1994). We used the multiscale maximum entropy method to deconvolve this image. The algorithm took about thirty iterations to converge. The deconvolved image (Fig. 3.10) shows that the disk is extended at 10 μm and asymmetrical (the right side is more extended than the left side). We compared our method to the standard Richardson-Lucy algorithm which shows poor regularization (see Fig. 3.10, upper right) and an inability to restore faint structures; and also to the classical maximum entropy method. The deconvolved image using the multiscale maximum entropy method proves to be more effective for regularizing than the other standard methods, and leads to a good reconstruction of the faintest structures of the dust disk.

Figure 3.10 Beta Pictoris: raw image (upper left) and deconvolved images using: Richardson-Lucy's method (upper right), classical maximum entropy (lower left), and multiscale maximum entropy (lower right).

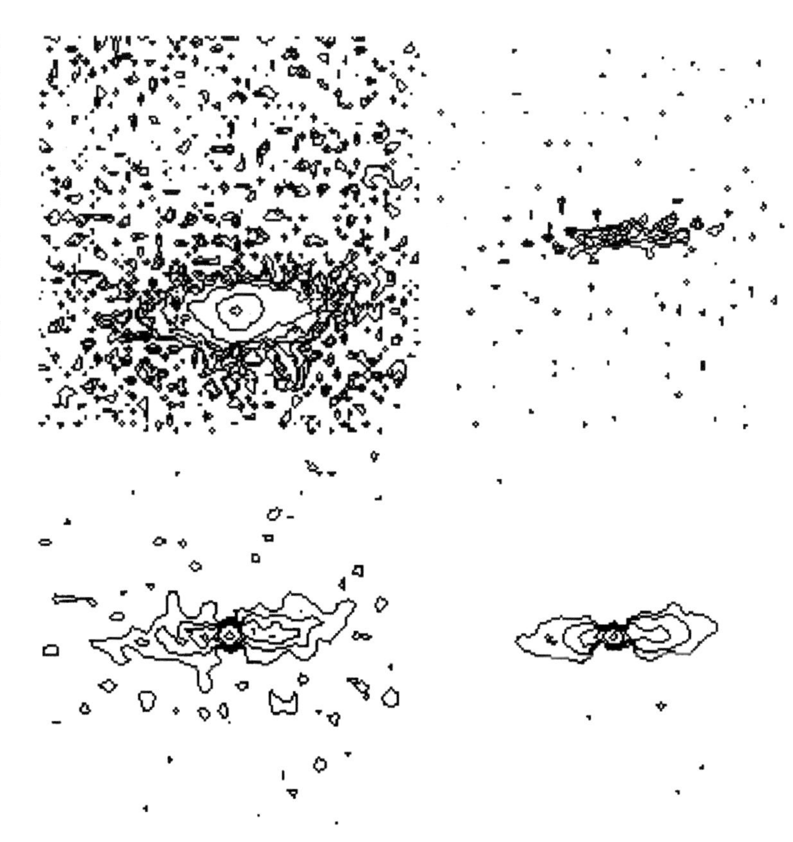

3.3.5 Another application of multiscale entropy: filtering

Multiscale entropy filtering. Filtering using multiscale entropy leads to the minimization of:

$$J(\tilde{I}) = \sum_{\text{pixels}} \frac{(I - \tilde{I})^2}{2\sigma_I^2} - \alpha S_{ms}(\tilde{I}) \tag{3.38}$$

where I and $\tilde{I}$ are the noisy and filtered image, α fixed experimentally to 0.5, and σ_I the standard deviation of the noise in the data I.

Then the gradient of the functional with respect to $\tilde{I}$ must be calculated:

$$\begin{aligned} \nabla(J(\tilde{I})) = & -\frac{(I - \tilde{I})}{\sigma_I^2} \\ & + \frac{\alpha}{\sigma_I^2} \sum_{\text{scales } j} \left[A(j)\sigma_j \operatorname{sgn}(w_j^{(\tilde{I})}) \log\left(\frac{\mid w_j^{(\tilde{I})} \mid}{m_j} \right) \right] * \psi_j^* \end{aligned} \tag{3.39}$$

where $\psi_j(x, y) = \frac{1}{2^j}\psi(\frac{x}{2^j}, \frac{y}{2^j})$, and ψ is the wavelet function corresponding to the à trous algorithm.

The 'one-step gradient' algorithm gives the iterative scheme:

$$\tilde{I}^{n+1} = \tilde{I}^n - \gamma \nabla(J(\tilde{I}^n)) \tag{3.40}$$

where γ is the step.

Experiments. We tested our filtering method on a mid-infrared image of the β Pictoris dust disk described above in the section on deconvolution experiments, but obtained with only one hour of integration time (see Fig. 3.11). The peak signal-to-noise ratio is around 30. After filtering, the disk appears clearly. For detection of faint structures (the disk here), one can calculate that the application of such a filtering method on this image provides a gain of observing time of a factor of around 60 (in the case of Gaussian additive noise leading to a signal-to-noise ratio varying like the square root of the integration time).

Fig. 3.12 shows a profile of the object (crosses). The profiles of the filtered images of the dust disk using multiscale maximum entropy filtering (plain line) and Wiener filtering (dots) are superimposed. Contrary to the Wiener filtering, the multiscale maximum entropy algorithm does not degrade the resolution, while carrying out the filtering effectively.

3.3.6 Conclusion on multiscale entropy and restoration

In the field of signal deconvolution, the maximum entropy method provided an attractive approach to regularization, and considerably improved on existing techniques. However several difficulties remained: the most important is perhaps the inability to find an optimal regularizing parameter (α) to reconstruct effectively the high and low spatial frequencies at the same time, while having good regularization.

Compared to classical maximum entropy, our method has a fixed α parameter and there is no need to determine it: it is the same for every image with normalized (σ=1) Gaussian noise. Furthermore this new method is flux-conservative and thus reliable photometry can be done

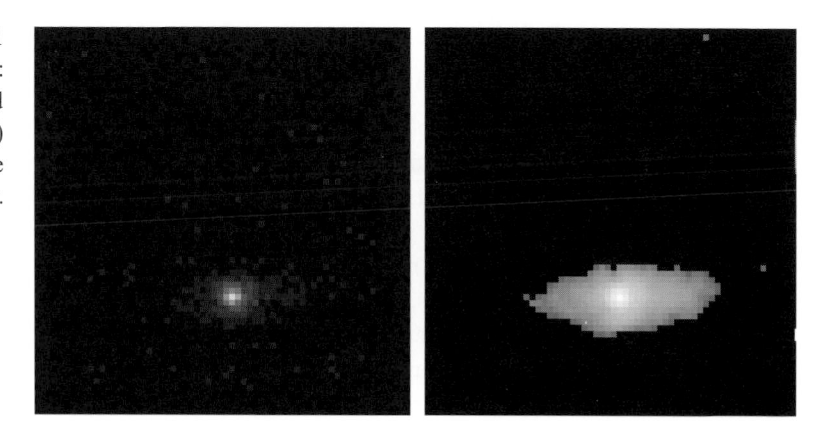

Figure 3.11 β Pictoris dust disk: Raw image (left) and filtered image (right) using multiscale maximum entropy.

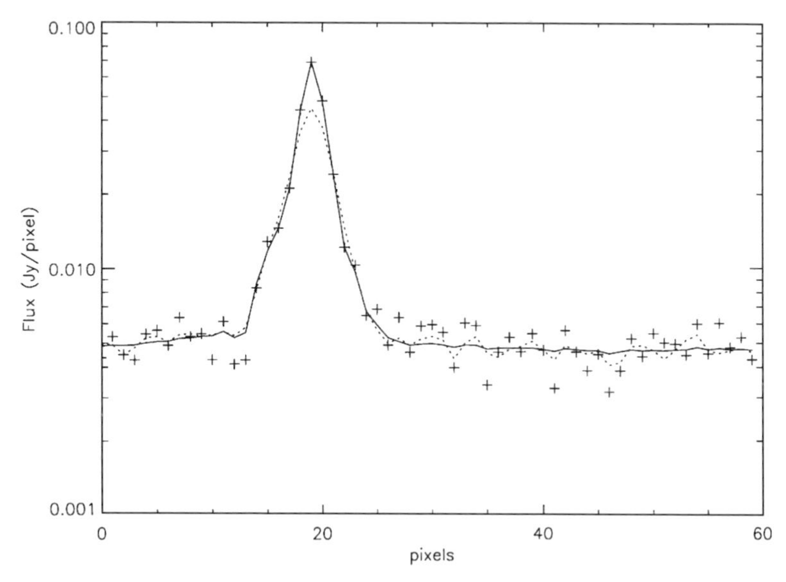

Figure 3.12 Raw image profile of β Pictoris dust disk (crosses). The profiles of the filtered images using multiscale maximum entropy filtering (plain line) and Wiener filtering (dots) are superimposed.

on the deconvolved image. In Bontekoe *et al.* (1994), it was noticed that the 'models' in multi-channel maximum entropy deconvolution should be linked to a physical quantity. We have shown here that this is the case since it is a fraction of the standard deviation of the noise at a given scale of resolution. Bontekoe *et al.* opened up a new way of thinking in terms of multiresolution decomposition, but they did not use the appropriate mathematical tool which is the wavelet decomposition. Using such an approach, we have proven that many problems they encountered are naturally solved. The result is an effective 'easy to use' algorithm since the user has no parameter to supply.

3.4 Image restoration for aperture synthesis

3.4.1 Introduction to deconvolution in aperture synthesis

Frequency holes. The principle of imaging by aperture synthesis lies in the determination of fringe contrasts and phases from different configurations (Labeyrie, 1978). The number of spatial frequencies after observations would be ideally the same as the number of independent pixels in the restored images. This is difficult to achieve with radio interferometers, and this cannot be seriously imagined today for optical interferometers (Beckers, 1991). Frequency holes are the consequence of this incomplete coverage. The point spread function, the inverse Fourier transform of the transfer function, is an irregular function with large rings and wings.

The relation between the object and the image in the same coordinate frame is a convolution:

$$I(x,y) = \int_{-\infty}^{+\infty} O(X,Y)P(x-X,y-Y)dXdY \tag{3.41}$$

where $P(x,y)$ is the point spread function of the imaging system. We want to determine $O(X,Y)$ knowing $I(x,y)$ and $P(x,y)$. With the Fourier transform we get the relation:

$$\hat{I}(u,v) = \hat{O}(u,v)\hat{P}(u,v) \tag{3.42}$$

This equation can be solved only if $|\hat{P}(u,v)| \neq 0$. Evidently this is not the case for aperture synthesis, so a perfect solution does not exist.

An image restoration is needed to fill the frequency holes. This operation is not available from a classical linear operation (Frieden, 1975). Image superresolution, which generalizes this problem, has been examined for many years (Fried, 1992; Harris, 1964; etc.). One of the most popular methods used in radio-interferometric imaging is the CLEAN algorithm (Högbom, 1974).

CLEAN. This approach assumes the object is composed of point sources. It tries to decompose the image (called the dirty map), obtained by inverse Fourier transform of the calibrated *uv* data, into a set of δ-functions. This is done iteratively by finding the point with the largest absolute brightness and subtracting the point spread function (dirty beam) scaled with the product of the loop gain and the intensity at that point. The resulting residual map is then used to repeat the process. The process is stopped when some prespecified limit is reached. The convolution of the δ-functions with an ideal point spread function (clean beam) plus the residual equals the restored image (clean map). This solution is only possible if the image does not contain large-scale structures. The algorithm is:

1 Compute the dirty map $I^{(0)}(x, y)$ and the dirty beam $A(x, y)$.
2 Find the maximum value, and the coordinate $(x_{\max}, y_{\max})$ of the corresponding pixel in $I^{(i)}(x, y)$.
3 Compute $I^{(i+1)}(x, y) = I^{(i)}(x, y) - \gamma I_{\max} A_{\mathrm{m}}(x, y)$ with $A_{\mathrm{m}}(x, y) = A(x - x_{\max}, y - y_{\max})$ and the loop gain γ inside [0,1].
4 If the residual map is at the noise level, then go to step 5.
Else $i \longleftarrow i + 1$ and go to step 2.
5 The clean map is the convolution of the list of maxima with the clean beam (which is generally a Gaussian).
6 Addition of the clean map and the residual map produces the deconvolved image.

Multiresolution CLEAN. The CLEAN solution is only available if the image does not contain large-scale structures. Wakker and Schwarz (1988) introduced the concept of Multiresolution CLEAN (MRC) in order to alleviate the difficulties occurring in CLEAN for extended sources. The MRC approach consists of building two intermediate images, the first one (called the smooth map) by smoothing the data to a lower resolution with a Gaussian function, and the second one (called the difference map) by subtracting the smoothed image from the original data. Both these images are then processed separately. By using a standard CLEAN algorithm on them, the smoothed clean map and difference clean map are obtained. The recombination of these two maps gives the clean map at the full resolution.

In order to describe how the clean map at the full resolution is obtained from the smoothed and difference clean map, a number of symbols must be defined:

- $G =$ the normalized ($\int G(x)dx = 1$) smoothing function; the width of the function is chosen such that the full-width at half maximum of the smoothed dirty beam is f times larger than the full-width at half maximum of the original dirty beam.
- $A =$ dirty beam
- $D =$ dirty map
- $\delta = \delta$-functions
- $R =$ residual after using CLEAN on the map
- $B =$ clean beam with peak value 1
- $C =$ clean map
- $s =$ the scale factor of the dirty beam needed to rescale the smooth dirty beam back to a peak value 1
- $r =$ the scale factor of the dirty beam needed to rescale the smooth clean beam back to a peak value 1
- $A_s =$ normalized smooth dirty beam $= sA * G$
- $A_d =$ normalized difference dirty beam $= 1/(1 - \frac{1}{s})(A - \frac{A_s}{s})$
- $B_s =$ normalized smooth clean beam $= rB * G$
- $B_d =$ normalized difference clean beam $= 1/(1 - \frac{1}{r})(B - \frac{B_s}{r})$

From the δ-functions found by the CLEAN algorithm, one can restore the dirty map by convolving with the dirty beam and adding the residuals:

$$D = D_s + D_d = \delta_s * A_s + R_s + \delta_d * A_d + R_d \tag{3.43}$$

which can be written also as:

$$D = \left[s\delta_s * G + \frac{s}{s-1}\delta_d * (1 - G) \right] * A + R_s + R_d \tag{3.44}$$

If we replace the dirty beam by the clean beam, we obtain the clean map:

$$C = \frac{s}{r}\delta_s * B_s + \frac{s(r-1)}{r(s-1)}\delta_d * B_d + R_s + R_d \tag{3.45}$$

The MRC algorithm needs three parameters. The first fixes the smoothing function G, and the other two are the loop gain and the extra loop gain which are used by CLEAN respectively on the smooth dirty map and difference dirty map.

This algorithm may be viewed as an artificial recipe, but we have shown (Starck and Bijaoui, 1991, 1992) that it is linked to multiresolution analysis as defined by Mallat (1989). Mallat's theory provides a new representation where a function is a sum of detail structures obtained with the same pattern, the wavelet, with suitable translations and dilations. Wavelet analysis leads to a generalization of MRC from a set of scales.

Our approach allows MRC algorithms to be harmonized with the classical theory of deconvolution.

3.4.2 CLEAN and wavelets

The wavelet transform chosen. We have seen that there are many wavelet transforms. For interferometric deconvolution, we choose the wavelet transform based on the FFT for the following reasons:

- The convolution product is kept at each scale.
- The data are already in Fourier space, so this decomposition is natural.
- There is a pyramidal implementation available which does not take much memory.

Hence until the end of this chapter, we will consider the use of the pyramidal transform based on the FFT.

Deconvolution by CLEAN in wavelet space. If $w_j^{(I)}$ are the wavelet coefficients of the image I at the scale j, we get:

$$\hat{w}_j^{(I)}(u,v) = \hat{w}_j^{(P)}\hat{O}(u,v) \tag{3.46}$$

where $w_j^{(P)}$ are the wavelet coefficients of the point spread function at the scale j. The wavelet coefficients of the image I are the convolution product of the object O by the wavelet coefficients of the point spread function.

At each scale j, the wavelet plane $w_j^{(I)}$ can be decomposed by CLEAN ($w_j^{(I)}$ represents the dirty map and $w_j^{(P)}$ the dirty beam) into a set, noted δ_j, of weighted δ-functions.

$$\delta_j = \{A_{j,1}\delta(x - x_{j,1}, y - y_{j,1}), A_{j,2}\delta(x - x_{j,2}, y - y_{j,2}), \ldots, A_{j,n_j}\delta(x - x_{j,n_j}, y - y_{j,n_j})\} \tag{3.47}$$

where n_j is the number of δ-functions at the scale j and $A_{j,k}$ represents the height of the peak k at the scale j.

By repeating this operation at each scale, we get a set $\mathcal{W}_\delta$ composed of weighted δ-functions found by CLEAN ($\mathcal{W}_\delta = \{\delta_1, \delta_2, \ldots\}$). If B is the ideal point spread function (clean beam), the estimation of the wavelet coefficients of the object at the scale j is given by:

$$\begin{aligned} w_j^{(E)}(x,y) &= \delta_j * w_j^{(B)}(x,y) + w_j^{(R)}(x,y) \\ &= \sum_k A_{j,k} w_j^{(B)}(x - x_{j,k}, y - y_{j,k}) + w_j^{(R)}(x,y) \end{aligned} \tag{3.48}$$

where $w_j^{(R)}$ is the residual map. The clean map at the full resolution is obtained by the reconstruction algorithm. If we take a Gaussian function as the scaling function, and the difference between two resolutions as the wavelet ($\frac{1}{2}\psi(\frac{x}{2}, \frac{y}{2}) = \phi(x,y) - \frac{1}{2}\phi(\frac{x}{2}, \frac{y}{2})$), we find the algorithm proposed by Wakker and Schwarz (1988). The MRC algorithm in the wavelet space is:

1 We compute the wavelet transforms of the dirty map, the dirty beam and the clean beam.
2 For each scale j, we decompose by CLEAN the wavelet coefficients of the dirty map into a list of weighted δ-functions δ_j.
3 For each scale j, we convolve δ_j by the wavelet coefficients of the clean beam and we add the residual map $w_j^{(R)}$ to the result in order to obtain the wavelet coefficients of the clean map.
4 We compute the clean map at the full resolution by using the reconstruction algorithm.

In Fig. 3.13a, we can see a simulated object containing two point sources lying on top of an extended Gaussian. Such a source is notoriously difficult to restore using classical deconvolution methods. Figures 3.13c, 3.13d, 3.13e, 3.13f show the results of restoring using

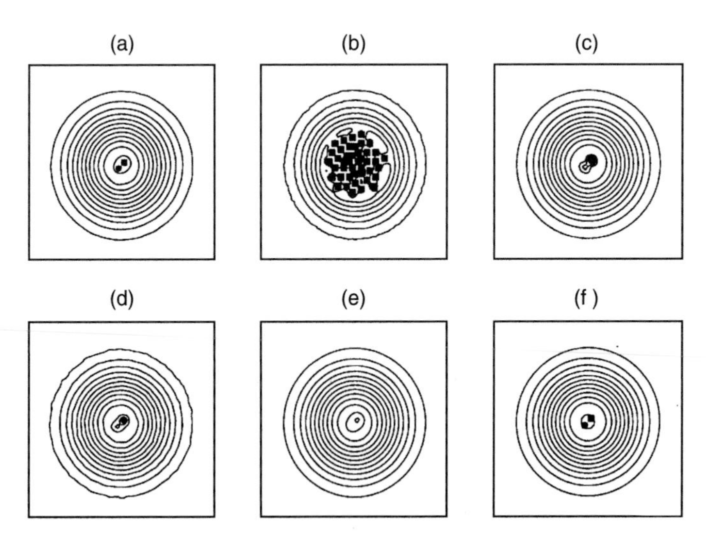

Figure 3.13 Restoration of two point sources on an extended background. (a) The original object. (b) The simulated image obtained by convolving (a) with a Gaussian point spread function of full-width at half maximum equal to 3 pixels and adding noise. (c), (d), (e), (f) Restorations obtained by maximum entropy method, Richardson-Lucy, CLEAN, and multiresolution CLEAN with regularization.

respectively the method of maximum entropy (MEM) (Frieden, 1975), Lucy (1974), CLEAN (Högbom, 1974) and Multiresolution CLEAN (MRC). For this kind of image, the MRC method is very effective and the point sources are well detected. The difficulties occurring in CLEAN for extended sources are resolved by the multiresolution approach.

Improvements to multiresolution CLEAN. We apply CLEAN to each plane of the wavelet transform. This allows us to detect at each scale the significant structure. The reconstructed image gives the estimation $\tilde{O}$ found by MRC of the object. But MRC does not assume that this estimation is compatible with the measured visibilities. We want:

$$| \hat{\tilde{O}}(u,v) - V_{\mathrm{m}}(u,v) | < \Delta_{\mathrm{m}}(u,v) \tag{3.49}$$

where $\Delta_{\mathrm{m}}(u,v)$ is the error associated with the measure V_{m}.

To achieve this, we use the position of the peaks determined by the MRC algorithm. We have seen that after the use of CLEAN, we get a list of positions δ_j on each plane j, with approximate heights A_j. In fact, we get a nice description of the significant structures in the wavelet space (see Fig. 3.14). The height values are not sufficiently accurate, but CLEAN enhances these structures. So we have to determine heights which reduce the error. We do so using Van Cittert's algorithm (1931) which converges, even in the presence of noise, because our system is

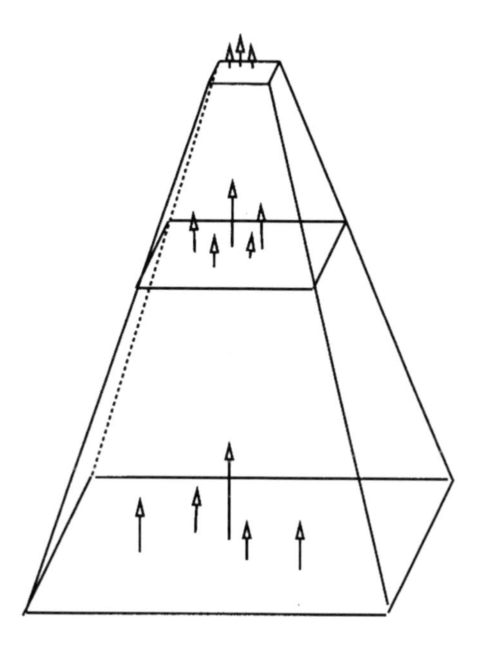

Figure 3.14 Example of detection of peaks by CLEAN at each scale.

well regularized. Then, heights of the peaks contained in $\mathcal{W}_\delta$ will be modified by the following iterative algorithm:

1 Set $n = 0$ and $\mathcal{W}_\delta^{(0)} = \mathcal{W}_\delta$.
2 Compute $A_{j,l}^{(n+1)} = A_{j,l}^{(n)} + \mathcal{Q}_{j,l}.\mathcal{W}_\delta^{(n)}$ so that we then have:

$$\delta_j^{(n+1)} = \{A_{j,1}^{(n+1)}\delta(x - x_{j,1}, y - y_{j,1}),$$

and:

$$\mathcal{W}_\delta^{(n+1)} = \{\delta_1^{(n+1)}, \delta_2^{(n+1)}, \ldots\}$$

3 $n = n + 1$ and go to step 1.

$\mathcal{Q}$ is the operator that:

- computes the wavelet coefficients of the clean map $w^{(C)}$ by convolving at each scale $\delta_j^{(n)}$ by the clean beam wavelet $w_j^{(B)}$

$$w_j^{(C)} = \delta_j^{(n)} * w_j^{(B)}$$

- reconstructs the estimated object $O^{(n)}$ at full resolution from $w^{(C)}$
- thresholds the negative values of $O^{(n)}$
- computes the residual $r^{(n)}$ by:

$$\hat{r}^{(n)} = p(V - \hat{O}^{(n)})$$

where p is a weight function which depends on the quality of the measurement V (error bars). A possible choice for p is:

  - ∘ $p(u, v) = 0$ if we do not have any information at this frequency (i.e. a frequency hole).
  - ∘ $p(u, v) = 1 - 2\frac{\Delta_{\rm m}(u,v)}{V_{\rm m}(0,0)}$ if $\Delta_{\rm m}(u, v)$ is the error associated with the measurement $V_{\rm m}(u, v)$.

- computes the wavelet transform $w^{(r^{(n)})}$ of $r^{(n)}$
- extracts the wavelet coefficient of $w^{(r^{(n)})}$ which is at the position of the peak $A_{j,l}\delta(x - x_l, y - y_l)$.

The final deconvolution algorithm is:

1 Convolution of the dirty map and the dirty beam by the scaling function.
2 Computation of the wavelet transform of the dirty map which yields $w^{(I)}$.
3 Computation of the wavelet transform of the dirty beam which yields $w^{(D)}$.

4 Estimation of the standard deviation of the noise N_0 of the first plane from the histogram of w_0. Since we process oversampled images, the values of the wavelet image corresponding to the first scale ($w_0^{(I)}$) are nearly always due to the noise. The histogram shows a Gaussian peak around 0. We compute the standard deviation of this Gaussian function, with a 3-sigma clipping, rejecting pixels where the signal could be significant.
5 Computation of the wavelet transform of the clean beam. We get $w^{(B)}$. If the clean beam is a Dirac delta, then $\hat{w}_j^{(B)}(u,v) = \frac{\psi(2^j u, 2^j v)}{\phi(u,v)}$.
6 Set j to 0.
7 Estimation of the standard deviation of the noise N_j from N_0. This is done from the study of the variation of the noise between two scales, with the hypothesis of a white Gaussian noise.
8 Detection of significant structures by CLEAN: we get δ_j from $w_j^{(I)}$ and $w_j^{(D)}$. The CLEAN algorithm is very sensitive to the noise. Step 1 of this algorithm offers more robustness. CLEAN can be modified in order to optimize the detection.
9 $j = j + 1$ and go to step 7.
10 Reconstruction of the clean map from $\mathcal{W}_\delta = \{\delta_1, \delta_2, \cdots\}$ by the iterative algorithm using Van Cittert's method.

The limited support constraint is implicit because we put information only at the position of the peaks, and the positivity constraint is introduced in the iterative algorithm. We have made the hypothesis that MRC, by providing the coordinates of the peaks, gives the exact position of the information in the wavelet space and we limited the deconvolution problem by looking for the height of the peaks which give the best results. It is a very strong limited support constraint which allows our problem to be regularized. CLEAN is not used as a deconvolution algorithm, but only as a tool to detect the position of structures.

3.4.3 Experiment

In Fig. 3.15, we can see on the left a simulated object containing a point source with an envelope and, on the right, the Fourier space or uv plane, coverage used in the simulation of an interferometric image.

Figure 3.16 (left) shows the simulated image, computed from 271 visibilities with a precision of 5%, and at right the restored image by our algorithm. In spite of the power coverage, the envelope of the star has been found. Before iterating, 108 visibilities were outside the error bars, and after 50 iterations, only five visibilities.

By our method, the images of two evolved stars surrounded by dust shells have been reconstructed from speckle interferometry observations (Starck *et al.*, 1994).

3.4.4 Observations of two evolved stars

Presentation of the sources. We observed the two sources OH 26.5+0.6 and Hen 1379 in the course of a high angular resolution study by using speckle interferometry. OH 26.5+0.6 is a typical OH/IR star. It exhibits a strong OH maser emission at 1612 MHz associated with a luminous infrared source. The OH/IR stars belong to an evolved stellar population located on the Asymptotic Giant Branch of the Hertzsprung-Russell diagram. Most of these sources appear to be very long period variables and their mass loss rate is the highest known for late type stars. In particular, OH 26.5+0.6 has a period of about 1600 days and a mass loss rate of 5.10^{-5} solar masses per year (Le Bertre, 1984). This object had already been observed by speckle interferometry (Perrier, 1982). High resolution imaging of the OH emission shows

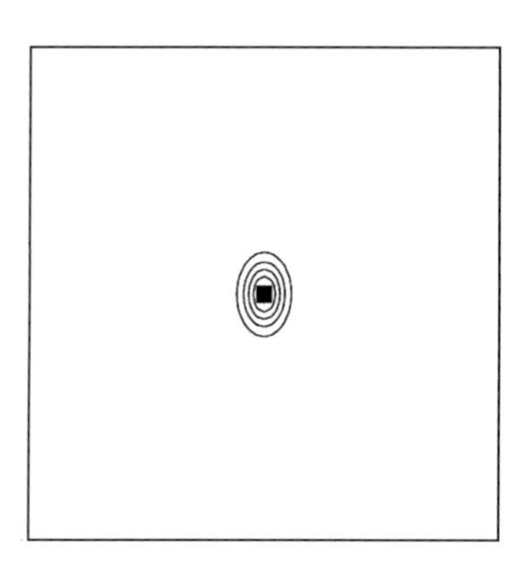
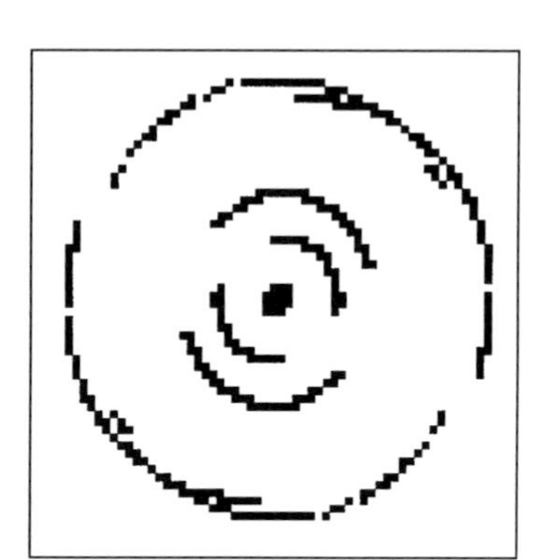

Figure 3.15 Simulated object and *uv* plane coverage.

Figure 3.16 Restoration of a simulated object.

that the structure of its circumstellar shell is spherical (Herman *et al.*, 1985).

Hen 1379 is a post Asymptotic Giant Branch star in a phase of intense mass loss (Le Bertre *et al.*, 1989). The circumstellar dust distribution for post-AGB stars generally departs from spherical geometry. This is the case for Hen 1379, the high polarization measured at visible and near-infrared wavelengths indicating that the envelope is strongly non-spherical (Le Bertre *et al.*, 1989).

Both kinds of stars, OH 26.5+0.6 and Hen 1379, contribute significantly to the enrichment of the interstellar medium in refractory material formed from the gas and dust expelled. Since dust forms in the shells of these stars, it absorbs visible and near-infrared light from the star photosphere and then reradiates it at longer wavelengths. These stars are by nature strong infrared sources.

High spatial resolution near-infrared observations are sensitive to dust emission coming primarily from the inner region of the shell where grain temperature is of the order of $500 - 1000$ K. Resolving the circumstellar shell geometry is of importance for understanding the physical mechanisms responsible for mass loss. In practice, mass loss rates are generally computed under the assumption of spherical symmetry while shell structure can be strongly different. Accurate mass loss rates should be determined only by considering the shell geometry as revealed by means of high resolution observations.

Speckle interferometry observations. The high angular resolution observations of these sources were performed using the ESO one-dimensional (1D) slit-scanning near-infrared specklegraph attached to the ESO 3.6m telescope Cassegrain focus. The instrument makes use of the telescope wobbling secondary $f/35$ mirror operated in a saw-tooth mode (Perrier, 1986). Rotating the secondary mirror mount together with the Cassegrain instrumental flange allows the scanning direction to be changed. This direction defines the actual position angle (PA) of the direction of exploration, counted counter-clockwise with a zero point corresponding to a scan vector going from South to North. A measurement consists of taking a series of typically a few thousand scans, temporally short enough to freeze the atmospheric turbulence, alternately on the source and a nearby point-like source (IRC-10450 for OH 26.5+0.6 and SRS 14119 for Hen 1379) providing the average modulated transfer function reference. The technique of fast alternated acquisition on the two objects aims at minimizing the discrepancies between seeing histograms that could otherwise prevent the calibration of the atmospheric average modulated transfer function at low spatial frequencies (Perrier, 1989).

A classical speckle 1D data treatment, using the power spectrum densities, was applied to provide the 1D visibility modulus $V(\mathrm{u})$ of the source in each band, defined as the 1D Fourier transform modulus of the spatial energy distribution of the object $O(\alpha, \beta)$ integrated along the slit direction (see Figs. 3.17 and 3.18):

$$V(\mathrm{u}) = \left| \frac{\int_\alpha \int_\beta O(\alpha, \beta) e^{-2\pi i \mathrm{u}\alpha} d\beta d\alpha}{\int_\alpha \int_\beta O(\alpha, \beta) d\beta d\alpha} \right| \tag{3.50}$$

$V(\mathrm{u})$ is the cut, along the axis u parallel to the scan direction, of the two-dimensional Fourier transform of $O(\alpha, \beta)$:

$$V(u) = \left| \tilde{O}(\mathrm{u}, \mathrm{v} = 0) \right| \tag{3.51}$$

with:

$$\tilde{O}(u, v) = \frac{\int_\alpha \int_\beta O(\alpha, \beta) e^{-2\pi i (\mathrm{u}\alpha + \mathrm{v}\beta)} d\beta d\alpha}{\int_\alpha \int_\beta O(\alpha, \beta) d\beta d\alpha} \tag{3.52}$$

Here α and β are angular coordinates on the sky and u, v their conjugates in Fourier space. The vertical lines $|\ |$ represent the modulus of the complex function. $\int_\alpha \int_\beta O(\alpha, \beta) d\beta\, d\alpha$ is the normalization factor

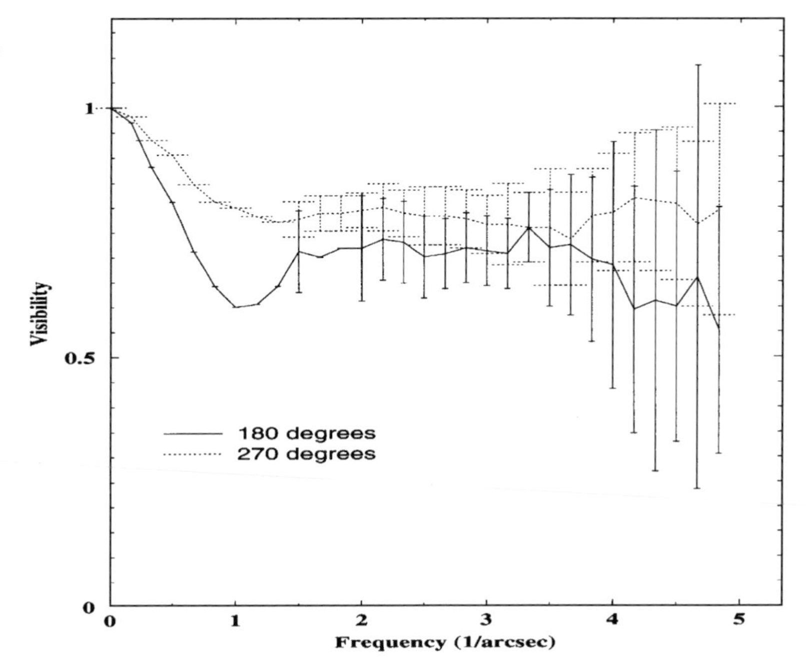

Figure 3.17 Example of visibility curves of Hen 1379 in two directions at band L (3.6 μm).

taken as the flux received in the spectral band. The fact that our 1D measurements yield a cut of the 2D visibility theoretically allows a full recovery of the 2D object distribution $O(\alpha, \beta)$ by sampling the *uv* space with enough scan directions, a technique referred to as tomographic imaging. These multi-position angle observations aim at studying the structure of the circumstellar envelopes.

We obtained measurements at K (2.2 μm), L (3.6 μm) and M (5.0 μm) bands. At L, the circumstellar envelopes are resolved for the two sources which were observed at 8 different position angles. The theoretical cut-off frequency of the telescope is 4.9 arcsec^{-1} at this wavelength. Due to atmospheric turbulence, the effective cut-off is slightly lower than the telescope diffraction limit. This effective cut-off is variable from one visibility curve to another due to temporal evolution of seeing conditions. As each visibility curve samples the modulus of the two-dimensional Fourier transform of the source, and the phase is unavailable, only the autocorrelation of the images of OH 26.5+0.6 and Hen 1379 can be reconstructed.

Image reconstruction. Figures 3.19 and 3.20 show the *uv* plane coverage and the reconstruction of the stars Hen 1379 and OH 26.5+0.6 by the wavelet transform.

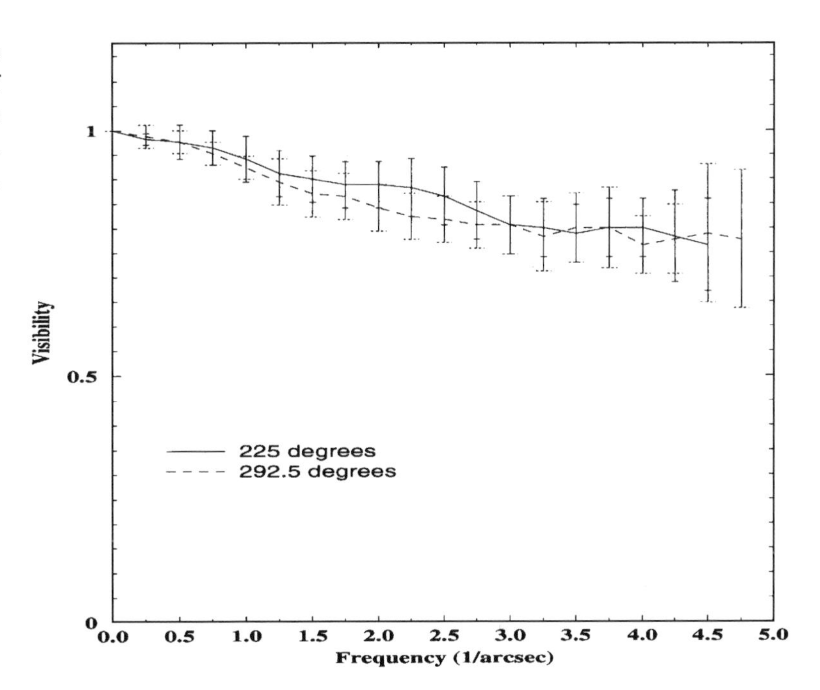

Figure 3.18 Example of visibility curves of OH 26.5+0.6 in two directions at band L (3.6 μm).

The reconstructed images give us information on the shell structure of the two sources. But they must be analyzed with care:

1 Only the modulus of the Fourier transform of the sources is available from the visibility information. The deconvolution algorithm therefore leads to the restoration of the autocorrelation of the object, i.e. a symmetrical form of OH 26.5+0.6 and Hen 1379 actual object distribution.
2 The images must be interpreted with the help of the visibility curves. The most complete visibility function is the one of Hen 1379. The shell is clearly resolved up to the plateau attributed to the stellar component. Moreover a second inverted lobe is seen on the curves. This could be created by the inner boundary of the Hen 1379 dust shell which clearly appears on the reconstructed frames. As the plateau of the stellar component is present, the flux ratio of dust shell emission to stellar radiation is known. Through the

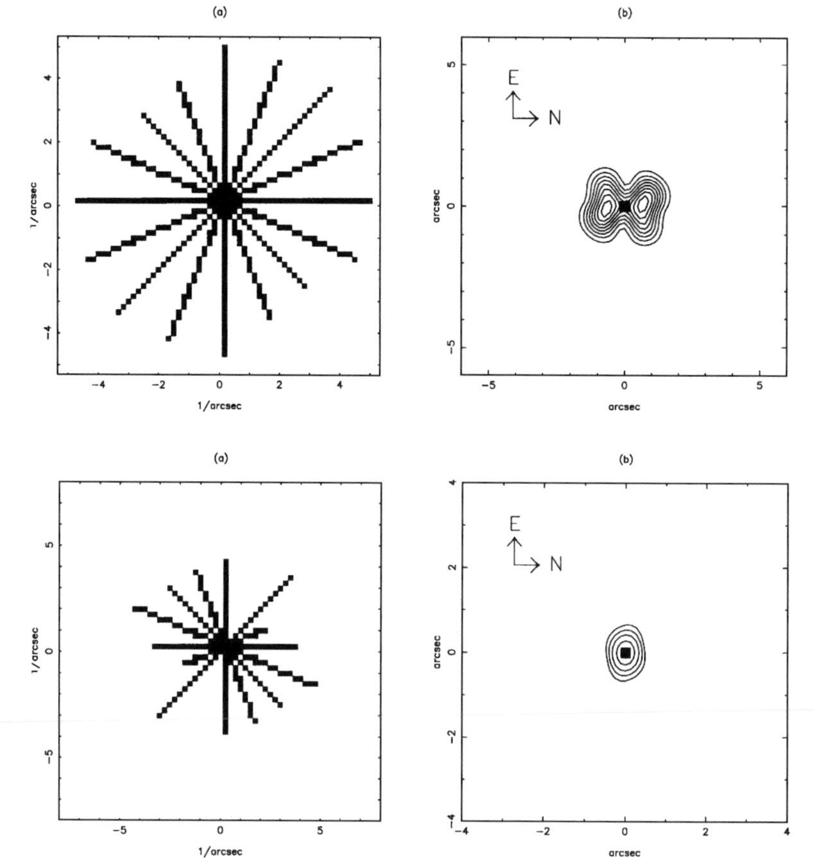

Figure 3.19 The *uv* plane coverage of Hen 1379 and reconstructed image. The ratio point-source to the maximum amplitude of the envelope is 290. Contour levels are 12, 24, 36,..., 96% of the maximum amplitude of the envelope.

Figure 3.20 The *uv* plane coverage of OH 26.5+0.6 and reconstructed image. The ratio point-source to the maximum amplitude of the envelope is 165. Contour levels are 20, 40,..., 100% of the maximum amplitude of the envelope.

deconvolution process, this ratio constrains the relative flux level between the circumstellar shell and the non-resolved central star in the image (see Fig. 3.21).

Obviously this is not the case of OH 26.5+0.6 which has a smaller angular size. The dust shell is partially resolved on visibility functions. Although the flux level between shell and star is undetermined, the deconvolution algorithm gives an image solution which is not constrained by the magnitude of the ratio. The relative level of the dust shell flux and the non-resolved central star flux is therefore not relevant in the reconstruction of the image of OH 26.5+0.6.

3.4.5 Conclusion on interferometric data deconvolution

Our approach using the MRC algorithm corresponds to a scheme in which the image is a set of atomic objects (Meyer, 1993). There are many ways of decomposing a signal into independent atoms. The

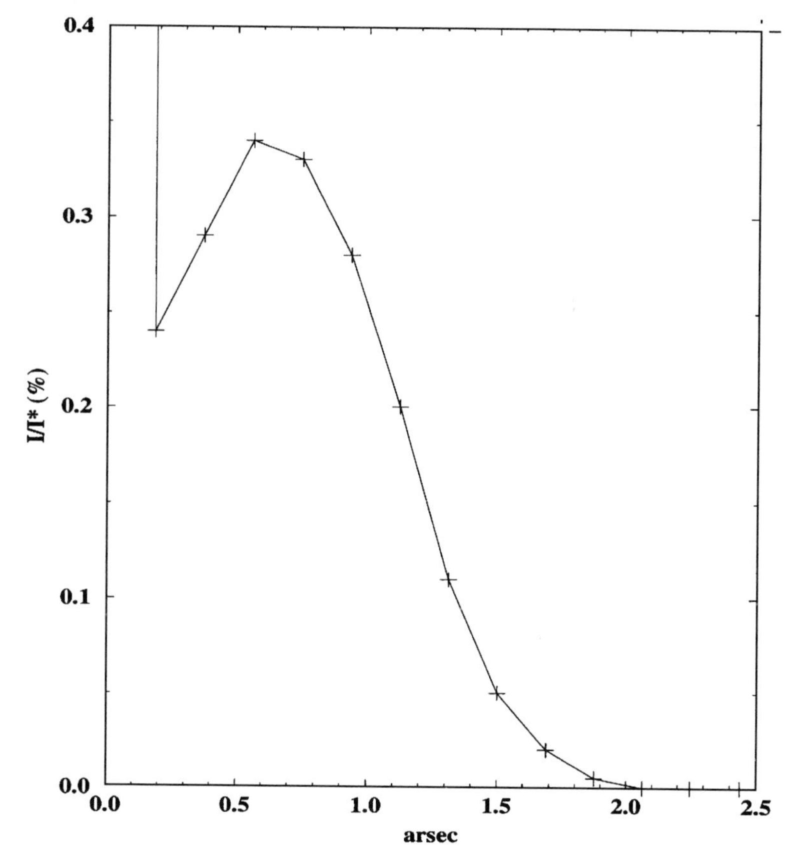

Figure 3.21 Intensity profile I of the star Hen 1379 for a position angle of 180 degrees normalized to the stellar intensity I^*.

orthogonal wavelet transform belongs to this class of linear decomposition. In our method, the deconvolution problem is reduced to a pattern recognition one: we search for the minimum atoms which permit the building up of an image, in agreement with the observation.

The classical CLEAN algorithm is a pattern recognition method, where the model is the simplest one: the image is a set of Dirac peaks convolved with the point spread function. Taking into account the clean beam, the atoms are Gaussian peaks. If the Fourier space is correctly sampled in a limited frequency band, the natural atoms are sine cardinal functions (resulting from the sampling theorem). In the case of a frequency hole, we cannot have duality between the frequency coverage and the sampled sine cardinals. We have to limit the number of significant peaks in order to get an equivalence between the number of visibilities and the number of peaks (Schwarz, 1978). If the image is too rich in structural elements, artifacts appear during the restoration process. We need a method to reduce the number of atoms without reducing the covered image. This is exactly what is provided by the wavelet transform. A set of close peaks is reduced to fewer wavelet peaks. This reduction of the number of atoms leads to a regularization of the restoration method. The best choice is the one which permits the use of all visibilities in an optimal manner. This is the case for the CLEAN algorithm for which each visibility contributes equally for each peak. If large structures are present, CLEAN fails and we cannot use all the visibilities for each atom. In our method, we decompose the Fourier space into rings having a radius reduced by a factor 2 from one step to the following one. Other atoms could be possible, if we want to use a priori information on the structure of the image in a better manner.

Our method is model- and parameter-free, and this is an essential advantage for a quick-look tool. The choice of the wavelet function plays an important part. We can constrain it:

- to be compact in Fourier space, in order to have a readily-available pyramidal algorithm using the FFT;
- to be as compact as possible in the direct space.

Compactness in Fourier space prevents compactness in the direct space, and that leads to rings in the direct space. The intensity of these rings is reduced if the wavelet function has a high degree of regularity. Then the wavelet must be:

- Unimodal in the Fourier space. Rings in Fourier space would lead to artificial high frequency patterns.

- Quite unimodal in the direct space.
- Isotropic in order to avoid any favored axis.

A nice choice lies in B-spline functions in Fourier space. We use a cubic function. The 5th degree B-spline is similar to a Gaussian. Other compact regular functions can be built (Jaffard, 1989).

An important characteristic of our restoring procedure lies in the possibility to compute the probability for each atom to be real using classical statistical theory. This is an important feature which keeps the restoration close to the needs of astrophysical interpretation.

Speckle interferometry observations providing an incomplete Fourier plane coverage were successfully processed and the method appears to be well-adapted to the analysis of future long baseline interferometry data which are expected to fill the Fourier plane irregularly.

4 1D signals and Euclidean data analysis

4.1 Analysis of 1D signals: spectral analysis

4.1.1 Spectral analysis

We present an application of the wavelet transform to the analysis of spectra. The wavelet transform of a signal by the à trous algorithm produces at each scale j a set $\{w_j\}$. The original signal S can be expressed as the sum of all the wavelet scales and the smoothed array c_p:

$$S(\lambda) = c_p(\lambda) + \sum_{j=1}^{p} w_j(\lambda)$$

4.1.2 Noise determination and detection criteria

The appropriate value of the noise σ_j in the succession of wavelet scales is difficult to estimate because the noise is varying with wavelength. To resolve this problem, we use the fact that often the final spectrum is obtained from several individual scans. We compute the root mean square error (RMS) for each wavelength, which gives a good estimation of the noise $N(\lambda)$ at the particular wavelength λ. We now generate Gaussian noise with $\sigma = 1$, take its wavelet transform, and compute the standard deviation σ_j of each scale j. As i wavelengths have been used to calculate the coefficient $w_j(\lambda)$ $(S(\lambda - i/2), ..., S(\lambda), ..., S(\lambda + i/2))$, the upper limit of the standard deviation of the noise at scale j and at position λ is computed by:

$$N_j(\lambda) = \sigma_j \cdot \max\{N(\lambda - i/2), ..., N(\lambda), ..., N(\lambda + i/2)\} \qquad (4.1)$$

If it is not possible to determine an RMS at each λ, that is, if only one scan is available, $N(\lambda)$ is obtained as follows: we consider that the noise is locally constant, and define $N(\lambda)$ as the local standard deviation around $S(\lambda)$.

Note that this approach is most conservative, N_j being an upper limit while the actual noise is lower.

An emission band is detected if $w_j(\lambda) > kN_j(\lambda)$, and an absorption is detected if $w_j(\lambda) < -kN_j(\lambda)$. As is the case in more conventional methods, a value of $k = 3$ implies a confidence level for the detection greater than 99.9%. Considering uncertainties in the noise estimation, and calibration problems etc., $k = 5$ is more robust. For constant noise, this corresponds to a 5σ detection.

Two important points should be stressed. Firstly, the only a priori knowledge we need is information about the width of the bands we want to detect. If we do not use this information, all the scales should be analyzed. Secondly, the detection criterion is independent of the actual shape of the continuum. This point is particularly important. Indeed, the continuum is often estimated 'by hand' or by fitting a low order polynomial. Furthermore, the fit is sometimes done without taking into account all the data. But it is clear that fitting by excluding data, where we assume we have an emission or an absorption line, is equivalent to introducing a priori information on the wavelength where we have signal. If the emission or absorption bands are weak, then this a priori information could force a detection at this wavelength. In other words, excluding data at a given wavelength from the fit could lead us to find a detected line at this wavelength. This continuum estimation method should be used with great care, and only for cases where the detection is clear, without any ambiguity. In the wavelet method, no a priori information is added, and the detection results only from the comparison between a wavelet coefficient and the noise.

4.1.3 Simulations

Simulation 1: featureless continuum with noise. In order to illustrate our new method and the detection criteria adopted, we present a few examples using simulated spectra. In the first case shown in Fig. 4.1a, variable Gaussian noise is superimposed on a smooth continuum, represented by the dashed line. All wavelet scales 1–5 are shown in Figs. 4.1b–f, respectively, together with the corresponding noise levels given by the dashed lines. None of the wavelet scales indicates the presence of an absorption or an emission band, as expected.

Simulation 2: continuum with noise and a strong emission band. We now add a strong emission band at 3.50 μm to the simulated spectrum of Fig. 4.1 (top), with a maximum of five times the local noise standard deviation and a width of FWHM = 0.01 μm. Figure 4.2 (top) contains the simulated spectrum. Figure 4.2 (bottom) shows wavelet scale 4 and the 3σ noise limits, indicated by the two dashed lines. It can be seen that at 3.50 μm, the wavelet exceeds the 3σ noise level. The wavelet analysis results in the detection of an emission band at 3.50 μm above 3σ.

Simulation 3: continuum with noise and a weak absorption band. A third example is shown in Fig. 4.3 (top), where a weak absorption band at 3.50 μm is superimposed on the featureless continuum given in Fig. 4.1 (top). The minimum of the absorption band is twice the local noise standard deviation, and its width is FWHM = 0.015 μm.

Figure 4.3 (top) shows the original simulated spectrum, and Fig. 4.3 (bottom) shows scale 5, and the 3σ limit. As we can see, the band is undetectable in the original simulated spectrum, but it is detected at scale 5 using the 3σ criterion. It must be clear that the standard deviation at scale 5 is not equal to the standard deviation in the original spectrum (see section 4.1.2). In practice, the larger the band, then the better we can detect it, even if the maximum of the band is low compared to the standard deviation of the noise in the original

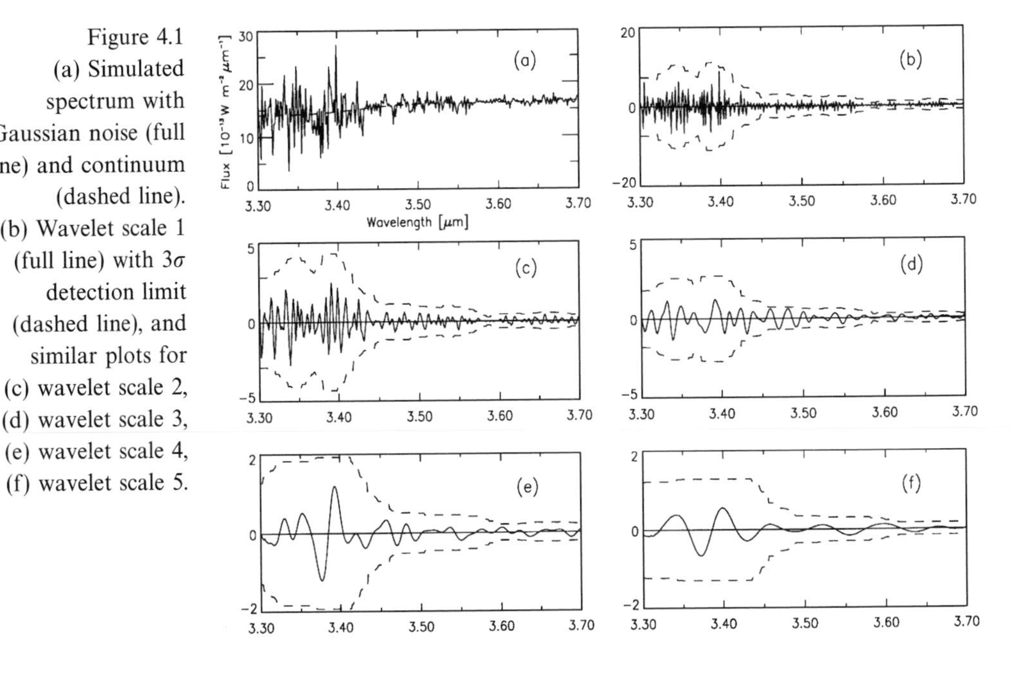

Figure 4.1
(a) Simulated spectrum with Gaussian noise (full line) and continuum (dashed line). (b) Wavelet scale 1 (full line) with 3σ detection limit (dashed line), and similar plots for (c) wavelet scale 2, (d) wavelet scale 3, (e) wavelet scale 4, (f) wavelet scale 5.

spectrum. What is important is the level of the noise at the detection scale, and this level is decreasing with scale.

4.1.4 Problems related to detection using the wavelet transform

Border problems. The most general way to handle the boundaries is to consider that $S(k+N) = S(N-k)$, where S represents our data, N is the number of samples (or pixels), and k is a positive integer value. Other methods can be used such as periodicity ($S(k+N) = S(k)$), or

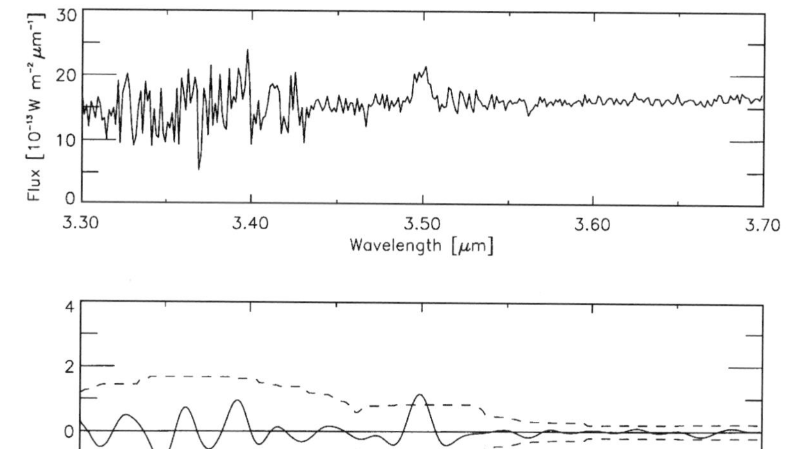

Figure 4.2 Top: simulated spectrum. Bottom: wavelet scale 4 (full line) and 3σ detection limits (dashed).

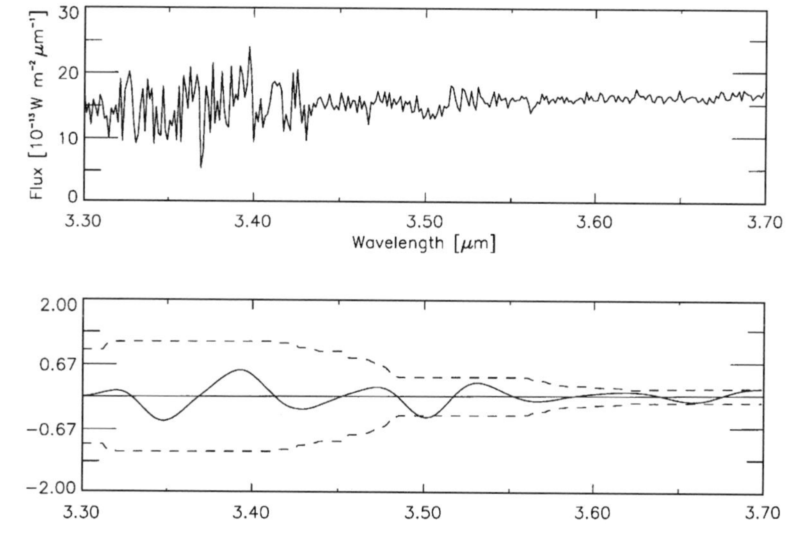

Figure 4.3 Top: simulated spectrum. Bottom: wavelet scale 5 (full line) and 3σ detection limits (dashed).

continuity ($c(k+N) = c(N)$). We used the first method. Choosing one of these methods has little influence on our general analysis strategy. But all detections on the border should be checked carefully because they are obtained from wavelet coefficients which are calculated from extrapolated data.

Bump problems (and simulation 4). Each scale corresponds to the convolution of the input signal with the (dilated) wavelet function. This means that if the signal contains a Dirac function (or sharp jump at a point), we will find this at each scale of the wavelet function. In the case of the à trous wavelet transform algorithm, the wavelet function is derived from a B_3-spline, and looks like the difference between two Gaussians (the second having a FWHM twice the value of the first). Then, the wavelet has a positive main beam, and two negative parts around it. Hence if a structure is very strong, it is possible that we detect not only the structure, but also the bumps around it.

To illustrate this problem, we have added to the simulated spectrum of Fig. 4.1 (top) a strong emission band at 3.5 μm, with a maximum equal to ten times the local noise standard deviation, and with a width (FWHM) equal to 0.02 μm.

Figure 4.4 contains the simulated spectrum and the wavelet scale 5. The wavelet analysis detects an emission band at 3.50 μm above 3σ (3σ noise given by the dashed lines). Also detected at 3σ are two absorption bands at 3.47 μm and 3.53 μm. These two bands are not contained in the simulated spectrum of Fig. 4.4 (top) and are artifacts of the wavelet transform analysis. Every time there is a very strong detection of an emission (absorption) band, detections of weaker absorption (emission) bands symmetrically centered around the main peak may be found which ought to be investigated carefully. This problem can be resolved by subtraction of the strong band from the original signal.

4.1.5 Band extraction

Once a band is detected in the wavelet space, the determination of parameters such as its position, its FWHM, or its optical depth, requires the reconstruction of the band. This is because the wavelet coefficients are the result of the convolution of the signal by the wavelet function. The problem of reconstruction (Bijaoui and Rué, 1995) consists of searching for a signal B such that its wavelet coefficients are the same as those of the detected band. By denoting $\mathcal{T}$ as the wavelet transform operator, and P_b the projection operator in the subspace of the detected coefficients (i.e. all coefficients set to zero at scales and positions where nothing was detected), the solution is found by minimization of

$$J(B) = \| W - (P_b \circ \mathscr{T})B \| \tag{4.2}$$

where W represents the detected wavelet coefficients of the signal S.

A description of algorithms for minimization of such a functional are discussed in Chapter 9 and also later in this chapter (section 4.2 onwards).

Figure 4.5 (top) shows the reconstruction of the detected band in the simulated spectrum presented in Section 4.1.4 (see Fig. 4.4). The real feature is over-plotted as a dashed line. Fig. 4.5 (bottom) contains the original simulation shown in Fig. 4.5 with the reconstructed band subtracted. It can be seen that there are no strong residuals near the location of the band, which indicates that the band is well reconstructed. The center position of the band, its FWHM, and its maximum, are then estimated via a Gaussian fit. The results obtained in the three simulations are presented in Table 4.1.

4.1.6 Continuum estimation

The continuum is the 'background' component of the spectrum. The determination of the shape of the local continuum in noisy astronomical spectra is generally a difficult task. Often, polynomials of low degree are fit to determine its shape, or it is determined by smoothing

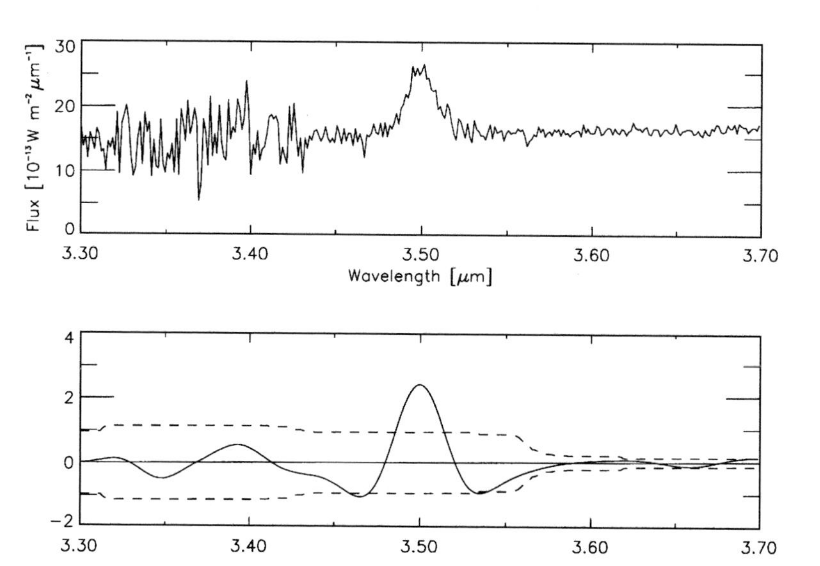

Figure 4.4 Top: simulated spectrum. Bottom: wavelet scale 5 (full line) and 3σ detection limits (dashed).

the spectrum using a broad band-pass. The wavelet transform provides a powerful tool to extract the shape of the continuum. The smoothed array c_p introduced above is a good representation of the continuum, because it contains all the information at a very low spectral resolution. It is equivalent to what we get by convolving the data with a wide filter.

Table 4.1 *Parameter estimation of the simulated detected bands.*

	Position	Maximum	σ	FWHM	Flux
Simulation 2					
Simulated band	3.5	5	$4.3e^{-3}$	0.01	40.00
Reconstructed	3.501	4.587	$3.9e^{-3}$	0.009	31.84
Error (%)	0.026	3.17	9.18	9.18	20.38
Simulation 3					
Simulated band	3.5	−2	$6.4e^{-3}$	0.015	24.00
Reconstructed	3.498	−2.68	$6.3e^{-3}$	0.0148	31.89
Error (%)	0.059	34.00	1.07	1.07	32.88
Simulation 4					
Simulated band	3.5	10	$8.5e^{-3}$	0.02	160.00
Reconstructed	3.499	10.31	$9.1e^{-3}$	0.021	165.50
Error (%)	0.011	3.18	6.59	6.59	3.44

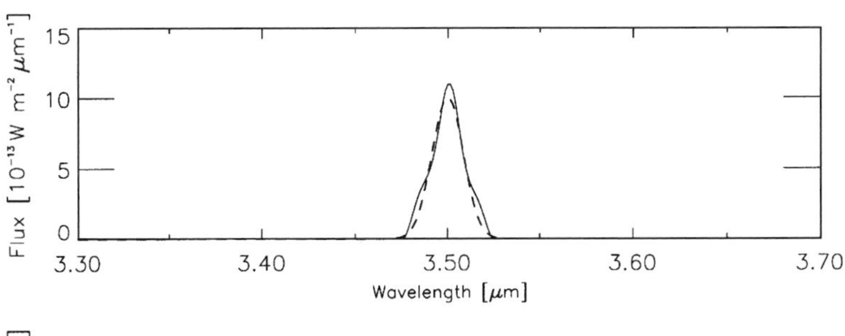

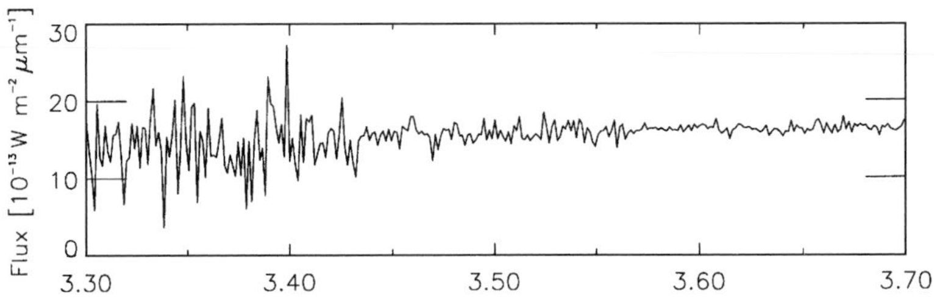

Figure 4.5 Top: reconstructed simulated band (full line) and original band (dashed line). Bottom: simulated spectrum minus the reconstructed band.

Continuum: $C(\lambda) = c_p(\lambda)$. The only decision that has to be made concerns the scale p, which requires a priori knowledge that we can have either after looking at the wavelet scales of the signal, or by fixing the maximum width of the features we expect in our spectra.

The signal $\tilde{S}(\lambda) = S(\lambda) - C(\lambda)$ corresponds to our data without any continuum. This means that $\tilde{S}$ should not contain any information after the scale p. This is not completely true in practice due to border problems when we compute the wavelet transform. The estimate of the continuum can be improved by the following iterative algorithm:

1 Set $C(\lambda)$ to 0, i to 0, and $\tilde{S}$ to S.
2 Compute the wavelet transform of $\tilde{S}$.
 $c_p(\lambda)$ is the last scale (smoothed array).
3 The new continuum estimation is: $C(\lambda) = C(\lambda) + c_p(\lambda)$.
4 Recompute $\tilde{S}$ by: $\tilde{S}(\lambda) = S(\lambda) - C(\lambda)$.
5 $i = i + 1$, and $i < N$ go to step 2.

N represents the number of iterations. In practice, the algorithm converges quickly, and only a few iterations are needed (3 or 4).

Figure 4.6 shows the result of this algorithm applied on real data. We see the original spectrum (upper left), the estimated continuum (upper right), spectrum (full line) and continuum overplotted (bottom left), and the spectrum minus the continuum (upper right). The only parameter used for this continuum estimation is the p parameter, and it was fixed at 7.

Because the continuum estimation involves a convolution of the sig-

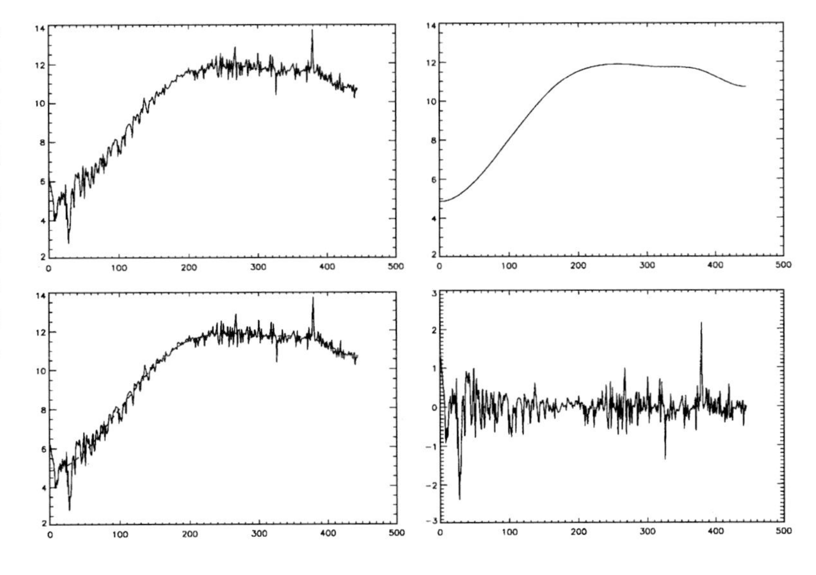

Figure 4.6 Example of continuum estimation. Upper left: spectrum. Upper right: estimated continuum. Bottom left: spectrum (full line) and continuum (dashed line). Bottom right: spectrum minus continuum.

nal with a very wide filter, very strong emission or absorption bands may cause problems. In such a case it is generally better to first extract the bands and then determine the location of the continuum. This is illustrated in Fig. 4.7a, which contains a spectrum with strong absorption features. Figure 4.7b shows the result after the absorption bands have been extracted using the method described in subsection 4.1.5. The continuum is now estimated from this spectrum with the result shown in Fig. 4.7c, where the continuum is represented by the dashed line. A second continuum, represented by the dotted line, is also plotted in Fig. 4.7c. The second continuum is obtained from the original spectrum, without extracting the bands in the first place. The strong spectral features influence the location of the continuum and produce a continuum level that is not correct.

4.1.7 Optical depth

The optical depth τ defines the flux absorbed by a medium. Once the continuum $C(\lambda)$ is estimated, the optical depth τ_s is calculated by:

$$\tau_s(\lambda) = -\ln(S(\lambda)/C(\lambda)) \quad (4.3)$$

As the data S are noisy, τ_s is noisy too. A smoother version τ_b of τ_s can be obtained using the extracted bands $B(\lambda)$:

$$\tau_b(\lambda) = -\ln((C(\lambda) + B(\lambda))/C(\lambda)) \quad (4.4)$$

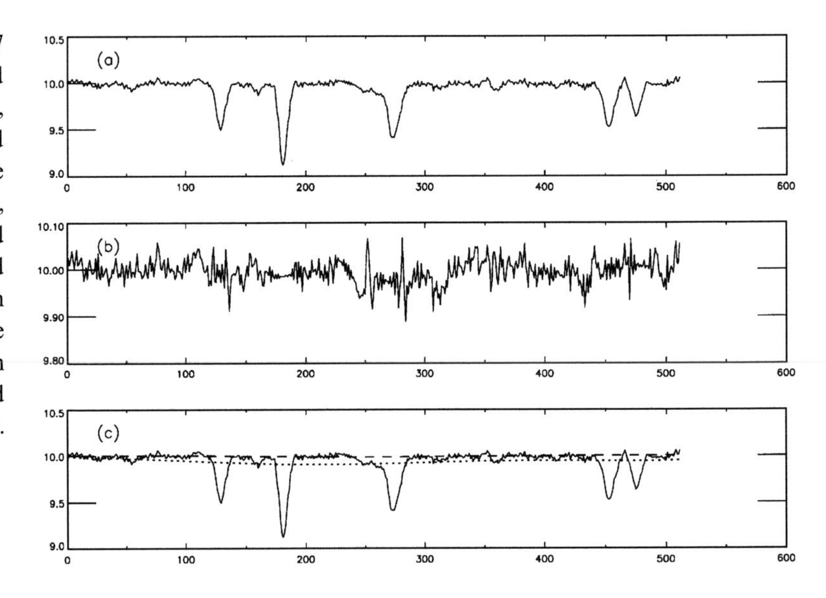

Figure 4.7 (a) Simulated spectrum, (b) simulated spectrum minus the extracted bands, (c) simulated spectrum and continuum over-plotted, before band extraction (dotted line), and after (dashed line).

and the column density of an element can be determined from the optical depth (Allamandola *et al.*, 1993) by

$$N = \frac{\Delta\nu\tau_{\max}}{A} \tag{4.5}$$

where $\Delta\nu$ is the FWHM of the band (in cm^{-1}), $\tau_{\max}$ is the peak optical depth, and A is the integrated absorption cross-section per molecule (cm mol^{-1}).

Figure 4.8 shows the comparison between the true optical depth of the simulated spectrum of Fig. 4.2, and the calculated one by eqns. (4.3) and (4.4). The upper curve diagram contains the true optical depth, the middle diagram is the calculated optical depth using eqn. (4.3) (full line) and eqn. (4.4) (dashed line), and the bottom diagram shows the difference between the true and the calculated optical depth, with the full line representing τ obtained from eqn. (4.3), and the dashed line representing τ obtained from eqn. (4.4). The results obtained from the simulation of the weak absorption band (Fig. 4.3) are given in Fig. 4.9. The two figures demonstrate the reliability of the method. The residuals between the true and the calculated optical depths (cf. Fig. 4.8 and Fig. 4.9) contain only noise.

We may want to compare our new analyzing technique with strategies employed elsewhere. As an example, we have processed the data on the protostar GL2136, recently obtained by Schutte (1995) using the European Southern Observatory near infrared spectrograph IRSPEC. In order to emphasize the weaker absorption features, Schutte

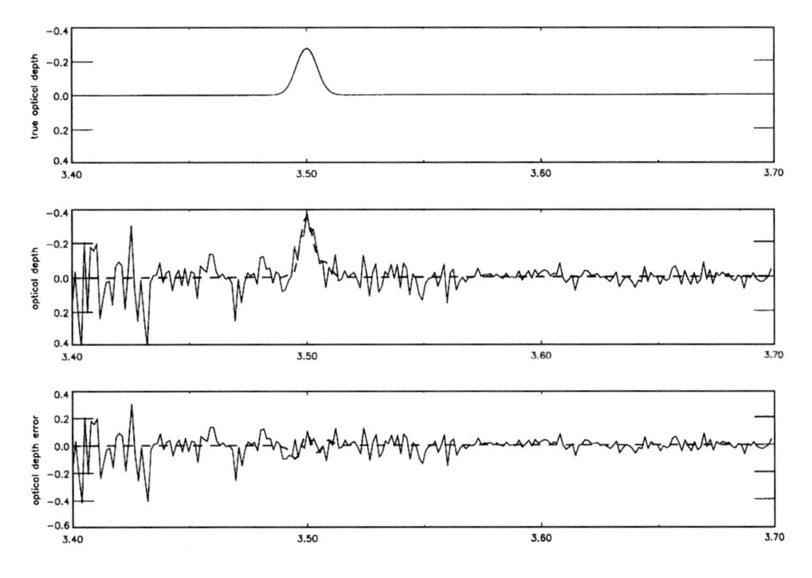

Figure 4.8 Top: true optical depth in the simulated spectrum. Middle: optical depth calculated from the signal (full line), and from the extracted bands (dashed line). Bottom: difference between the true and the calculated optical depth by the two methods. Note that the difference between both methods is within the noise.

et al. (1996) fit a baseline in the $\log(S(\lambda)) - \lambda$ plane and subtract it from the original spectrum. The authors note that their procedure does not represent the 'true' continuum of the source. The fit was done in the $\log(S(\lambda)) - \lambda$ plane rather than in the $S(\lambda) - \lambda$ plane to avoid spurious features in the optical depth plots. Their analysis resulted in the detection of an absorption band at 3.54 μm.

Figure 4.10 (up) contains the spectrum of Schutte *et al.* (1996) which was kindly provided by the authors, together with the continuum determined here. The corresponding optical depth is given below,

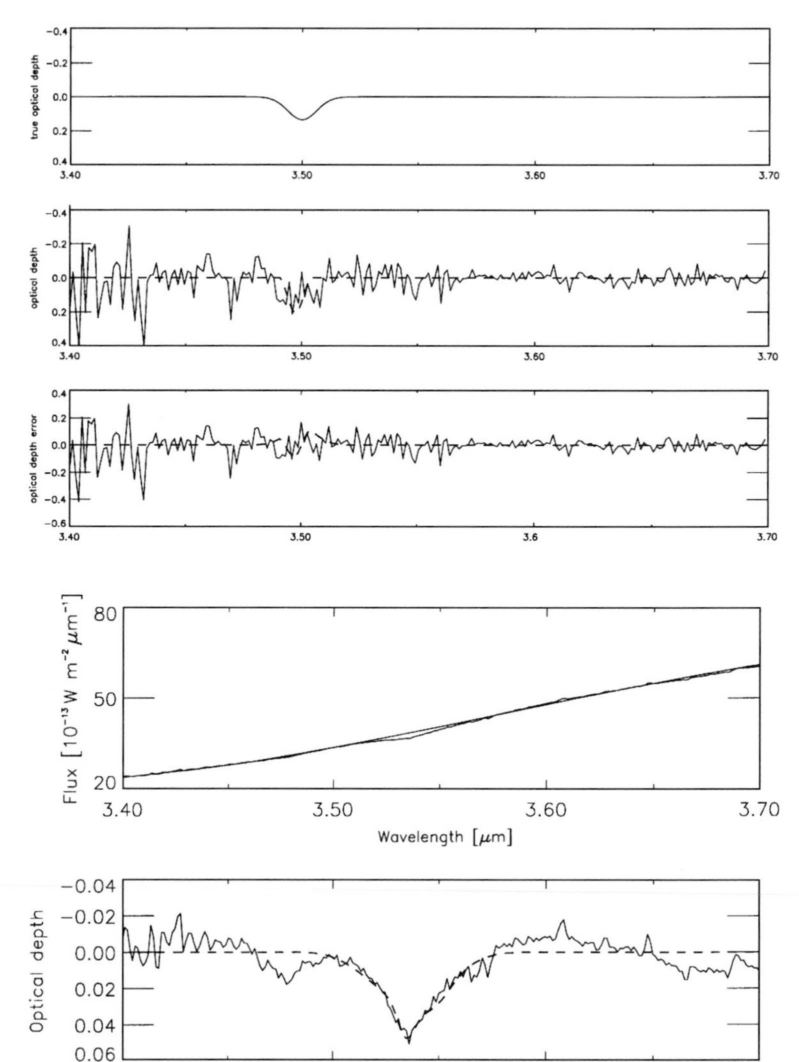

Figure 4.9 Top: true optical depth in the simulated spectrum. Middle: optical depth calculated from the signal (full line), and from the extracted bands (dashed line). Bottom: difference between the true and the calculated optical depth by the two methods. Note that the difference between both methods is within the noise limit.

Figure 4.10 Results from GL2136 (Schutte *et al.*, 1996). Top: spectrum and continuum. Bottom: optical depth (full line) and its smooth version (dashed).

together with a smoothed version (see eqn. (4.4)). A feature is detected near 3.54 μm with a maximum of the optical depth of 0.047. Schutte *et al.* (1996) derived an optical depth of 0.037 μm for the absorption band near 3.54 μm. It can be seen that there are some differences in the derived optical depth. In the wavelet transform method, the determination of the continuum is done fully automatically, and no assumption is made about the position or the wavelength of the bands we are searching for. The only parameter that has to be fixed is the expected width of the bands.

4.1.8 The multiresolution spectral analysis algorithm

The previous sections have shown that absorption or emission bands can be extracted from the spectrum without continuum estimation. For strong bands, the continuum estimation is better done by extracting the bands first. This approach may however fail if the continuum has a strong curvature. In this case, the band extraction may cause problems because of border problems. In such a case, it is generally better to subtract the continuum first. This leads to the following iterative spectral analysis technique:

1 Set i to 0, and estimate the continuum C_i.
2 Compute the difference between the signal and the continuum: $d_i(\lambda) = S(\lambda) - C_i(\lambda)$.
3 Extract the bands in d_i. We obtain $B_i(\lambda)$.
4 Suppress the bands in the data $n_i(\lambda) = S(\lambda) - B_i(\lambda)$.
5 Estimate the continuum C_{i+1} from n_i.
6 Iterate steps 2–5.
7 Compute the optical depth from $B_i(\lambda)$ and C_{i+1}, and the parameters relative to the bands.

A few iterations (2–3) should normally suffice. The procedure aims to suppress false detections which may occur near the border by subtracting a continuum first. The continuum is then determined again from the spectrum once the bands have been detected and extracted.

4.2 Wedding the wavelet transform and multivariate data analysis

In this and the following sections of this chapter, we turn our attention to the processing of 1D data in the form of time series or spectra or vectors. The objective is data exploration – e.g. clustering – or in later sections time series prediction. For descriptive data analysis, we will switch attention to use of an orthogonal wavelet transform. For prediction, we will return to use of the à trous method.

Data analysis, for exploratory purposes, or prediction, is usually preceded by various data transformations and recoding. In fact, we would guess that 90% of the work involved in analyzing data lies in this initial stage of data preprocessing. This includes: problem demarcation and data capture; selecting non-missing data of fairly homogeneous quality; data coding; and a range of preliminary data transformations.

The wavelet transform offers a particularly appealing data transformation, as a preliminary to data analysis. It offers additionally the possibility of close integration into the analysis procedure. The wavelet transform may be used to 'open up' the data to de-noising, smoothing, etc., in a natural and integrated way.

4.2.1 Wavelet regression in multivariate data analysis

Our task is to consider the approximation of a vector x at finer and finer scales. The finest scale provides the original data, $x_N = x$, and the approximation at scale m is x_m where usually $m = 2^0, 2^1, \ldots, 2^N$. The incremental detail added in going from x_m to x_{m+1}, the detail signal, is yielded by the wavelet transform. If ξ_m is this detail signal, then the following holds:

$$x_{m+1} = H^T(m)x_m + G^T(m)\xi_m \tag{4.6}$$

where $G(m)$ and $H(m)$ are matrices (linear transformations) depending on the wavelet chosen, and T denotes transpose (adjoint). This description is similar to that used in Strang (1989) and Bhatia, Karl and Willsky (1996).

An intermediate approximation of the original signal is immediately possible by setting detail components $\xi_{m'}$ to zero for $m' \geq m$ (thus, for example, to obtain x_2, we use only x_0, ξ_0 and ξ_1). Alternatively we can de-noise the detail signals before reconstituting x and this has been termed wavelet regression (Bruce and Gao, 1994).

Define ξ as the row-wise juxtaposition of all detail components, $\{\xi_m\}$, and the final smoothed signal, x_0, and consider the wavelet transform $\mathcal{T}$ given by

$$\mathcal{T}x = \xi = \begin{bmatrix} \xi_{N-1} \\ \cdot \\ \cdot \\ \xi_0 \\ x_0 \end{bmatrix} \tag{4.7}$$

Taking $\mathcal{T}^T\mathcal{T} = I$ (the identity matrix) is a strong condition for exact reconstruction of the input data, and is satisfied by the orthogonal wavelet transform. This permits use of the 'prism' (or decomposition in terms of scale and location) of the wavelet transform.

Examples of these orthogonal wavelets, i.e. the operators G and H, are the Daubechies family, and the Haar wavelet transform (Daubechies, 1992; Press *et al.*, 1992). For the Daubechies D_4 wavelet transform, H is given by

$$(0.482\,962\,913, 0.836\,516\,304, 0.224\,143\,868, -0.129\,409\,523)$$

and G is given by

$$(-0.129\,409\,523, -0.224\,143\,868, 0.836\,516\,304, -0.482\,962\,913).$$

Implementation is by decimating the signal by two at each level and convolving with G and H: therefore the number of operations is proportional to $n+n/2+n/4+\ldots = O(n)$. Wrap-around (or 'mirroring') is used by the convolution at the extremities of the signal.

We consider the wavelet transform of x, $\mathcal{T}x$. Consider two vectors, x and y. The squared Euclidean distance between these is $\|x-y\|^2 = (x-y)^T(x-y)$. The squared Euclidean distance between the wavelet transformed vectors is $\|\mathcal{T}x - \mathcal{T}y\|^2 = (x-y)^T\mathcal{T}^T\mathcal{T}(x-y)$, and hence identical to the distance squared between the original vectors. For use of the Euclidean distance, the wavelet transform can replace the original data in the data analysis. The analysis can be carried out in wavelet space rather than in direct space. This in turn allows us to directly manipulate the wavelet transform values, using any of the approaches found useful in other areas. The results based on the orthogonal wavelet transform exclusively imply use of the Euclidean metric, which nonetheless covers a considerable area of current data analysis practice. Future work will investigate extensions to other metrics.

Note that the wavelet basis is an orthogonal one, but is not a principal axis one (which is orthogonal, but also optimal in terms of least squares projections). Wickerhauser (1994) proposed a method to find an approximate principal component basis by determining a large number of (efficiently-calculated) wavelet bases, and keeping the one closest to the desired Karhunen-Loève basis. If we keep, say, an approximate representation allowing reconstitution of the original n components by n' components (due to the dyadic analysis, $n' \in \{n/2, n/4, \ldots\}$), then we see that the space spanned by these n' components will not be the same as that spanned by the n' first principal components.

Filtering or nonlinear regression of the data can be carried out by deleting insignificant wavelet coefficients at each resolution level (noise filtering), or by 'shrinking' them (data smoothing). Reconstitution of the data then provides a cleaned data set. A practical overview of such approaches to data filtering (arising from work by Donoho and Johnstone at Stanford University) can be found in Bruce and Gao (1994, chapter 7).

4.2.2 Degradation of distance through wavelet approximation

The input data vector of most interest to us in this chapter is associated with an ordered set of values: e.g. a time series, or a spectrum ordered by the wavelength at which the flux value was observed. But nothing prevents us taking any arbitrary vector as an ordered sequence. This just implies that, if we wavelet-filter such a vector, an equally valid result could have been obtained on the basis of a different sequence of the input values. Subject to such non-uniqueness, there is nothing wrong with analyzing unordered input vectors in this way and we will in fact begin with such an investigation.

An example is shown in Figs. 4.11–4.13. A 64-point, 64-dimensional data set was simulated to contain some structure and additive Gaussian noise. (This was carried out by creating a 2D Gaussian shape, and superimposing lower-level noise; and then reading off each row as a separate vector.) Figure 4.11 shows a principal coordinates (met-

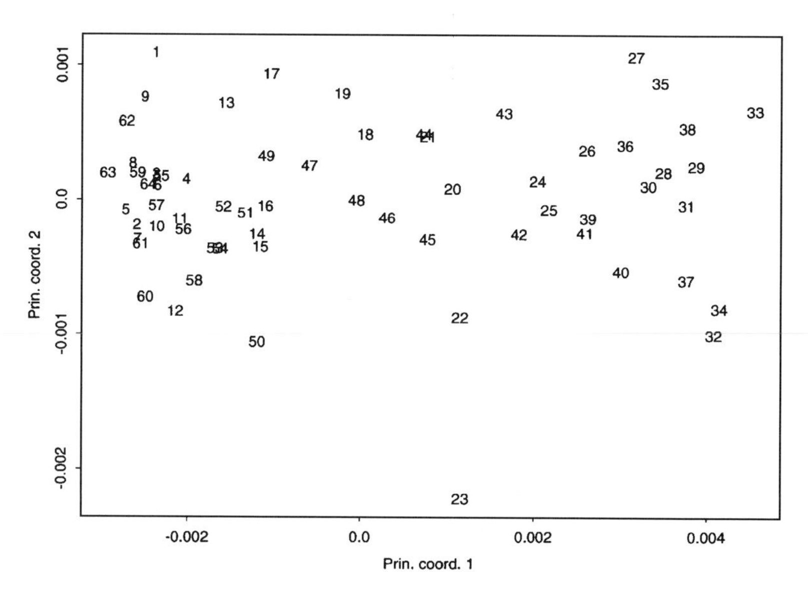

Figure 4.11 Planar projection of original data (64 points).

ric multidimensional scaling; Torgerson, 1958) plot of the original data, projected onto the plane. The principal coordinates analysis was carried out on Euclidean pairwise distances between the 64 vectors (as was done for Figs. 4.12 and 4.13 also). A D_4 Daubechies wavelet transform was made of this same dataset and a principal coordinate plot constructed (not shown here). This principal coordinate plot of the wavelet transformed data gave (for reasons explained in previous

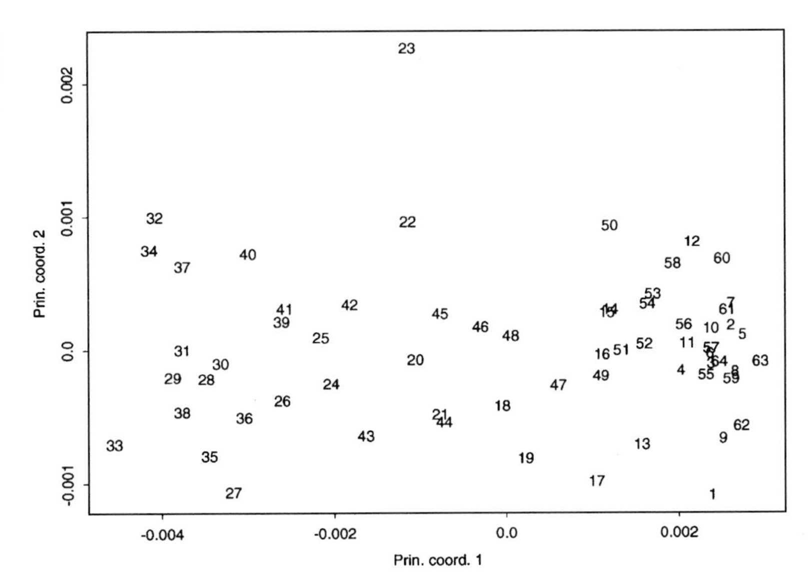

Figure 4.12 Planar projection of filtered data.

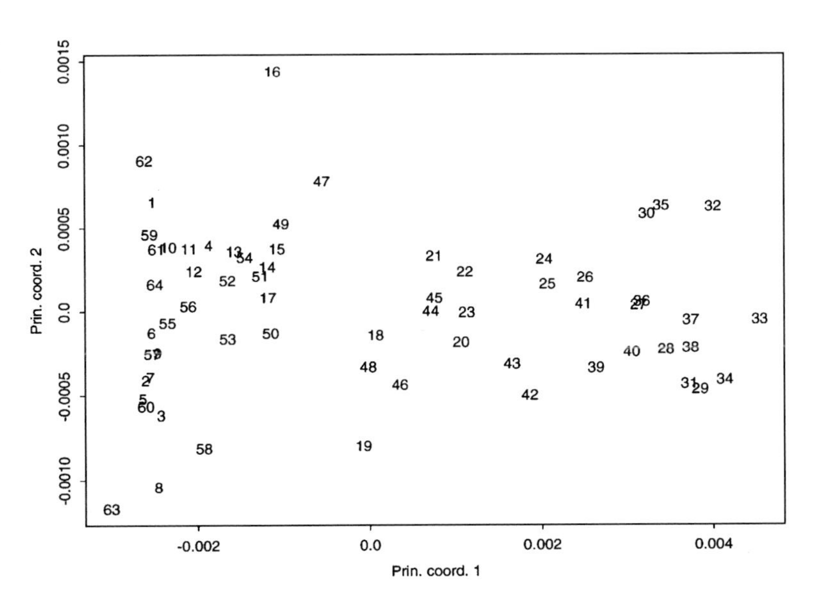

Figure 4.13 Planar projection of filtered data.

sections) an identical result to the principal coordinate analysis of the original data.

Figure 4.12 shows a similar principal coordinate planar projection output following suppression of wavelet transform coefficients which were less than 0.1 times the overall standard deviation. In this way, 12.2% of the coefficients were set to zero. The result (Fig. 4.12) is extremely similar to the result on the original data (Fig. 4.11). Therefore 'cleaning' the data by removing 12.2% of the values had an imperceptible effect.

Increasing the suppression of wavelet coefficient values to one standard deviation (Fig. 4.13) leads to appreciable change in the points' positions. Bear in mind, of course, that the figures show planar projections of (in principle) a 64-dimensional input space. With the wavelet thresholding used for Fig. 4.13, 78.5% of the values were set to zero.

Setting wavelet coefficients to zero, as has been stated, amounts to 'cleaning' the data. This can be furthered also by data quantization. This in turn can lead to economies of storage, and of calculation of distances. Jacobs, Finkelstein and Salesin (1995) discuss the use of the Haar wavelet transform, wavelet coefficient clipping, and then quantization of remaining wavelet coefficients to −1 and +1. They proceed to discuss a very efficient dissimilarity calculation method.

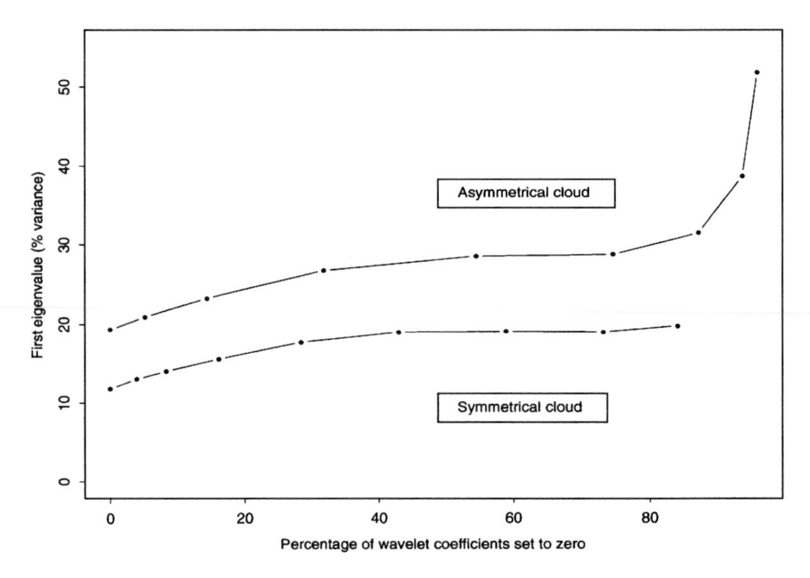

Figure 4.14 First eigenvalue (percentage of variance) as a function of zeroing of wavelet coefficient values.

4.2.3 Degradation of first eigenvalue through wavelet filtering

Linearity of data plays an important role in a number of areas such as regression and using local linearity for pattern recognition (an example of the latter is Banfield and Raftery, (1993)). Zeroing wavelet coefficients and then reconstituting the data help to clean it. We checked in a number of cases what effect this had on the first few eigenvalues. The data clouds consisted of 150 points in 64D space. The 'symmetric cloud' consisted of eight Gaussian (mean 4, standard deviation 4) subclouds each in 8D spaces. The 'asymmetric cloud' was constructed similarly but had standard deviations of 2 or 4, and means of 4, 6, or 8 (in the different 8D subspaces). For varying thresholds, wavelet coefficients were set to zero. The data was then reconstituted from these 'cleaned' wavelet coefficients. Figure 4.14 shows the outcome, when the eigenvalues were determined on the cleaned and original (i.e. corresponding to no cleaning: the zero point on the horizontal axis) data. We see that, with up to 75% of the wavelet coefficients zeroed, the change in first eigenvector is not very large.

4.3 The Kohonen self-organizing map in wavelet space

We now turn attention to a small sample of real data. We will look at clustering and dimensionality reduction methods, in direct and in wavelet space. We begin with a method which combines both of these goals.

The Kohonen self-organizing feature map (SOFM) with a regular grid output representational or display space, involves determining vectors w_k, such that inputs x_i are parsimoniously summarized (clustering objective); and in addition the vectors w_k are positioned in representational space so that similar vectors are close (low-dimensional projection objective) in *representation space*: $k, k', k'' \in \{(r,s) \mid r = 1,\ldots,R; s = 1\ldots,S\}$.

Clustering: Associate each x_i with some one w_k such that

$$k = \text{argmin} \parallel x_i - w_k \parallel$$

Low-Dimensional projection: $\parallel w_k - w'_k \parallel < \parallel w_k - w''_k \parallel \Longrightarrow \parallel k - k' \parallel \leq \parallel k - k'' \parallel$

By way of example, $R = S = 10$ and the output representation grid is a regular, square one. The metric chosen for norm $\parallel . \parallel$ is usually Euclidean, and this will be assumed here. Without loss of generality, we can consider the squared Euclidean distance whenever this is useful to us. Evidently $x_i \in \mathbb{R}^m$ and $w_k \in \mathbb{R}^m$ for some dimensionality, or cardinality of attribute-set, m.

Iterative algorithms for clustering are widely used, requiring an initial random choice of values for w_k to be updated on the basis of presentation of input vectors, x_i. At each such update, the low-dimensional projection objective is catered for by updating not just the so-called winner w_k, but also neighbors of w_k with respect to the representational space. The neighborhood is initially chosen to be quite large (e.g. a 4×4 zone) and as the epochs proceed, is reduced to 1×1 (i.e. no neighborhood). An epoch is the term used for a complete set of presentations, and consequent updates, of the N input vectors. The result obtained by the SOFM algorithm is sub-optimal, as also is the case usually for clustering algorithms of this sort (k-means, partitioning) and quite often for dimensionality-reduction methods (Kruskal, 1964; Sammon, 1969). A range of studies showing how well the SOFM method performs compared to these methods can be found in Murtagh and Hernández-Pajares (1995).

We have seen that $\| x_i - w_k \|^2 = \| \mathcal{T} x_i - \mathcal{T} w_k \|^2$ where $\mathcal{T}$ is an orthogonal wavelet transform. Thus the idea presents itself to (i) transform the x_is with linear computational cost to the $\mathcal{T} x_i$s; and (ii) use the SOFM iterative training method to determine the $\mathcal{T} w_k$s. In doing this, we expect (i) the final assignments of each x_i to a 'winner' w_k to provide the same results as heretofore; and (ii) if required, the final values of $\mathcal{T} w_k$ to be inverse-transformable to provide the values of w_k. Performing the SOFM in direct or in wavelet space are equivalent since either (i) definitions and operations are identical, or (ii) all relevant update operations can be expressed in terms of the Euclidean metric. Differences between the two analyses can arise only due to the different starting conditions, i.e. the random initial values of all w_k. Clearly, stable, robust convergence to a quasi-optimal solution precludes such differences.

4.3.1 Example of SOFM in direct and in wavelet spaces

We used a set of 45 astronomical spectra of the object examined in Mittaz, Penston and Snijders (1990). These were of the complex AGN (active galactic nucleus) object, NGC 4151, and were taken with the small but very successful IUE (International Ultraviolet Explorer) satellite which ceased observing in 1996 after nearly two decades of operation. We chose a set of 45 spectra observed with the SWP spectral camera, with wavelengths from 1191.2 Å to approximately 1794.4 Å, with values at 512 interval steps. There were some minor discrepancies in the wavelength values, which we discounted: an alternative would have been to interpolate flux values (vertical axis, y) in order to have values at identical wavelength values (horizontal axis, x), but we did not do this since the infrequent discrepancies were fractional parts of the most common regular interval widths. Figure 4.15 shows a sample of 20 of these spectra. A wavelet transform (Daubechies 4 wavelet used) version of these spectra was generated, with a number of scales generated which was allowed by dyadic decomposition. An overall 0.1σ (standard deviation, calculated on all wavelet coefficients) was used as a threshold, and coefficient values below this were set to zero. Spectra which were apparently more noisy had relatively few coefficient values set to zero, e.g. 31%. More smooth spectra had up to over 91% of their coefficients set to zero. On average, 76% of the wavelet coefficients were zeroed in this way. Figure 4.16 shows the relatively high quality spectra re-formed, following zeroing of wavelet coefficient values.

Figures 4.17 and 4.18 show SOFM outputs using 5×6 output representational grids. When a number of spectra were associated with a representational node, one of these is shown here, together with an indication of how many spectra are clustered at this node. Hatched nodes indicate no assignment of a spectrum. Each spectrum was normalized to unit maximum value. While some differences can be noted between Figs. 4.17 and 4.18, it is clear that Fig. 4.18 (based on 76% zeroing of wavelet coefficients, and then reconstitution of the data) is very similar to Fig. 4.17.

We then constructed the SOFM on the wavelet coefficients (following zeroing of 76% of them). The assignments of the 45 spectra were identical to the assignments associated with Fig. 4.18. The values associated with output representational nodes were, in this case, the values $\mathcal{T} w_k$, which can be converted back to w_k values (with linear computational cost).

This approach to SOFM construction leads to the following possibilities:

1 Efficient implementation: a good approximation can be obtained by zeroing most wavelet coefficients, which opens the way to more appropriate storage (e.g. offsets of non-zero values) and distance calculations (e.g. implementation loops driven by the stored non-zero values). Similarly, compression of large datasets can be carried out. Finally, calculations in a high-dimensional space, $\mathbb{R}^m$, can be carried out more efficiently since, as seen above, the number of

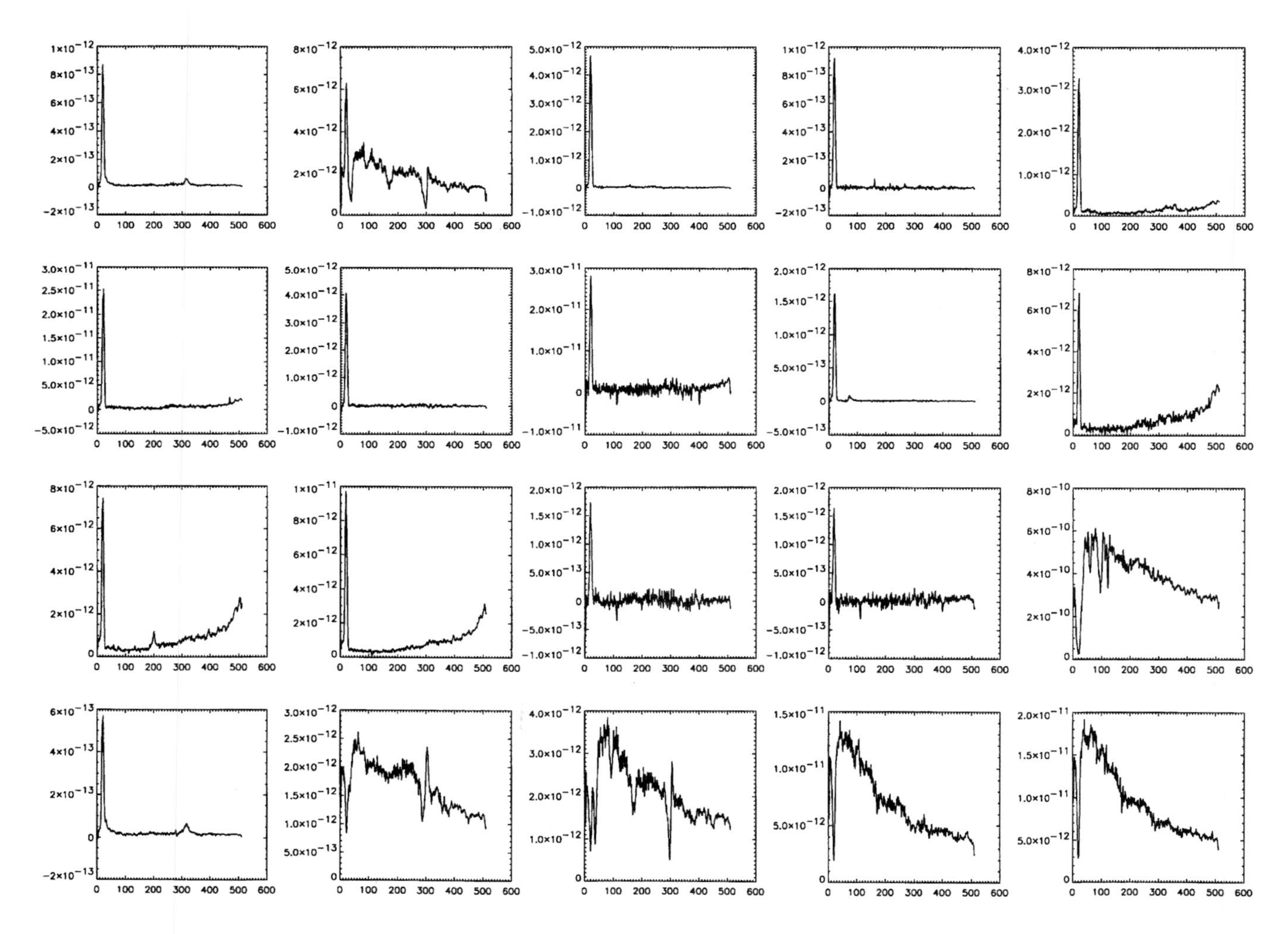

Figure 4.15 Sample of 20 spectra (from 45 used) with original flux measurements plotted on the y-axis.

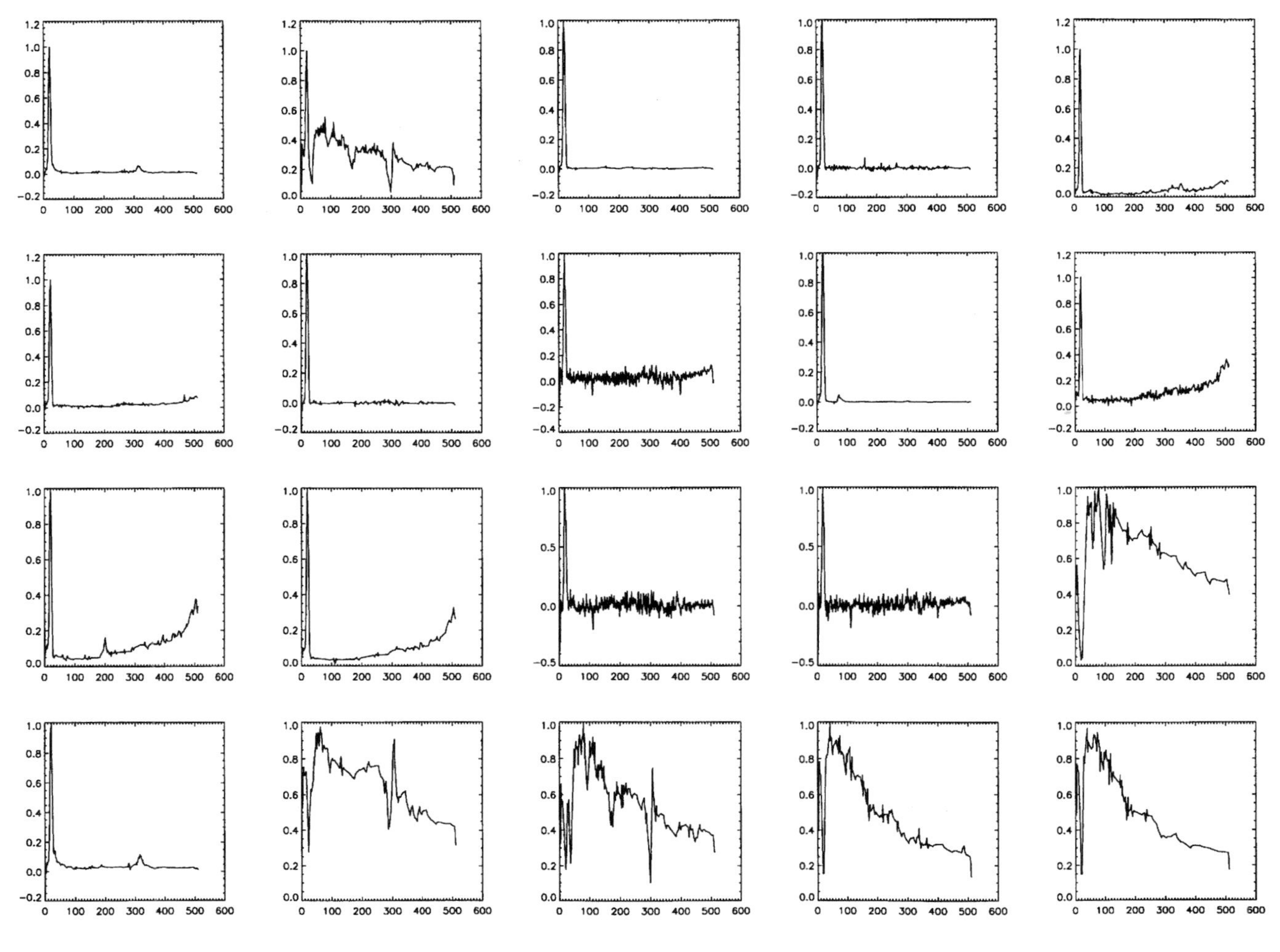

Figure 4.16 Sample of 20 spectra (as in previous figure), each normalized to unit maximum value, then wavelet transformed, approximately 75% of wavelet coefficients set to zero, and reconstituted.

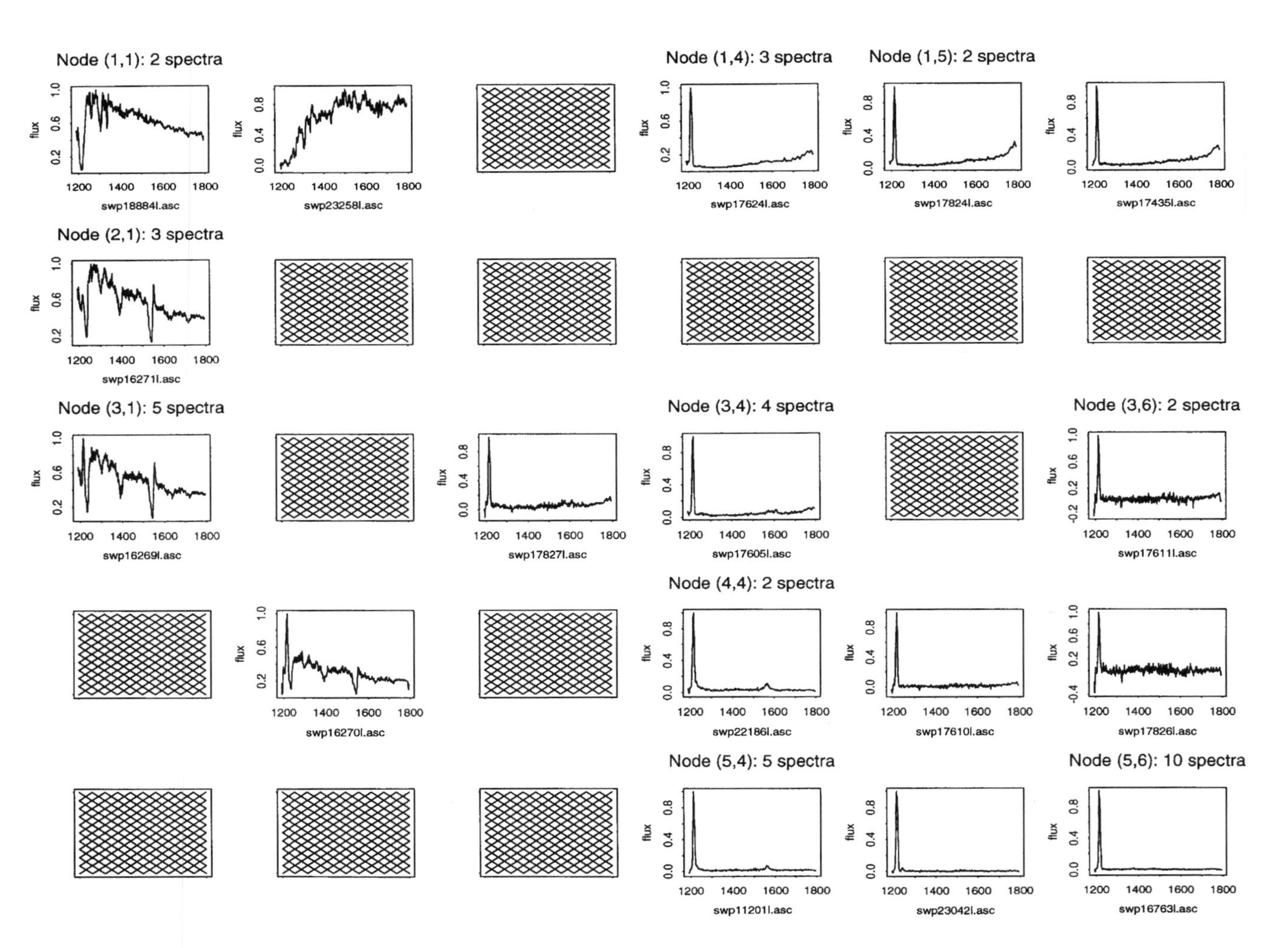

Figure 4.17 Kohonen SOFM of 45 spectra: original data.

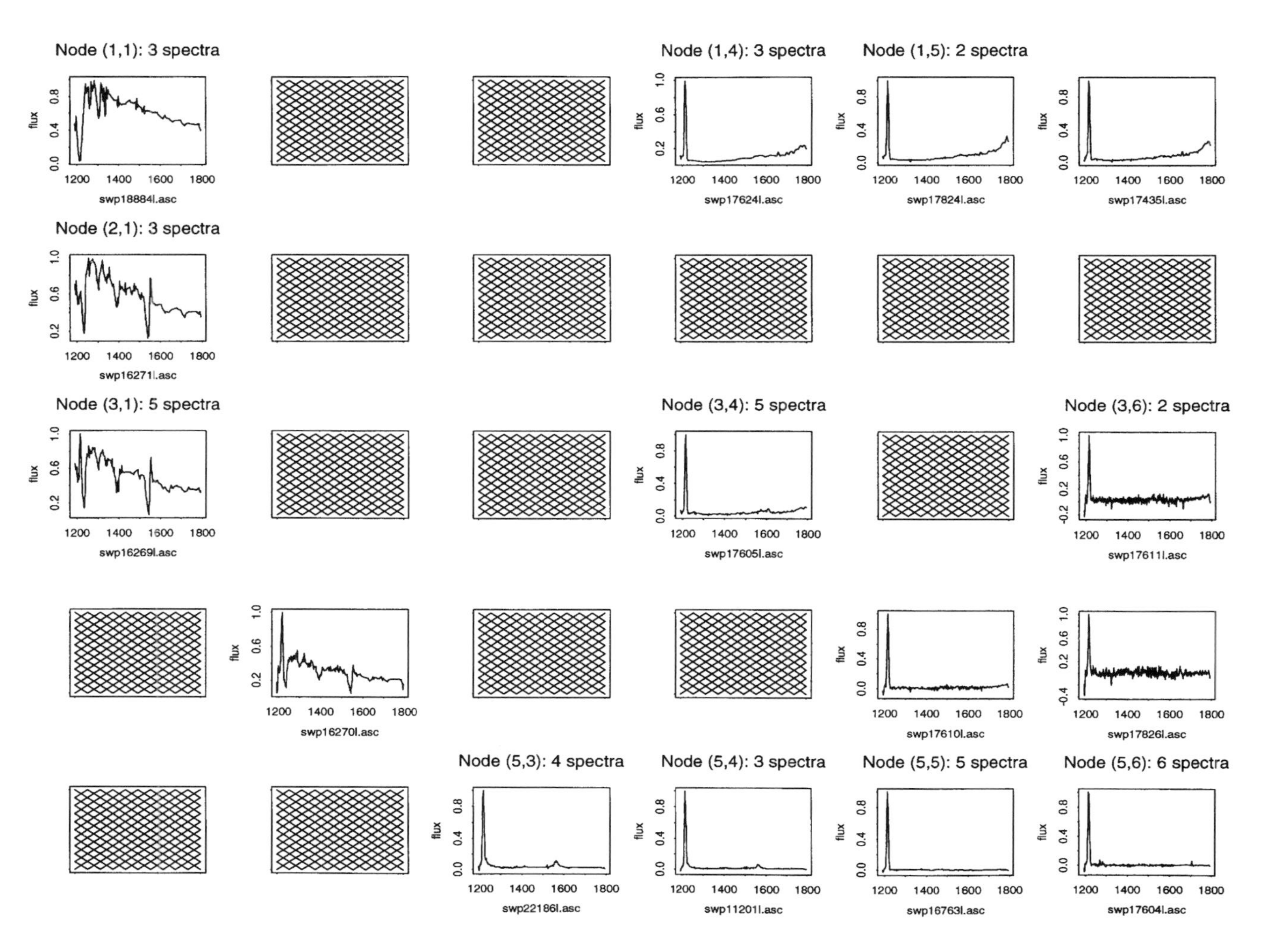

Figure 4.18 Kohonen SOFM of 45 spectra: data reconstituted following zeroing of 76% of wavelet coefficients.

non-zero coefficients may well be $m'' << m$ with very little loss of useful information.

2 Data 'cleaning' or filtering is a much more integral part of the data analysis processing. If a noise model is available for the input data, then the data can be de-noised at multiple scales. By suppressing wavelet coefficients at certain scales, high-frequency (perhaps stochastic or instrumental noise) or low-frequency (perhaps 'background') information can be removed. Part of the data coding phase, prior to the analysis phase, can be dealt with more naturally in this new integrated approach.

This powerful, integrated wavelet transform and neural network method allows some major questions relating to input data pre-treatment (filtering, coding, preliminary transformation) to be addressed. The various uses of the wavelet transform can be availed of.

Separately, there may be a considerable computational gain. If 75% of the input data (let us take this as n m-valued vectors as in the example above) is set to zero, we can note that $m/4$ offset values are required to indicate the $m/4$ non-zero values on average in each vector. Thus there are $m/2$ values to be considered in each such vector. The scalar product, or Euclidean distance, is then 'driven' by these values. This implies $m/2$ calculations of Euclidean metric-related components. Thus we have, in this case, a potential two-fold computational gain immediately. The greater the value of m, the more interest we have in using this result.

The integrated wavelet/neural net approach proposed is of clear practical relevance, in different application fields.

4.3.2 K-means and principal components analysis in wavelet space

We used the set of 45 spectra which were 'cleaned' in wavelet space, by putting 76% of the coefficients to zero (cf. above). This was therefore a set of cleaned wavelet-transformed vectors. We also reconstructed these cleaned spectra using the inverse wavelet transform. Due to design, therefore, we had an input data array of dimensions 45×512, with 76% of values equal to zero; and an input data array of dimensions 45×512, with no values exactly equal to zero.

A number of runs of the k-means partitioning algorithm were made. The exchange method, described in Späth (1985) was used. Four, or two, clusters were requested. Identical results were obtained for both data sets, which is not surprising given that this partitioning method is based on the Euclidean distance. For the 4-cluster, and 2-cluster, solutions we obtained respectively these assignments:

```
123213114441114311343133141121412222222121114
```

```
122211111111111111111111111121112222222121111
```

The case of principal components analysis (PCA) was very interesting. We know that the basic PCA method uses Euclidean scalar products to define the new set of axes. Often PCA is used on a variance-covariance input matrix (i.e. the input vectors are centered); or on a correlation input matrix (i.e. the input vectors are rescaled to zero mean and unit variance). These two transformations destroy the Euclidean metric properties vis-à-vis the raw data. Therefore we used PCA on the unprocessed input data. We obtained identical eigenvalues and eigenvectors for the two input data sets.

The eigenvalues are similar up to numerical precision:

```
1911.217  210.355  92.042  13.909   7.482   2.722   2.305
```

```
1911.221  210.355  92.042  13.909   7.482   2.722   2.305
```

The eigenvectors are similarly identical. The actual projection values are entirely different. This is simply due to the fact that the principal components in wavelet space are themselves inverse-transformable to provide principal components of the initial data.

Various aspects of this relationship between original and wavelet space remain to be investigated. We have argued for the importance of this, in the framework of data coding and preliminary processing. We have also noted that if most values can be set to zero with limited (and maybe beneficial) effect, then there is considerable scope for computational gain also. The processing of sparse data can be based on an 'inverted file' data-structure which maps non-zero data entries to their values. The inverted file data-structure is then used to drive the distance and other calculations. Murtagh (1985, pp. 51–54 in particular) discusses various algorithms of this sort.

4.4 Multiresolution regression and forecasting

4.4.1 Meteorological prediction using the à trous method

We used a set of 96 monthly values of the jetstream (m s^{-1}), measured over central Chile. We used $p = 5$ wavelet resolution levels. Figure 4.19 shows (i) the original data – the most volatile of the curves to be seen in this figure; (ii) the curve yielded by $c_5 + w_5 + w_4 + w_3$; and (iii) the near-'background' curve yielded by $c_5 + w_5$.

For predictability, we used a one-step-ahead 'carbon-copy' approach (the simplest autoregressive approach: for example, predictor of $c_0(k) = c_0(k-1)$). The MSE (mean square error) predictions for c_5, $c_5 + w_5$, $c_5 + w_5 + w_4$, ..., c_0, were respectively: 0.01, 0.03, 0.03, 9.0, 18.0, 149.3, 235.3. Not surprisingly the more coarse the approximation to the original data, the greater the ease of making a good forecast. The wavelet transform data-structure provides a disciplined output – an ordered set of increasingly smoothed versions of the input data.

Although one-step-ahead forecasts were carried out, it must be noted in this case that the 'future' values of the data did go into the making of the wavelet coefficients at the various resolution scales. This was not the case in a further set of forecasting experiments, in which the predictions were made on entirely withheld data.

Figure 4.20 shows the jetstream values and a set of 90 consecutive values. The lower part of this figure, and Fig. 4.21, show the wavelet transform. The wavelet coefficients, plus the residual, can be added to reconstruct the input data. The residual shows an interesting secular trend.

Forecasts were carried out on the input 90-valued data; and on each wavelet resolution level, with the sum of values being used to give an overall prediction. Six time-steps ahead were used in all cases. The withheld values were of course available, for comparison with the forecasts. A simple autoregressive forecast was carried out, which was sufficient for this study. For an autoregressive forecast using a parameter of 12 (plausible, based on a 12-month period) the mean square error of the 6 time-step prediction using the given data was

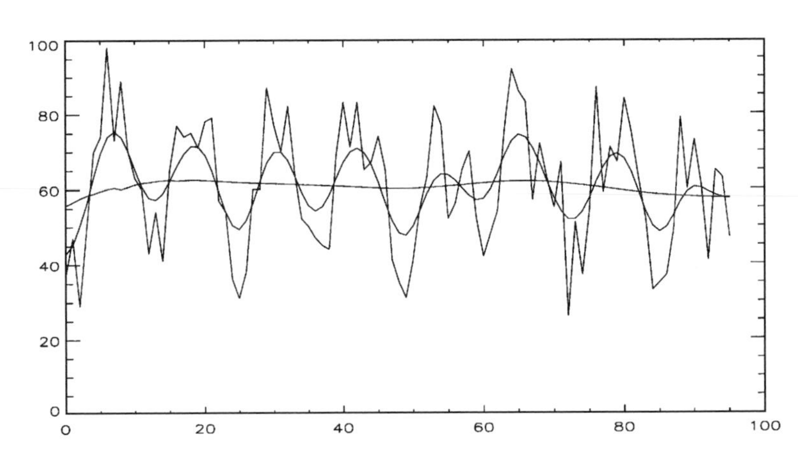

Figure 4.19 Ninety-six monthly values of jetstream, overplotted with two smoothed versions, provided by the à trous wavelet transform.

1240.17. The mean square error of the 6 time-step prediction using the cumulated multiple scale predictions was 584.49.

4.4.2 Sunspot prediction using à trous and neural networks

The sunspot series was the first time series studied with autoregressive models (Priestley, 1981; Tong, 1990; Yule, 1927), and thus has served as a benchmark in the forecasting literature. The sunspots are dark blotches on the sun that can be related to other solar activities such as the magnetic field cycles, which in turn influence, by indirect and intangible means, the meteorological conditions on earth. The solar cycle with which sunspots occur varies over a 7 to 15 year period. Although the data exhibit strong regularities, attempts to understand the underlying features of the series have failed because the amount of available data was insufficient. Thus the sunspots provide an interesting data-set to test our wavelet decomposition method.

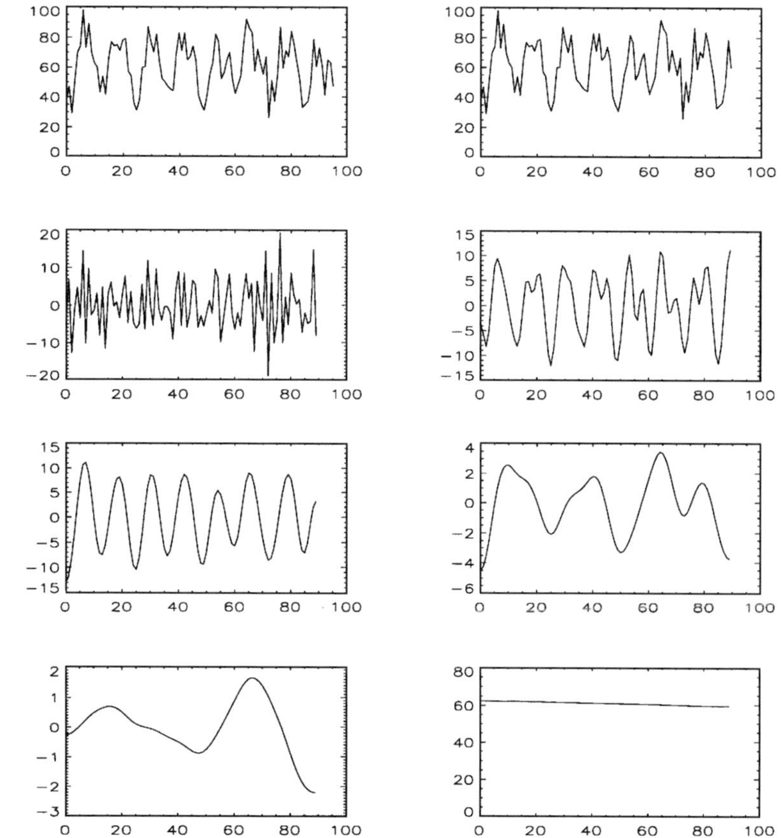

Figure 4.20 Upper left: ninety-six monthly values of jetstream. Upper right: ninety values used for forecasting. Lower left and right: first two sets of wavelet coefficients.

Figure 4.21 Upper left and right: third and fourth sets of wavelet coefficients. Lower left: fifth set of wavelet coefficients. Lower right: residual.

Consistent with previous appraisals (Tong, 1990; Weigend, Rumelhart and Huberman, 1990), we use range-normalized yearly averages of the sunspot data tabulated from 1720 to 1979. Since quantitative performance measures are expressed in terms of one-step ahead errors in the literature, the latter is used as a performance criterion. The single-step prediction error is monitored on the 59 withheld sunspot values ranging from 1921 to 1979, while the remaining data is used for training. The performance measure is expressed in terms of one-step ahead *normalized mean square errors* $\sum_j (x_j - \hat{x}_j)^2 / N\sigma^2$ where the sum is taken over the N patterns in the test set, x_j denotes the true value of the sequence, $\hat{x}_j$ is the forecast, and σ^2 is the variance of the true sequence.

4.4.3 Dynamic recurrent neural network architecture

A very general class of neural network was investigated. This was a dynamic recurrent neural network (DRNN), a fully connected connectionist network with N nodes which, when solved, obeys the following:

$$\mathbf{v}_k = g\left(\sum_{d=0}^{D} \mathbf{W}_k^{d\,T} \mathbf{v}_{k-d}\right) + \mathbf{i}_k \tag{4.8}$$

where k is a time-point, d is a delay, $\mathbf{i}$ is a clamped (fixed) input vector, D is the maximum connection delay, g is a sigmoid function, $\mathbf{W}$ are weights, and $\mathbf{v}$ is an m-dimensional vector. Relative to the more traditional neural network models, here each synaptic weight is replaced by a finite impulse response (FIR) filter. The DRNN method is studied in detail in Aussem (1995) and Aussem, Murtagh and Sarazin (1995).

A natural manner to assess the performance of the wavelet-based strategy is to take a given network architecture and compare the mean square error obtained by the direct prediction with the error obtained by the decomposition method.

With this in mind, it is wise to select an efficient architecture which balances pure performance against complexity, given some constraints on the topology, the number of neurons and synaptic connections. For the sake of simplicity, we restrict ourselves to a network having one input unit, two hidden units fully connected with FIR filters of order 2, and one output unit: see Fig. 4.22. Both input and output units are connected with static links so that the memory of the system is purposely confined to the core of the model. In particular, a single sunspot number is presented to the input and the network is trained to output the next value, thus forcing the DRNN to encode some past information on the sequence. A time lagged vector is no longer required thanks to the internal dynamics of the model.

The network was trained up to complete convergence. This tiny architecture yielded quite acceptable performance. The resulting normalized mean square error for single step predictions over the prediction set is 0.154. Of course, better performance is achieved for larger networks but the aim of this experiment was not to achieve pure performance but instead to compare two methodologies.

4.4.4 Combining neural network outputs

The same DRNN configuration was used to forecast each wavelet resolution. Each network was fed with the prior series value for the series $w_j(t)$, at some time t, to provide an estimate of the next value for $w_j(t+1)$. So the prediction engine was called into operation 6 times, here, entirely independently. Once the univariate predictions were obtained for the series $w_j(t)$ for j between 1 and 6, the error was afterwards compared to other indexes of performance, namely the MLP (multilayer perceptron, more on which below) and a very basic index which we term the 'carbon-copy' error, i.e. the error based on the most simple autoregressive forecast. This carbon-copy error

Table 4.2 *Single-step prediction errors on the test set. The more negative the outcome, the more accurate the prediction.*

Model	w_1	w_2	w_3	w_4	w_5	*res*
DRNN	−0.55	−1.04	−1.46	−0.65	−1.03	−1.24
MLP	−0.38	−1.00	−0.95	−0.38	−0.66	−1.12
Carbon copy	+0.06	−0.13	−0.21	−0.49	−1.12	−1.66

Table 4.3 *Single-step prediction errors on the test set.*

Model	data type	nMSE
DRNN $1 \times 2 \times 1$ with 0 :2 :0 order filters	coarse data	0.154
Hybrid MLP $6 \times 5 \times 1$ + Carbon copy	wavelet coeff.	0.209
Hybrid DRNN + Carbon copy	wavelet coeff.	0.123

was more favorable for the last coefficient and the residual (w_5 and *res* in Table 4.2). Note that these are the more smoothly-varying series resulting from the wavelet decomposition. This led us to benefit from improved performance of this simple autoregressive forecaster: we retained predictions by it, for w_5 and *res*; and we retained predictions by the very much more sophisticated and long-memory DRNN for w_1, w_2, w_3 and w_4.

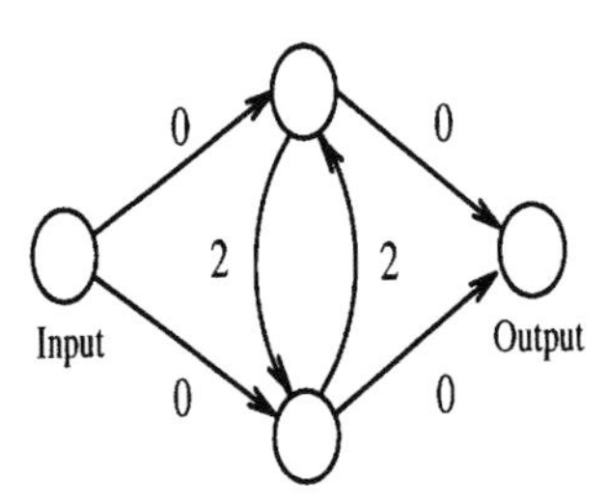

Figure 4.22 Schematic representation of DRNN model used for prediction. The figures associated with the arrows indicate the order of the FIR filter used to model the synaptic link. The first layer has 1 input unit; the second layer has 2 hidden units; and the output layer has 1 unit.

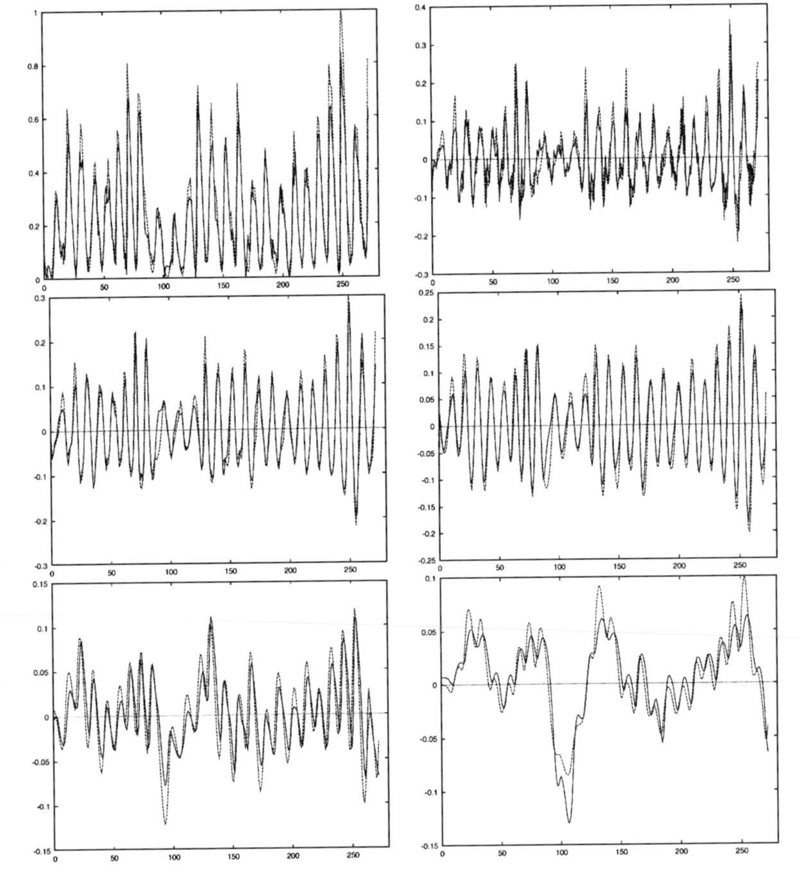

Figure 4.23 Single step prediction with several DRNNs over the whole data set (training + test sets). Top left: original and predicted sunspot series overplotted; the prediction was obtained from the combination of the individual wavelet resolution scale forecasts w_i for $i = 1, \ldots, 5$ and the residual. The w_i are shown from the top right to the bottom. The dashed curve corresponds to the actual series and the plain curve is the prediction.

To illustrate the predictive ability of such a small DRNN model, an MLP of size $(6 \times 5 \times 1)$ having 41 weights including bias and fed with the delay vector $[w_i(t-1), w_i(t-2), \ldots, w_i(t-6)]$ for j between 1 and 5, was trained to provide comparative predictions. Selection of these dimensions was based mostly on trial and error. We remark that the MLP has twice as many parameters as the DRNN although better predictions are obtained for all wavelet scales with the DRNN as can be seen in Table 4.2.

The ability of the network to capture dynamical behavior over higher resolution levels deteriorates quite fast. This is intuitively clear in view of the plots in Fig. 4.23. The higher the order of the resolution scale, the smoother the curve, and thus the less information the network can retrieve. This is also the reason why we stopped the wavelet decomposition process at order 6. From lag 5 onwards, the network outputs were replaced by the carbon-copy estimates in these cases. The individual resolution coefficient predictions were afterwards summed together to form an estimate of the sunspot value, $x(t+1)$. The resulting normalized mean square error for single step predictions over the prediction set is 0.123 with the hybrid DRNN and carbon-copy method, instead of 0.209 with the hybrid MLP and carbon copy (Table 4.3). Finally, this compares with 0.154 with use of the direct DRNN approach on the original time series (not wavelet-decomposed). Consequently, the combination of the DRNN on the wavelet decomposition provides improvement of the prediction accuracy.

5 Geometric registration

Image registration of remotely sensed data is a procedure that determines the best spatial fit between two or more images that overlap the same scene, acquired at the same time or at different times, by identical or different sensors. This is an important step, as it is frequently necessary to compare data taken at different times on a point-to-point basis, for many applications such as the study of temporal changes for example. Therefore we need to obtain a new dataset in such a way that its image under an appropriate transform is registered, geometrically, with previous datasets.

The inventory of natural resources and the management of the environment requires appropriate and complex perception of the objects to be studied. Often a multiresolution approach is essential for the identification of the phenomena studied, as well as for the understanding of the dynamical processes underlying them. In this case, the processing of data taken with different ground resolutions by different or identical sensors is necessary.

Another important situation where the need for different images acquired with a different ground resolution sensor arises is when the generalization to larger surface areas of an identification or an interpretation model, based on small areas, is required (Achard and Blasco, 1990). This is the case for studies on a continental scale. Examples of this application can be found in Justice and Hiernaux (1986), Hiernaux and Justice (1986) and Prince, Tucker and Justice (1986). Therefore, the data must be geometrically registered with the best possible accuracy.

Several digital techniques have been used for automatic registration of images such as cross-correlation, normal cross-correlation and minimum distance criteria; e.g. Barneau and Silverman (1972), Jeansoulin (1982) and Pratt (1978). After a brief description of these methods, we will present a procedure for automatic registration of remotely sensed data based on multiresolution decomposition of the data with the use of the wavelet transform. The advantage of the wavelet transform is that it produces both spatial and frequency domain information which allows the study of the image by frequency band. We will illustrate this study with different sets of data obtained with identical sensors as well as with different sensors. In particular, we will illustrate registration from among the registration of SPOT versus SPOT data, MSS versus MSS data, SPOT versus MSS data, and SPOT versus TM data.

5.1 Image distortions

Images acquired by on-board satellite sensors are affected by a number of distortions which, if left uncorrected, would affect the accuracy of the information extracted and thereby reduce the usefulness of the data. These distortions can be characterized globally by two categories: geometrical distortions; and radiometrical distortions.

5.1.1 Geometrical distortions

Geometrical distortions can be separated into two groups: systematic or predictable, and random or non-predictable. Each of these groups can also be separated into two types:

- Internal distortions: these are sensor-related distortions, systematic and stationary and can be corrected by calibration.
- External distortions: these are due to platform perturbations and to the scene characteristics and are variable by nature. These distortions can be determined from a model of the altitude and in some cases from a digital terrain model (DTM) or from control points taken on the images.

The effect of these errors have been characterized in Silverman and Bernstein (1971) and are shown in Fig. 5.1.

5.1.2 Radiometrical distortions

Radiometric distortions are characterized by incorrect intensity distribution, spatial frequency filtering of the scene data, blemishes in the imagery, banding of the image data, etc. These distortions are caused by camera or scanner shading effects, detector gain variations, atmospheric and sensor induced filtering, sensor imperfections, sensor detector gain errors, etc. They can also be separated into two types:

systematic or predictable, and random or non-predictable (Lillesand and Kiefer, 1987). As in the case of geometric distortions, each of these groups can be separated into internal and external (Berstein, 1975, 1976).

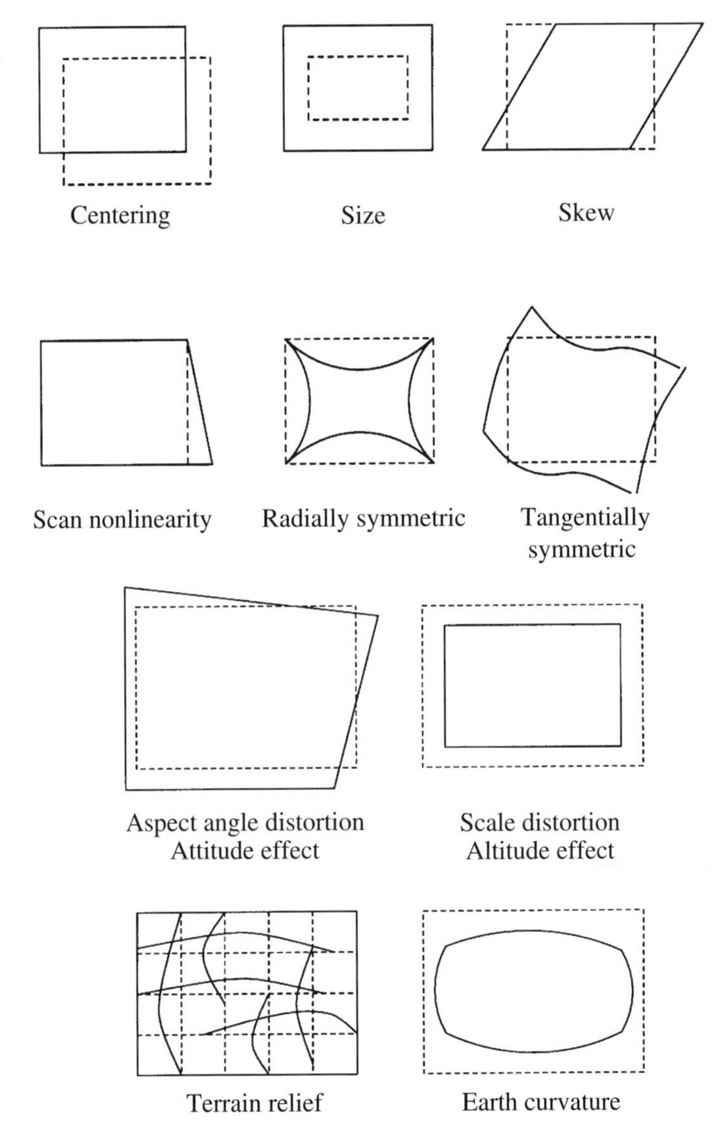

Figure 5.1 Typical external sensor distortions.

5.2 Geometrical image registration

Geometrical image registration is a procedure to determine the best spatial fit between two images of the same scene. It consists of geometrically matching two or more images from the same scene acquired with the same or different sensors and with or without the same ground resolution, at the same or at different times.

Let us define some of the terminology used. We shall call input image the image to be registered or warped, output image or working image the corrected or registered image, and reference image the image to which all the others will be registered (Castleman, 1979; Djamdji, 1993; Niblack, 1986).

The output image is then generated by defining a spatial transformation, which is a mapping function that establishes a spatial correspondence between all points of coordinates (X, Y) of the reference image and the coordinates (x, y) of the input image.

We can consider two kinds of transformation: global, which will be applied on the whole image, and local, which can take into account specific local deformations in particular areas. The latter will be discussed in the next chapter.

Global transformations impose a single mapping function on the whole image. They often do not account for local geometric distortions such as scene elevations, atmospheric turbulence and sensor nonlinearity. All the control points are considered in order to derive the unknown coefficients of the mapping function. Generally, coefficients computed from a global method will remain fixed across the entire image, i.e. the same transformation is applied over each pixel. Furthermore, the least squares technique used for the determination of coefficients averages a local geometric difference over the whole image independently of the position of the difference. As a result, local distortions cannot be handled as they instead contribute to errors at distant locations. In the case of polynomial functions, we may instead interpolate the surface with a global mapping by increasing the degree of the polynomial to match the number of control points. However the resulting polynomial is likely to exhibit excessive spatial undulations and thereby introduces further artifacts.

We can consider two major cases for the mapping: forward and inverse.

- Forward Mapping: Forward or input-to-output mapping consists of transferring the gray level of the input image onto the output image. The latter is constructed gradually from the input image in the following. For each pixel (x, y) of the input image we look for its corresponding (X, Y) in the output image by applying the following mappings: $X = f(x, y)$ and $Y = g(x, y)$, and the value of the pixel in (x, y) will be given to the pixel in (X, Y). Notice that the set of input pixels are mapped from a set of integers to a set of real numbers.

However, despite the fact that this method is the most intuitive or natural one, it gives rise to two major problems: holes and overlaps. Holes, or patches of undefined pixels, come about when there is an output pixel onto which no input pixel is mapped. Overlaps occur when several input pixels map onto one single output pixel. An alternative is to use inverse or output-to-input mapping.

- Inverse Mapping: Inverse or output-to-input mapping projects each output coordinate pixel into the input image by the following mapping functions: $x = Q(X, Y)$ and $y = R(X, Y)$. The value of the data input sample at (x, y) is copied onto the output pixel at (X, Y). All the pixels of the output image will then have a value, and holes are avoided. For this reason, this method is the most practical one for general use (Castleman, 1979) and the most widely used. We will therefore use this method for registration.

The main steps. The main steps of the geometrical correction are usually (Pratt, 1978):

- We measure a set of well-defined ground control points (GCPs), which are well-located features both in the input image and in the reference image.
- We determine the warping or deformation model, by specifying a mathematical deformation model defining the relation between the coordinates (x, y) and (X, Y) in the reference and input image respectively.
- We construct the corrected image by output-to-input mapping.

The main difficulty lies in the automated localization of the corresponding GCPs, since the accuracy of their determination will affect the overall quality of the registration. In fact, there are always ambiguities in matching two sets of points, as a given point corresponds to a small region D, which takes into account the prior geometric uncertainty between the two images and many objects could be contained in this region.

One property of the wavelet transform is having a sampling step proportional to the scale. When we compare the images in the wavelet transform space, we can choose a scale corresponding to the size of the region D, so that no more than one object can be detected in this area, and the matching is done automatically.

5.2.1 Deformation model

Geometric correction requires a spatial transformation to invert an unknown distortion function. A general model for characterizing misregistration between two sets of remotely sensed data is a pair of bivariate polynomials of the form:

$$x_i = \sum_{p=0}^{N} \sum_{q=0}^{N-p} a_{pq} X_i^p Y_i^q = Q(X_i, Y_j) \tag{5.1}$$

$$y_i = \sum_{p=0}^{N} \sum_{q=0}^{N-p} b_{pq} X_i^p Y_i^q = R(X_i, Y_j) \tag{5.2}$$

where (X_i, Y_i) are the coordinates of the ith GCP in the reference image, (x_i, y_i) the corresponding GCP in the input image and N is the degree of the polynomial. Usually, for images taken under the same *imaging direction*, polynomials of degree one or two are sufficient as they can model most of the usual deformations like shift, scale, skew, perspective and rotation (see Table 5.1). We then compute the unknown parameters ($(N+1)(N+2)/2$ for each polynomial) using the least mean square estimator.

5.2.2 Image correction

Image correction is considered as follows. One may consider three cases for geometric registration (Djamdji, Bijaoui and Manière, 1993a):

Table 5.1 *Some common deformations.*

Shift	$x = a_0 + X$ $y = b_0 + Y$
Scale	$x = a_1 X$ $y = b_2 Y$
Skew	$x = X + a_2 Y$ $y = Y$
Perspective	$x = a_3 XY$ $y = Y$
Rotation	$x = \cos\theta X + \sin\theta Y$ $x = -\sin\theta X + \cos\theta Y$

1 The registration of images acquired with the same sensor and having the same ground resolution and imaging direction. The registration is then done in the pixel space.
2 The registration of images taken by different sensors and having different ground resolutions. The registration is then done in the real coordinate space.
3 The registration of images taken with the same sensors and having different imaging directions. This will be discussed in the next chapter.

We now look at pixel space. Once the coefficients of the polynomials have been determined, $Q(i,j)$ and $R(i,j)$ are computed, and the output image is generated as follows:

- For each output pixel location (i,j), we compute (k,l), $k = Q(i,j)$ and $l = R(i,j)$, record the pixel value at location (k,l) and assign it to the output pixel at (i,j). The process is iterated over the entire image and the image output is thus generated.
- The pixel values (k,l) are generally not integers, so an interpolation must be carried out to calculate the intensity value for the output pixel. Nearest-neighbors, bilinear or bicubic spline interpolations are the most widely used.

The real space is handled in the following way. We transform the coordinates of the GCPs from the pixel space (i,j) to the real space (i_r, j_r) and then compute the coefficients of the polynomials $Q(i_r, j_r)$ and $R(i_r, j_r)$. The output image is then generated as follows:

- Each pixel location (i,j) is transformed into its real coordinate value (i_r, j_r). Then $(k_r = Q(i_r, j_r)\ ,\ l_r = R(i_r, j_r))$ is computed. These values are transformed back into their pixel space value (k,l). We then record the pixel value at (k,l) and assign it to the output pixel at (i,j) as in the pixel space case. The image output is then generated.
- As in the pixel space case, the pixel values (k,l) are generally not integers, so an interpolation must be carried out.

5.3 Ground control points

A GCP or tiepoint is a physical feature detectable in a scene, whose characteristics (location and elevation) are known precisely (Berstein, 1976). Typical GCPs are airports, highway intersections, land-water interfaces, geological and field patterns and so on. Three fundamental methods of GCP selection may be distinguished (Niblack, 1986): manual; semi-automatic; and automatic.

Manual methods consist of locating and selecting manually the GCPs with the help of an operator. This is usually done interactively, the operator identifying the points on the image using a mouse. This method is very constraining, as it is time-consuming and highly dependent on the visual acuity of the operator, the quality of the warping depending on the good localization of the points.

In the semi-automatic method, the user again identifies corresponding points in each of the two images, but here the location in one image is taken as the exact location around which a window is extracted. The location in the other image is taken as the center of a larger search area. The exact location of the control point in the search area is computed as the point of maximum correlation with the window within the search area (Niblack, 1986).

In the fully automatic method, registration can be accomplished with some measure of similarity or dissimilarity between two or more images. This measure is a function of a relative shift between them. One similarity measure is the correlation between two overlapping image areas (Showengerdt, 1983). If the two image areas are registered, the correlation value is maximum.

Because correlations are computationally expensive for large areas, relatively small areas, distributed over the total overlapping region of the two images, are used (Showengerdt, 1983) (Fig. 5.2).

Areas in the reference image are the *search windows* containing the control points and those of the working image will be the *search areas*. Windows from the reference scene are stored in a file or library and automatically located by correlation in search areas in the input image.

The estimated locations of the search areas must be computed, and

Figure 5.2 Automatic correlation areas.

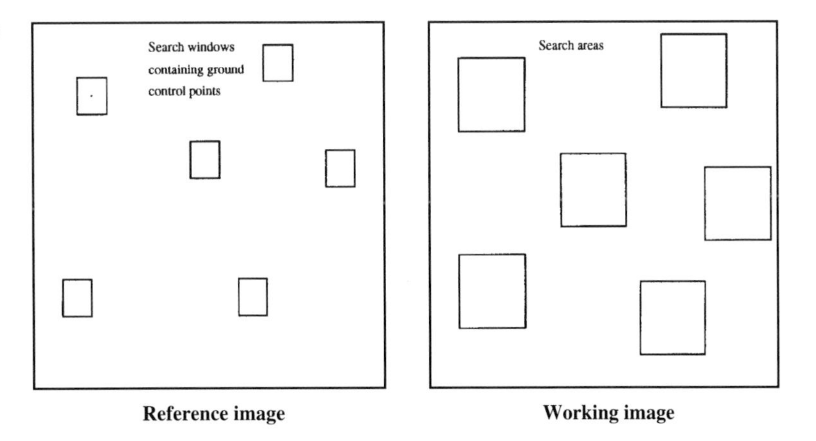

for the automatic method to be feasible, there must be some way to do this with reasonable accuracy. For satellite images, initial estimates of ground features, given their latitude and longitude can normally be made within 10 to 20 pixels, sometimes much better, so this method is often used in operational systems for correcting satellite images (Niblack, 1986). Typical window search areas range from 16×16 to 51×51 pixels. The size of the search area is chosen to guarantee that the feature is included and so depends on the uncertainty in the estimated feature location (Niblack, 1986). An efficient implementation for determining points of correspondence between images is the sequential similarity detection algorithm (SSDA) or template matching (Barnea and Silverman, 1972; Silverman and Bernstein, 1971).

After applying a registration model on the working image, the values of (x, y) do not necessarily occur exactly at the coordinates of an integer pixel. Interpolation is then necessary to compute the pixel value to be inserted into (X, Y). Many different interpolations exist, and the most widely used in remote sensing are (Bernstein, 1976; Park and Showengerdt, 1983; Schlien, 1991): zero order or nearest-neighbor interpolation; first order or bilinear interpolation; and second order or bicubic interpolation. More details can be found in Berstein (1976), Djamdji (1993), Keys (1981), Niblack (1986), Park and Showengerdt (1983), Rifman (1973), Rifman and McKinnon (1974) and Showengerdt (1983).

5.4 Image registration using the wavelet transform

5.4.1 Registration of images from the same satellite detector

Let I_n, $n \in (1, N)$, $N \geq 2$, be the images to be processed. Let I_1 be the reference image, and let M be the largest distance in the pixel space between two identical features. The matching will be first processed with the largest scale L, $2^{L-1} < M \leq 2^L$, in order to automatically match the identical features without errors (Bijaoui and Guidicelli, 1991; Djamdji *et al.*, 1993a,b).

On each image I_n, we compute the wavelet transform with the à trous algorithm, up to the scale L. We then obtain N smoothed images, $S_n(i, j)$, and $N \times L$ wavelet images, $W_{nl}(i, j)$, $n \in (1, N)$ and $l \in (1, L)$. The smoothed images are not used in the matching procedure. The reference image will be for $n = 1$.

Since L is the initial dyadic step, we specify a detection procedure on $W_{nl}(i, j)$ and we keep only the structures above a threshold of $(\theta \times \sigma_{n1})$, θ being a constant which increases when the resolution decreases, and σ_{n1} being the standard deviation of W_{n1}. We only retain from these structures their local maxima which will then play the role of GCP. These points correspond to significant image patterns, and we must find them in each wavelet image corresponding to the same area. Considering that the noise $n(x)$, which is located in the high frequencies, has a Gaussian distribution, with a standard deviation of σ, then 99.73% of the noise is located in the interval of $[-3\sigma, 3\sigma]$ (Stuart and Kendall, 1973). Therefore, the wavelet image W_{n1} for the first resolution contains the high frequencies of the image and thus contains the noise. By thresholding this image at 3σ, only the significant signal is retained, since 99.73% of the noise is eliminated. The algorithm being a dyadic one, the bandwidth is reduced by a factor 2 at each step, so the amount of noise in the signal decreases rapidly as the resolution increases.

For the step L, and for each image $n \in (2, N)$, we compare the positions of the objects detected to the ones found in the reference image W_{1L}. At this step, we can match identical features with confidence, and therefore determine the relationship between the coordinates of the different frames.

Let (ξ_{nl}, η_{nl}) be the position of a maximum for W_{nl}, the matching identifying it as the object m with a set of coordinates:

$$x_{nlm} = \xi_{nl} \tag{5.3}$$

$$y_{nlm} = \eta_{nl} \tag{5.4}$$

The deformation model is then calculated by:

$$x_{nlm} = Q(X_{1lm}, Y_{1lm}) \tag{5.5}$$

$$y_{nlm} = R(X_{1lm}, Y_{1lm}) \tag{5.6}$$

We now consider the wavelet images of order $L-1$ and detect a new set of maxima in each image. We then transform the coordinates of each maximum detected in the reference image using the previous parameters. That allows us to easily match the maxima and to determine the new parameters of the deformation model (Djamdji *et al.*, 1993a,b).

This process is iterated until the last scale corresponding to the wavelet of the best resolution is reached. The best geometrical correspondence is then established. A polynomial of degree one is used for the first steps, and may eventually be increased to two in the last few steps. The image is then warped using the final coefficients. The flowchart of this algorithm is given in Fig. 5.3.

5.4.2 Registration of images from different sensor detectors

The registration of images obtained from different sensors with a different ground resolution, is done in three steps (Djamdji *et al.*, 1993a,b):

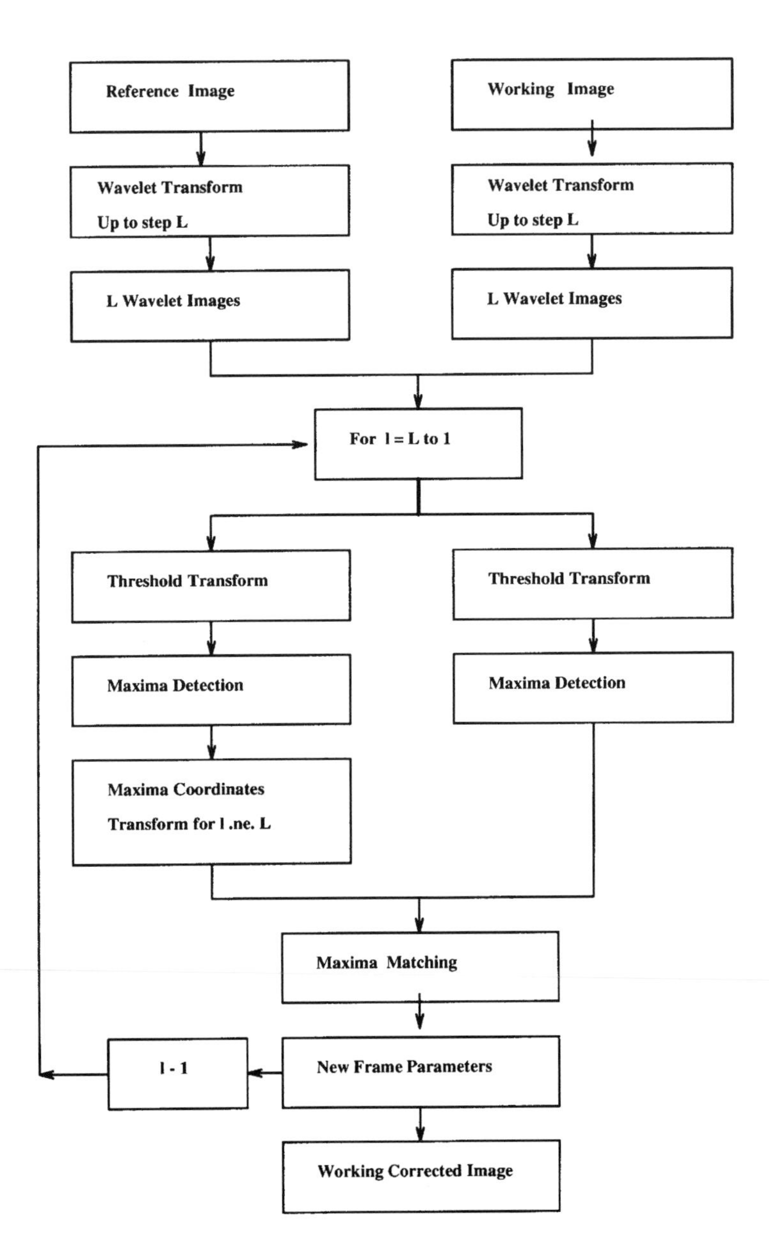

Figure 5.3 Flowchart of the geometrical registration algorithm.

- First the images are reduced to the same ground resolution, generally the lowest one.
- The matching is then carried out and the deformation model is calculated both in the real and the pixel space.
- The image of higher resolution is then registered in the real coordinate space.

Reduction to the same ground resolution. We studied the registration of SPOT versus MSS data, and SPOT versus TM data. In order to be able to perform the registration, the images have to be reduced to the same ground resolution. This is done by reducing the high resolution image to the lowest resolution in order to be able to match the two images. This reduction is done in the following:

- For SPOT to MSS data (20 m to 80 m), a pyramidal algorithm was used up to 2 scales. One sample out of two is retained at each scale, the data being then reduced by a factor of 2 at each step.
- For the SPOT to TM data (20 m to 30 m), a different approach was used: the sampled image was assumed to be the result of the scalar product of the continuous image with a scaling function $\phi(x)$, where $\phi(x)$ is the door function. Thus an analytical approach was used to compute the resulting transformation of SPOT data into TM data. We get:

for $k = 2l$

$$n_T(k) = \left\{ n_S(3l) + \frac{1}{4}[n_S(3l-1) + n_S(3l+1)] \right\} \frac{2}{3} \tag{5.7}$$

for $k = 2l + 1$

$$n_T(k) = \left\{ \frac{3}{4}[n_S(3l+1) + n_S(3l+2)] \right\} \frac{2}{3} \tag{5.8}$$

where n_T is the pixel in the TM image, n_S the pixel in the SPOT image and $\frac{2}{3}$ is a coefficient introduced to satisfy flux conservation.

The flowchart of the registration of images obtained from different sensors is given in Fig. 5.4.

5.4.3 Registration of images with a pyramidal algorithm

The method developed above is not well adapted to the processing of large images for two reasons: computation time becomes large in this case; and a lot of disk space is needed for the processing. One way to improve this situation is to use a pyramidal implementation of the à trous algorithm. The process is quite similar to the one above. The main difference lies in the matching procedure. The image being reduced by a factor of 4 at each step, due to the decimation procedure, the matching must be done in the real coordinate space (Djamdji *et al.*, 1993b). The flowchart of the pyramidal algorithm is given in Fig. 5.5.

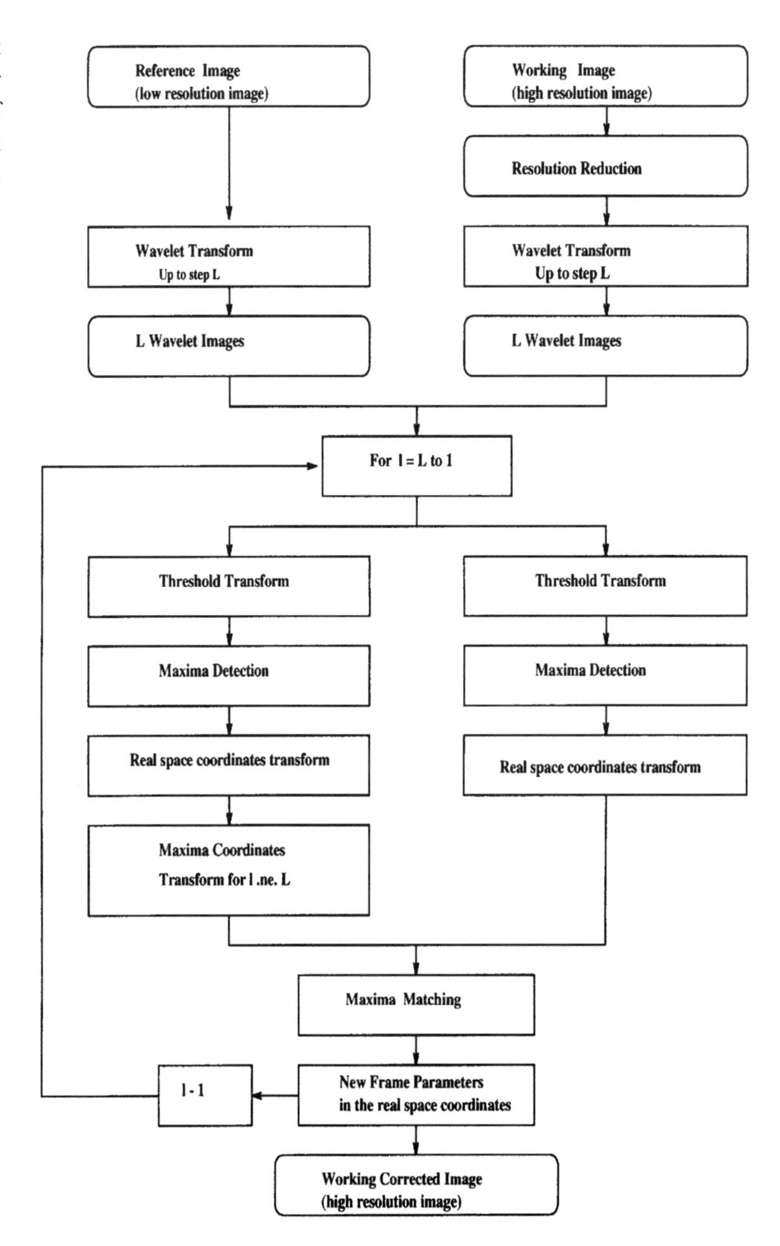

Figure 5.4 Flowchart of the algorithm for the registration of images obtained from different sensors.

5.5 Application

We applied our algorithms on different scenes from LANDSAT and SPOT satellites as well as on combinations of these, SPOT XS versus LANDSAT MSS. For all of these images, we extracted subregions of different sizes in order to avoid: (i) working on very large images since

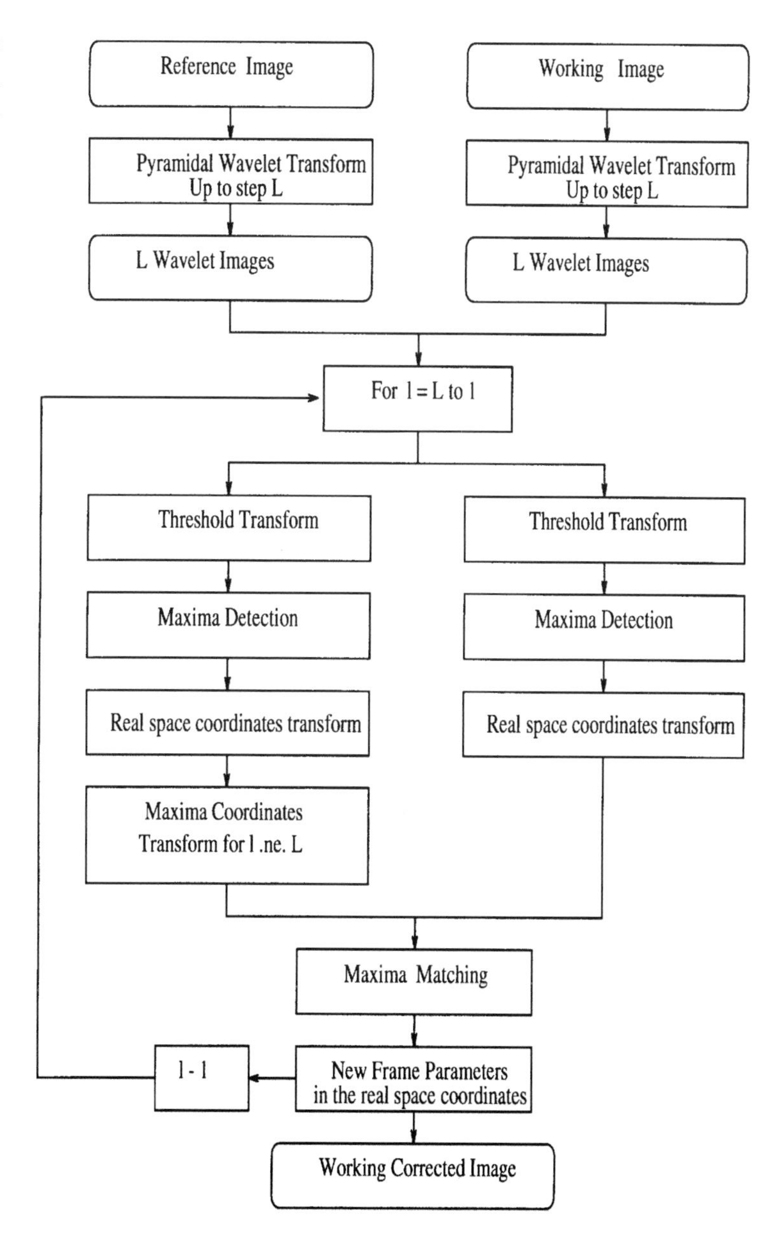

Figure 5.5 Flowchart of the pyramidal registration algorithm.

this is time-consuming and requires lots of storage space in the case of a non-pyramidal algorithm; (ii) dealing with regions full of clouds or snow. Sometimes, it is very difficult to have a scene totally free of clouds or snow.

For each of these images, we will present the results of the registration in the following detailed format:

- Threshold level:
  Value of the constant θ used to estimate the detection threshold.
- Matching distance:
  Distance in pixels used to match the maxima of identical structures in the wavelet space.
- Deformation model:
  Type of the polynomial model of degree n used.
- Number of detected maxima:
  Number of detected maxima above the threshold level in each of the processed wavelet images.
- Number of matched maxima before preprocessing:
  Number of matched maxima as a first approximation using only a distance criterion.
- Number of matched maxima after preprocessing:
  Number of matched maxima after preprocessing. Preprocessing consists of eliminating border points, multiple matching and GCP with a high residual.
- Number of maxima used for estimating the deformation model:
  Number of maxima (GCP) used for estimating the polynomial deformation model.

Bicubic interpolation was used for the registration.

5.5.1 SPOT data

The scenes we worked on are:

- Scene number 51–279 dated 18/05/1989 taken at 10h_48mn_10s, composed of 3005 rows and 3270 columns.
- Scene number 51–279 dated 12/05/1986 taken at 10h_41mn_28s, composed of 3003 rows and 3253 columns.

These scenes, from the region of Ain Ousseira in Algeria, were taken with a three-year interval time, in a region subject to desertification, and are therefore radiometrically very different as one can easily see. Two sub-scenes of 750×750 pixels were extracted in the XS1 (0.5 – 0.59 μm), XS2 (0.61 – 0.68 μm) and XS3 (0.79 – 0.89 μm). Our registration algorithm was then applied to the selected sub-images, using the scene of 1986 as the reference one (Fig. 5.6) and the one of 1989 as the image to be registered (Fig. 5.7). The processing was done using a six-level wavelet decomposition and the results are shown for the XS3 band. The resulting registration is given in Fig. 5.8, and the automatically selected GCPs can be seen in Figs. 5.9 and 5.10. A summary of the registration procedure is given in Tables 5.2, 5.3, 5.4.

5.5.2 MSS data

The scenes we worked on are:

- Scene number 195–29 dated 23/07/1984 composed of 2286 rows and 2551 columns.
- Scene number 195–29 dated 28/11/1984 composed of 2286 rows and 2551 columns.

These scenes, from the region of the Alpes Maritimes in France, were taken in the same year but in different seasons, and are therefore radiometrically different. Two subscenes of 600×700 pixels were extracted in the MSS3 band ($0.7 - 0.8\ \mu m$) which correspond to the near infrared wavelength. The scene of July 1984 was chosen as the reference image (Fig. 5.11) and the one from November 1984 as the working image (Fig. 5.12). The processing was done using a six-level wavelet decomposition. The final registration is given in Fig. 5.13. A summary of the registration procedure is given in Table 5.5.

5.5.3 SPOT versus MSS data

The scenes we worked on are:

- Scene number 53–261 dated 24/07/1986, acquired by HRV of SPOT satellite in its multispectral bands and composed of 3002 rows and 3140 columns.
- Scene number 195–29 dated 23/07/1984, acquired by MSS of LANDSAT satellite and composed of 2286 rows and 2551 columns.

These scenes are from the region of the Alpes Maritimes in France. The bands processed were the XS3 for SPOT and the MSS3 for LANDSAT. The SPOT scene was first reduced to an 80 m ground resolution, then two sub-scenes of 350×400 pixels were extracted from the 80 m SPOT and from the MSS3 scenes. Once these regions were selected, our algorithm was applied using the MSS3 scene as the reference one (Fig. 5.14), and the 80 m SPOT scene as the working

Figure 5.6 The SPOT XS3 reference image.

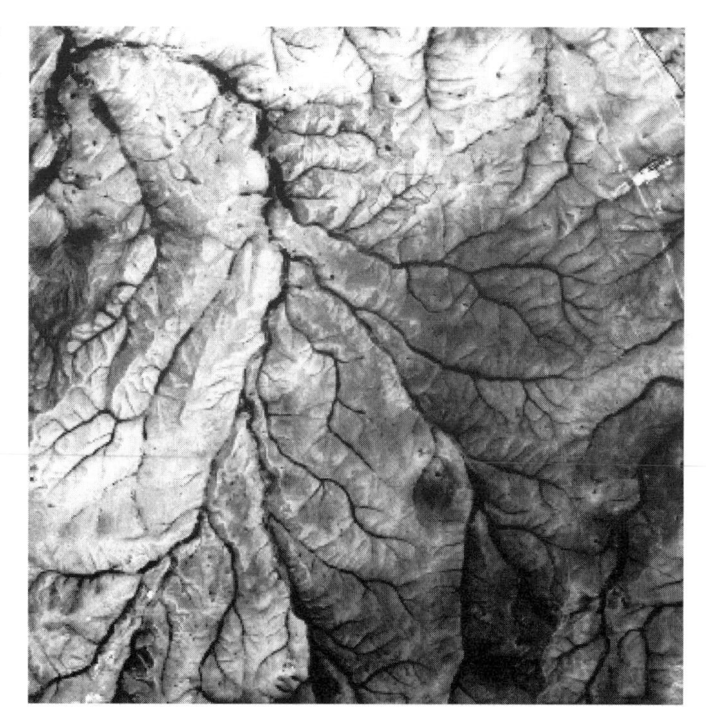

Figure 5.7 The SPOT XS3 working image.

Figure 5.8 The SPOT XS3 corrected image.

Figure 5.9 Selected GCPs in the 86 SPOT image – part of the original image.

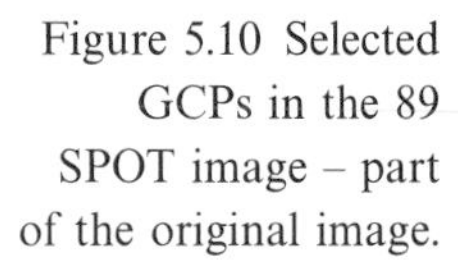

Figure 5.10 Selected GCPs in the 89 SPOT image – part of the original image.

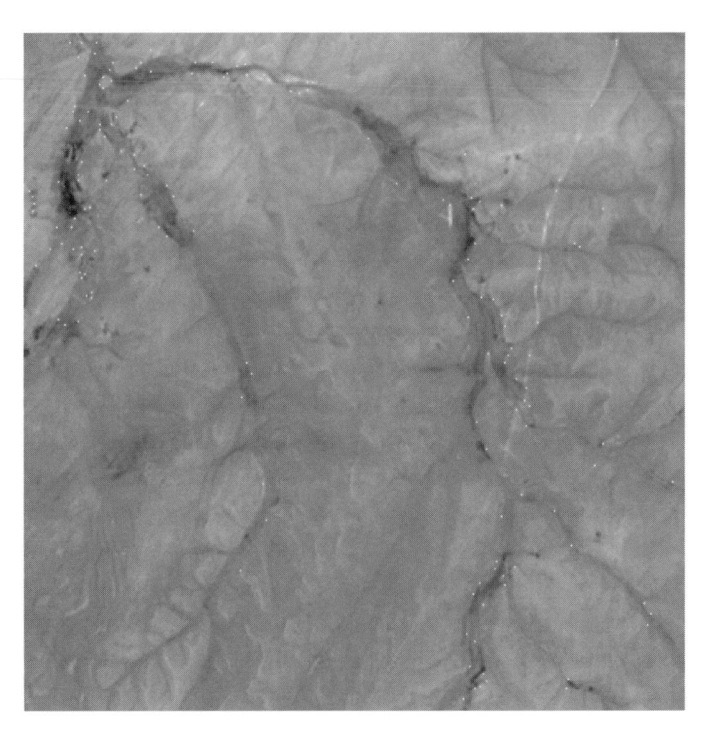

Figure 5.11 The LANDSAT MSS3 reference image.

one (Fig. 5.15). The processing was then done using a six-level wavelet decomposition. Finally the original SPOT scene was registered. Once the process was carried out, the rectified original SPOT scene was reduced to an 80m ground resolution and inlayed (encrusted) in the

Figure 5.12 The LANDSAT MSS3 working image.

Figure 5.13 The LANDSAT MSS3 corrected image.

MSS3 scene in order to have a visual evaluation of the accuracy of the procedure (Fig. 5.16). A summary of the registration procedure is given in Table 5.6.

Figure 5.14 The LANDSAT MSS3 reference image.

Figure 5.15 The SPOT XS3 reduced to an 80 m ground resolution working image.

Table 5.2 *XS1 band.*

Reference image	1986					
Working image	1989					
Image size	750 × 750					
Scale	6	5	4	3	2	1
Threshold level	1	1	1	1	2	3
Matching distance	32	16	8	4	2	2
Deformation model	1	1	1	1	1	2
No. detected maxima (reference image)	90	327	1117	3696	3919	2226
No. detected maxima (working image)	87	322	1146	3410	3097	1772
No. matched maxima before preprocessing	70	263	807	2189	1368	601
No. matched maxima after preprocessing	45	155	532	1342	791	331
No. maxima used for est. the deformation model	45	155	532	1342	791	281

Table 5.3 *XS2 band.*

Reference image	1986					
Working image	1989					
Image size	750 × 750					
Scale	6	5	4	3	2	1
Threshold level	1	1	1	1	2	3
Matching distance	32	16	8	4	2	2
Deformation model	1	1	1	1	1	2
No. detected maxima (reference image)	101	352	1228	4001	4173	2100
No. detected maxima (working image)	95	344	1191	3668	3437	1422
No. matched maxima before preprocessing	77	277	858	2436	1397	474
No. matched maxima after preprocessing	42	158	555	1380	755	230
No. maxima used for est. the deformation model	42	158	555	1380	755	180

Table 5.4 *XS3 band.*

Reference image	1986					
Working image	1989					
Image size	750x750					
Scale	6	5	4	3	2	1
Threshold level	1	1	1	1	2	3
Matching distance	32	16	8	4	2	2
Deformation model	1	1	1	1	1	2
No. detected maxima (reference image)	90	385	1343	4335	4338	1574
No. detected maxima (working image)	97	378	1337	4027	4182	1338
No. matched maxima before preprocessing	72	285	1053	2995	1992	338
No. matched maxima after preprocessing	33	130	742	1915	1129	169
No. maxima used for est. the deformation model	33	130	742	1915	1129	119

Table 5.5 *MSS3 band.*

Reference image	July					
Working image	November					
Image size	600 × 700					
Scale	6	5	4	3	2	1
Threshold level	1	1	1	1	2	3
Matching distance	32	16	8	4	2	2
Deformation model	1	1	1	1	1	2
No. detected maxima (reference image)	31	173	701	2538	2242	2213
No. detected maxima (working image)	39	158	609	2448	2531	2288
No. matched maxima before preprocessing	16	92	344	1318	475	242
No. matched maxima after preprocessing	14	38	156	654	240	100

Table 5.6 *MSS3 and XS3 bands.*

Reference image	MSS3 – 80 m					
Working image	XS3 – 80 m					
Image size	350 × 400					
Scale	6	5	4	3	2	1
Threshold level	1	1	1	1	2	3
Matching distance	32	16	8	4	2	2
Deformation model	1	1	1	1	1	2
No. detected maxima (reference image)	13	70	285	967	791	505
No. detected maxima (working image)	25	79	324	1043	1396	764
No. matched maxima before preprocessing	14	69	251	813	519	166
No. matched maxima after preprocessing	9	48	193	640	360	87

5.5.4 SPOT with different imaging directions

The scenes we worked on are:

- Scene number 148–319 dated 05/02/1991 taken at 07h_51mn_04s, composed of 3000 rows and 3000 columns, level 1a.
- Scene number 148–319 dated 02/04/1988 taken at 07h_34mn_40s, composed of 3003 rows and 3205 columns, level 1b.

This two scenes from the eastern region of Marib in the Republic of Yemen, were taken under different imaging directions. The level 1a (Spotimage, 1986) (Fig. 5.17) was taken with an incidence of 25.8 degrees left, while the level 1b (Spotimage, 1986) (Fig. 5.18) was taken with an incidence of 6.3 degrees right. Two subscenes of 512 × 512 pixels were extracted. An attempt to register these two images was made, using level 1b as the reference image and level 1a as the working image. The registration was globally not very good, especially in the elevation areas (Fig. 5.19). This was due to the polynomial model used which is a global model and therefore inadequate for the modeling of the local distortions introduced by the difference in the viewing angles. Another approach for the registration of such images will be seen in the next chapter.

Figure 5.16 The original registered SPOT image, reduced to an 80 m ground resolution and encrusted into the MSS scene.

Figure 5.17 The SPOT 1a input image.

Figure 5.18 The SPOT 1b reference image.

Figure 5.19 The SPOT 1a output image.

5.5.5 Astronomical image registration

In order to assess the robustness of the method for astronomical images (Djamdji, Starck and Claret, 1996), a strong distortion was applied to the galaxy NGC 2997 (see Fig. 1.6). A simulated image was made by shifting the previous one by 5 and 10 pixels in each axis direction. Then this image was rotated by 10 degrees and Gaussian noise was added. Figure 5.20 shows the simulated image (upper left panel), and also the difference between the original image and the simulated one (upper right panel). Figure 5.20 (bottom left and right) shows the corrected image and the residual between the original image and the corrected image. The two images have been correctly registered.

5.5.6 Field of view distortion estimation in ISOCAM images

The ISOCAM infrared camera is one the four instruments on board the ISO (Infrared Space Observatory) spacecraft which was launched

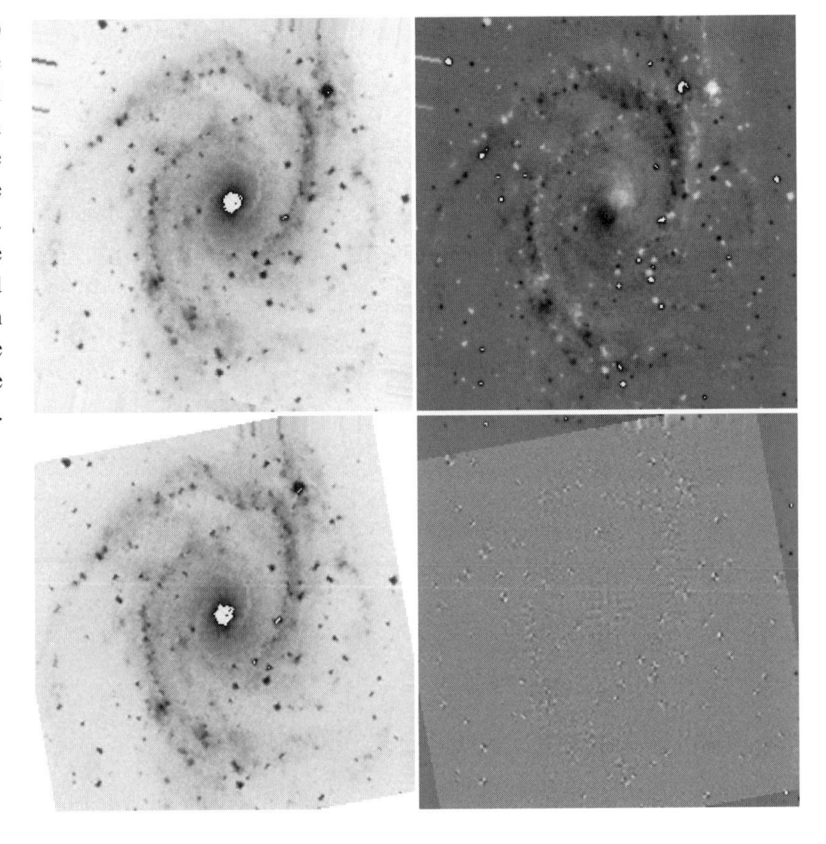

Figure 5.20 Simulated image (upper left) and difference between the original and the simulated image (upper right). Registered image (bottom left) and difference between NGC 2997 and the registered image (bottom right).

successfully on November 17, 1995. It operates in the 2.5–17 micron range, and was developed by the Service d'Astrophysique of CEA Saclay, France.

One way to assess the FOV (field of view) distortion consists of measuring the astrometry of an open star cluster from ISOCAM data, and comparing its astrometry with the theoretical values (as deduced from optical observations). The left panel of Fig. 5.21 represents a 9 arcmin × 9 arcmin field in the vicinity of the NGC 7790 star cluster as observed by ISOCAM. The right panel of Fig. 5.21 corresponds to the same FOV as observed in the optical range (Digitized Sky Survey). Provided that all visible stars of this FOV belong to the main stellar sequence, we can assume that the brightest ones in the optical range are also the brightest ones in the infrared range.

The task consists of identifying the optical counterpart of all detected sources in the ISOCAM image. Such an identification is easier once the optical image has been rebinned to the ISOCAM resolution (see left panel of Fig. 5.22). The registered ISOCAM image is presented in the right panel of Fig. 5.22. The difference between the theoretical star positions (from the optical image) and the measured ones (from the ISOCAM image) were automatically estimated. The polynomial transformation of the ISOCAM image into the optical image was also determined.

In order to visualize the distortion between the two images, this polynomial transformation was applied to a 32 × 32 rectangular grid (see Fig. 5.23). The distortion effect was also applied to a 300 × 300 grid in order to magnify it (Fig. 5.24).

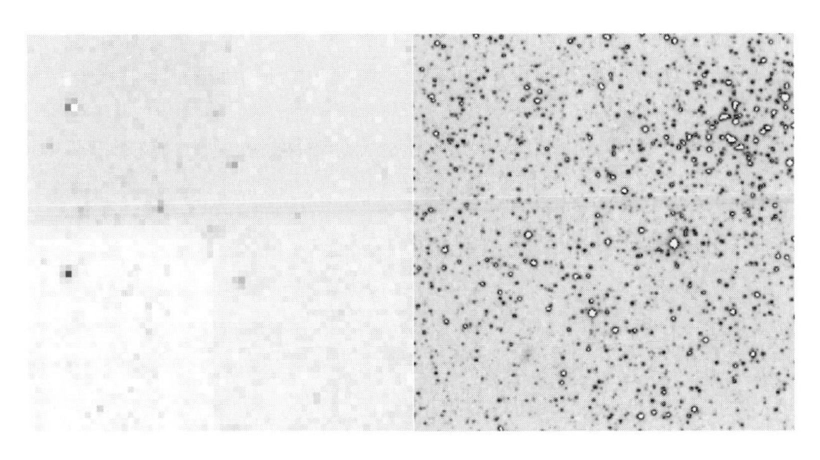

Figure 5.21 ISO image (left) and optical image (right).

5.6 Error analysis

One of the criteria used for the evaluation of the registration accuracy is the root mean square distance error (Djamdji *et al.*, 1993a; Ford and Zanelli, 1985), given by:

$$\text{RMSDE} = \sqrt{\frac{1}{N}\sum_{i=1}^{N} \text{line-residual}[i]^2 + \text{column-residual}[i]^2} \quad (5.9)$$

N being the total number of ground control points.

This criterion was used in order to evaluate the quality of the registration, and the results are given in Table 5.7. From these results, we can conclude that the final registration is of good quality, as we have reached subpixel accuracy. This is true for the case of images obtained with different sensors, the RMSDE in the real coordinate system being less than the largest ground resolution, as well as for images of the

Table 5.7 *Summary of the principal results for the geometric correction algorithm.*

Image	Number of GCP	RMSDE (pixel)	RMSDE (m) (real coordinate system)	Processing Time (mn)
SPOT	327	0.623		15
MSS	106	0.559		10
SPOT/MSS	108	0.547	69.870	20
SPOT 1a–1b	329	0.594		6

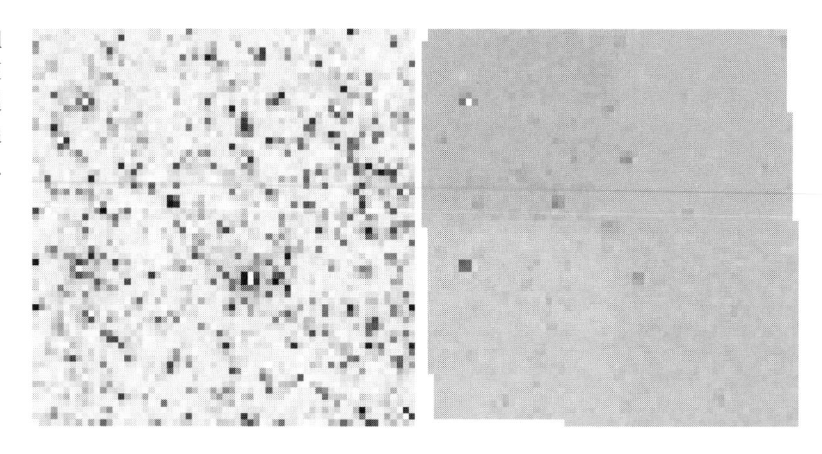

Figure 5.22 Optical image at ISOCAM resolution (left) and ISOCAM registered image (right).

Figure 5.23 Pattern 32×32.

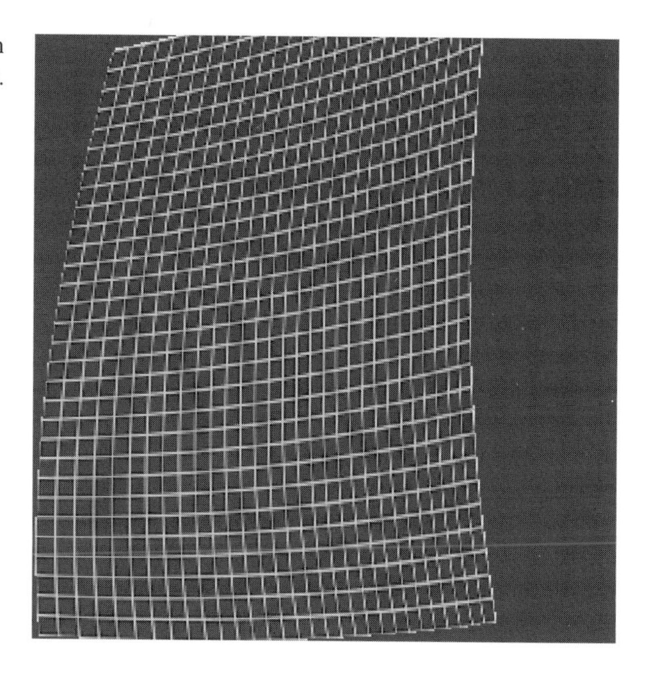
Figure 5.24 Pattern 300×300.

same sensor. But this is not true for images obtained with different imaging directions, despite the fact that the RMSDE is 0.594. This can be explained by the fact that the polynomial deformation model is inadequate to model local distortions introduced by the difference in the viewing angle and that the selected GCPs do not account for these distortions.

The processing time gives the time taken by the procedure on a Sparc IPC workstation. This includes the wavelet transform of the images, the matching procedure and the registration and interpolation of the working image and the image of higher resolution (in case of images with different ground resolution) with a bicubic convolution.

Since the RMSDE criterion is not reliable in all cases for the evaluation of the registration quality, we will now present a more accurate method for this purpose.

This method consists of evaluating the quality of the registration by comparing the transformation obtained with our method to the one obtained from a selection of a number of test GCPs. The test GCPs are different from the ones selected by our method and usually chosen manually. As manual choice was difficult in most of our images (difference in scene radiometry, difference in size for the characteristic structures), we chose an automatic selection with visual validation for the test GCPs. The method is the following: from the GCP file automatically detected by our method at the last scale of analysis, a number N of points, regularly spaced in order to have a global distribution over the entire image, is extracted. This will ensure that the two sets of points are totally different. The set of N points will be used to compute a test deformation model.

Let $(X_i, Y_i)_{i=1,\ldots,M}$ and $(x_i, y_i)_{i=1,\ldots,M}$ be the coordinates of the GCP obtained by our automatic method of registration, respectively in the reference and the working image, and used to compute the associated deformation models Q_o and R_o. Let $(Xt_i, Yt_i)_{i=1,N}$ and $(xt_i, yt_i)_{i=1,N}$ be the coordinates of the test GCP and Q_t and R_t be the deformation models evaluated on these points by a least mean square approximation. The quality of the registration is estimated from four measures which are the minimum (Min), the maximum (Max), the mean (E) and the standard deviation (σ) of the residues e_i between the point transformation (Xt_i, Yt_i) by the deformation models (Q_t, R_t) and (Q_o, R_o), for $i = 1, \ldots, N$.

$$e_i^2 = (Q_t(Xt_i, Yt_i) - Q_o(Xt_i, Yt_i))^2 + (R_t(Xt_i, Yt_i) - R_o(Xt_i, Yt_i))^2 \quad (5.10)$$

$$\text{Max}(e) = \max_{i=1,\ldots,N} (e_i) \tag{5.11}$$

$$\text{Min}(e) = \min_{i=1,\ldots,N} (e_i) \tag{5.12}$$

$$E(e) = \frac{1}{N}\sum_{i=1}^{N}(e_i) \tag{5.13}$$

$$\sigma(e) = \sqrt{\frac{1}{N}\sum_{i=1}^{N}(e_i^2) - \left(\frac{1}{N}\sum_{i=1}^{N}(e_i)\right)^2} \tag{5.14}$$

The following tables present the values of the quality measures for the processed cases.

The automatic selection of the GCPs (20 points) being located in the well-registered part of the images (low elevations), we manually added a number of points, located in the badly registered areas (high elevations). The statistical result is therefore different, as one can easily see.

5.7 Conclusion on multiscale geometric registration

We have described a new method for the geometric registration of images having the same, or a different, ground resolution. The procedure is fully automated, the algorithms are fast and the implementation easy.

The main drawback of the non-pyramidal algorithm is the large amount of disk space, as well as the large amount of processing time, needed to process large images due to non-decimation. This algorithm has an order of magnitude complexity of kN^2. The main advantage is the possibility to process small images up to a relatively high scale.

This drawback can be overcome by a pyramidal algorithm based on pyramidal implementation of the à trous algorithm, but it is not possible to process small images due to the reduction of the image size by a factor of 4 at each scale. The complexity of the algorithm is of $\frac{4}{3}N^2$ in this case.

The pyramidal à trous algorithm was preferred relative to Mallat's algorithm, due to the isotropy of the analyzing wavelet.

The final registration is of good quality, as we reach subpixel accuracy, in both cases as shown in section 5.5.

The polynomial model used for the registration is inadequate for processing images taken with different viewing angles as it cannot model local deformations. But this model is adequate for processing images acquired with the same viewing angles, since these images present only global deformations. In order to process local deformations, another approach has to be considered. This will be detailed in the next chapter.

The non-decimation algorithm is therefore optimal for small images and the pyramidal one for large images.

Table 5.8 *Registration accuracy for the SPOT data.*

Band	No. GCP N	Evaluation (pixel) Min(e)	Max(e)	$E(e)$	$\sigma(e)$
XS1	50	0.109	0.312	0.214	0.058
XS2	50	0.122	0.544	0.252	0.073
XS3	50	0.027	1.030	0.293	0.216

Table 5.9 *Registration accuracy for the MSS data.*

Band	No. GCP N	Evaluation (pixel) Min(e)	Max(e)	$E(e)$	$\sigma(e)$
MSS3	20	0.063	0.539	0.282	0.118

Table 5.10 *Registration accuracy for the SPOT – MSS data.*

Band	No. GCP N	Evaluation (pixel) Min(e)	Max(e)	$E(e)$	$\sigma(e)$
MSS3-XS3	20	0.068	0.525	0.272	0.142

Table 5.11 *Registration accuracy, SPOT data, different viewing angles.*

Band	No. GCP N	Evaluation (pixel) Min(e)	Max(e)	$E(e)$	$\sigma(e)$
XS1	20	0.138	0.929	0.383	0.225
XS1	31	0.152	7.551	2.025	2.008

6 Disparity analysis in remote sensing

6.1 Definitions

Differences in images of real world scenes may be induced by the relative motion of the camera and the scene, by the relative displacement of two cameras or by the motion of objects in the scene. These differences are important because they contain enough information to allow a partial reconstruction of the three-dimensional structure of the scene from its two-dimensional projections. When such differences occur between two images, we say that there is a disparity between them, which may be represented by a vector field mapping one image onto the other (Barnard and Thompson, 1980). The evaluation of the disparity field has been called the correspondence problem (Duda and Hart, 1973). Time-varying images of the real world can provide kinematical, dynamical and structural information (Weng, Huang and Ahuja, 1989). The disparity field can be interpreted into meaningful statements about the scene, such as depth, velocity and shape.

Disparity analysis, in the sense of stereovision, may be broadly defined as the evaluation of the existing geometrical differences, in a given reference frame, between two or more images of the same or similar scenes. The differences in remote sensing are mainly the result of different imaging directions. The goal of the analysis is to assign disparities, which are represented as two-dimensional vectors in the image plane, to a collection of well-defined points in one of the images. Disparity analysis is useful for image understanding in several ways. Since the images are generally not in the same geographical frame, a geometrical registration of the images is therefore necessary. There is information in a disparate pair of images that is difficult or even impossible to find in any single image. Disparity is therefore a very general property of images which may be used in a variety of situations.

Our purpose is to determine the disparities $\epsilon_x(i,j)$ and $\epsilon_y(i,j)$ respectively in the x and y direction, at each point (i,j) of the image. Our approach will rely on two main steps:

- The detection of a set of ground control points (GCPs) using a multiresolution approach (Djamdji *et al.*, 1993a) over which the disparities are computed.
- A mapping of the disparities over the entire image by the kriging method.

An example of a situation where the disparities are useful, the geometrical registration of a stereoscopic pair of images, will be presented.

6.1.1 Disparity

Let P be a point in the real world, and P_i^1 and P_i^2 be the images of this point in frames 1 and 2 respectively. These two points are similar in that they are the image plane projection of the same real world surface point. Consequently, matching P_i^1 with P_i^2 is the same as assigning to P_i^1 a disparity with respect to image 2 of:

$$D_i = (x_i^1 - x_i^2, y_i^1 - y_i^2) \tag{6.1}$$

We will modify this classical definition by getting rid of the deformation polynomial model underlying the geometrical registration. Instead, we will consider the disparity as being the divergence between two identical points with respect to the deformation model considered.

Figure 6.1
Stereoscopic system.

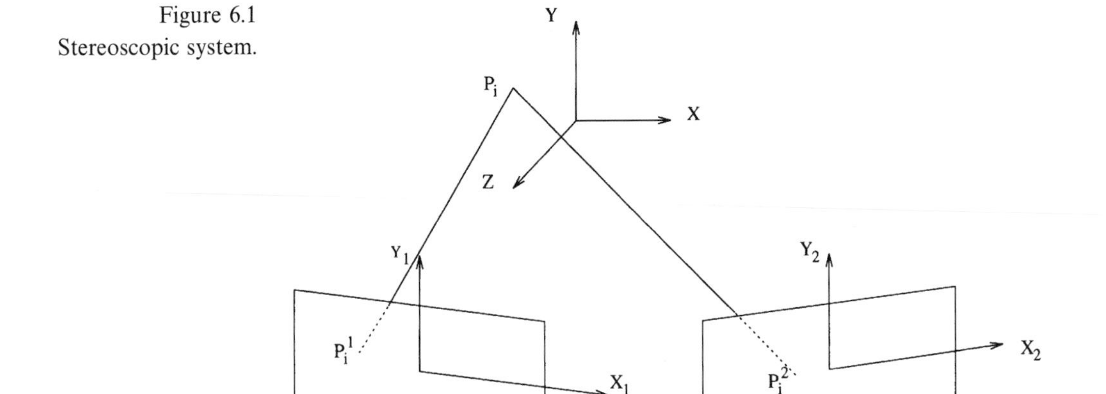

These distortions are produced by physical phenomena, and mainly in remote sensing from the differences in the viewing angle.

Let (X_i, Y_i) and (x_i, y_i) be the coordinates of an identical point in the reference and the working frames. Then:

$$(x_i^1, y_i^1) = (X_i, Y_i)$$
$$(x_i^2, y_i^2) = (x_i, y_i) \qquad (6.2)$$

If the viewing angles were the same, (X_i, Y_i) and (x_i, y_i) would be related by:

$$x_i = f(X_i, Y_i)$$
$$y_i = g(X_i, Y_i) \qquad (6.3)$$

where f and g are polynomials that take into account the global deformations between the two frames. But when the viewing angles are different, the model considered previously is no longer valid, and a correction term has to be introduced in order to take into account the local deformations introduced by the viewing angles. Thus the previous relationship can be rewritten as:

$$x_i = f(X_i, Y_i) + \epsilon_{x_i}(X_i, Y_i)$$
$$y_i = g(X_i, Y_i) + \epsilon_{y_i}(X_i, Y_i) \qquad (6.4)$$

where ϵ_{x_i} and ϵ_{y_i} describe the new local deformations. Then the disparity at the point (X_i, Y_i) along the X and Y axis is given by ϵ_{x_i} and ϵ_{y_i} respectively.

6.1.2 Matching

Matching is a natural approach to disparity analysis in an image (Barnard and Thompson, 1980). Assigning a disparity to points in a sequence of images is equivalent to finding a matching between sets of homologous points in each image. Let $S_1 = (s_1^x, s_1^y)$ and $S_2 = (s_2^x, s_2^y)$ be two points in images 1 and 2 respectively. These two points should be matched if and only if they are image plane projections of the same real world surface plane.

A matching approach for disparity analysis has to solve two major problems: how to select points for the matching; and how to quantify the quality of the matching.

It is obvious that all the points cannot be matched with the same precision, since some of these will necessarily be located in regions of weak detail. Some cannot be matched at all, as they are visible only in one image. In order to avoid any ambiguity in the matching, it is advantageous to try to match points that are easily distinguishable and have similar properties. Wavelet analysis will provide a perfect tool for this problem.

6.1.3 Extraction of ground control points

The extraction of the GCPs is achieved via a multiscale approach and its implementation is given in Djamdji *et al.* (1993a). The method is based on a special implementation of the discrete wavelet transform, the à trous algorithm (Holschneider *et al.*, 1989; Shensa, 1992). The matched GCPs are the maxima of the detected structures in the wavelet images, on a set of dyadic scales, the multiscale schemes starting from the coarsest resolution and proceeding to the finest one. In the case of scenes acquired with the same viewing angle, we have shown (Djamdji *et al.*, 1993a,b) that we were able to detect and correctly match the GCPs which, in turn, enables us, through this multiscale scheme, to register the images with subpixel accuracy.

For images taken with different viewing angles, a residual remains due to the disparity between these images. Nevertheless, we will use this multiscale procedure for the GCP extraction.

6.2 Disparity mapping

Let us introduce some of the mathematical tools used in the following, principally kriging and the variogram.

6.2.1 Kriging

The theory of regionalized variables was developed by G. Matheron in the late 1950s. Matheron demonstrated that spatially dependent variables can be estimated on the basis of their spatial structure and known samples (Matheron, 1970). A random variable distributed in space is said to be regionalized. These variables, because of their spatial aspect, possess both random and structured components. On a local scale, a regionalized variable is random and erratic. Two regionalized variables $F(x)$ and $F(x+h)$ separated by a distance vector h are not independent, but are related by a structured aspect. This structure function or variogram, $\gamma(h)$, is dependent on h (Carr and Myers, 1984),

$$\begin{aligned}\gamma(h) &= \frac{1}{2}\mathrm{Var}[F(x+h) - F(x)] \\ &= \frac{1}{2}E[(F(x+h) - F(x))^2] \qquad (6.5)\end{aligned}$$

Usually as the length of h increases, the similarity between two regionalized variables decreases.

At first glance, a regionalized variable appears to be a contradiction. In one sense, it is a random variable which locally has no relation to surrounding variables. On the other hand, there is a structured aspect to a regionalized variable which depends on the distance separating the variables. Both of these characteristics can however be described by a random function for which each regionalized variable is but a single realization. By incorporating both the random and structured aspects of a regionalized variable in a single function, spatial variability can be accommodated on the basis of the spatial structure shown by these variables (Carr and Myers, 1984).

An important technique used in geostatistics for estimation and interpolation purposes is kriging. Kriging is a regression method used on irregularly-spaced data in 1, 2 or 3 dimensions for the estimation of values at unsampled locations (Myers, 1991) using the available information. The value at unsampled points is estimated from a linear combination of all the available samples weighted by a certain coefficient. The estimator is considered as non-biased when the weight total equals one.

It is therefore possible to solve a kriging problem, which is to compute the optimal weights for each sample, only if we are given the covariance or variogram function (Matheron, 1965).

The kriging technique is optimal since it uses the spatial interdependence information represented by the variogram $\gamma(h)$ or by the covariance function (Chiles and Guillen, 1984). As will be seen later, the weights are obtained from a linear system of equations in which the coefficients are the values of the covariance or variogram function, values which quantify the correlation between two samples for a given distance. These equations are obtained by minimizing the variance of the estimation error. Estimation and modeling of the structure function is the most important and potentially the most difficult step in the process (Myers, 1991).

One important characteristic of the kriging estimator is that the weights (i.e. the kriging equations) do not depend on the data, but rather only on the variogram or covariance function and on the sample pattern (Myers, 1987).

6.2.2 Variogram

One way of examining the spatial structure of a regionalized variable is to analytically relate the change of the variables as a function of the separating distance h. The function which defines the spatial correlation or structure of a regionalized function is the variogram given by:

$$\gamma(h) = \frac{1}{2}E(\{f(x) - f(x+h)\}^2) \tag{6.6}$$
$$= C(0) - C(h) \tag{6.7}$$

where $C(h)$ is the covariance function, E the mathematical expectation and h the lag or separating distance. Equation (6.7) holds only if the covariance function is defined. The shape of the variogram reflects the degree of correlation between samples. Variogram functions that rise as h increases indicates that the spatial correlation decreases as more distant samples are chosen, until a separation distance is reached at which knowledge of one sample tells us nothing about the others (uncorrelated) (Glass *et al.*, 1987).

In order to use the variogram for kriging, a mathematical model must be fitted (McBratney, Webster and Burgess, 1981a,b). This model must meet certain criteria, and several so called 'authorized models' are available (Journel and Huijbregts, 1978; Myers, 1991).

Theoretical models. Different theoretical models of variograms exist. We will describe the most common ones. Let us first describe the principal characteristics of a stationary variogram which are (Journel and Huijbregts, 1978):

1 Its behavior at the origin (parabolic, linear and nugget effect).
2 The presence or absence of a sill in the increase of $\gamma(h)$, i.e., $\gamma(h) =$ constant when $|h| > a$.

The continuity and regularity of a random function $F(x)$ can be inferred from the behavior of the variogram at the origin (Matheron, 1970). By decreasing regularity order, we can distinguish four types of variogram behavior (Fig. 6.2):

– Parabolic:
  $\gamma(h)$ is twice differentiable at the origin. $F(x)$ is then differentiable

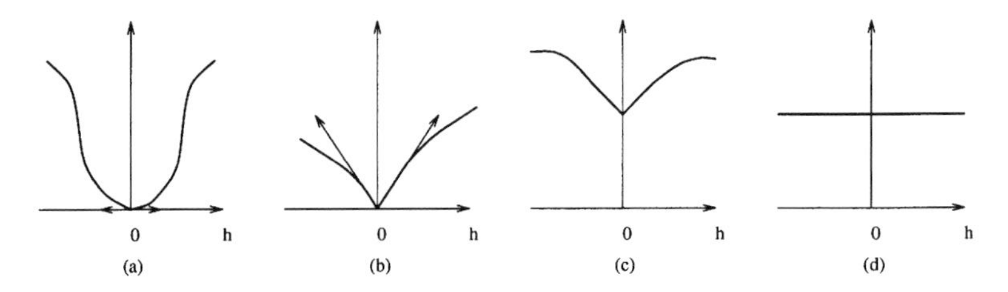

Figure 6.2 Behavior of a variogram at the origin: (a) parabolic (b) linear (c) nugget effect (d) pure nugget effect.

in the mean square sense and presents a highly regular spatial variability.

- Linear:
  $\gamma(h)$ is no longer differentiable at the origin but remains continuous at $h = 0$. $F(x)$ is mean-square continuous but no longer differentiable, consequently less regular.
- Microscopic correlation or nugget effect:
  $\gamma(h)$ does not tend towards 0 when h tends toward 0 (discontinuity at the origin), although by definition $\gamma(0) = 0$. $F(x)$ is no longer even mean-square continuous, consequently highly non-regular.
- Pure nugget effect:
  This is the limit case when $\gamma(h)$ appears solely as a discontinuity at the origin. $F(x)$ and $F(x')$ are uncorrelated (white noise). This corresponds to the total absence of autocorrelation.

The theoretical models can therefore be classified as:

- Models with a sill (or transition models):

  1 Linear behavior at the origin:

  (a) spherical model:

$$\gamma(r) = \begin{cases} \frac{3}{2}\frac{r}{a} - \frac{1}{2}\frac{r^2}{a^3} & \forall r \in [0, a] \\ 1 & \forall r \geq a \end{cases} \tag{6.8}$$

  (b) exponential model:

$$\gamma(r) = 1 - \exp\left(-\frac{r}{a}\right) \tag{6.9}$$

  The spherical model reaches its sill for a finite distance $r = a =$ range while the exponential one reaches it asymptotically. The difference between the spherical and the exponential model is the distance at which their tangents intersect the sill (Fig. 6.3):

  ∘ $r = \frac{2a}{3}$ for the spherical model,

  ∘ $r = a = \frac{a'}{3}$ for the exponential model.

  2 Parabolic behavior at the origin:

  (a) Gaussian model:

$$\gamma(r) = 1 - \exp\left(-\frac{r^2}{a^2}\right) \tag{6.10}$$

  The sill is reached asymptotically and a practical range can be considered for $a' = a\sqrt{3}$ value for which $\gamma(a') = 0.95 \simeq 1$.

– Models without sill:

These models correspond to random functions with unlimited capacity for spatial dispersion (Journel and Huijbregts, 1978) and therefore do not have variance or covariance.

1 Model in r^θ (Fig. 6.4):

$$\gamma(r) = r^\theta \quad \text{with } \theta \in]0, 2[\tag{6.11}$$

In practice only the linear model is used:

$$\gamma(r) = \omega r \tag{6.12}$$

where ω is the slope at the origin.

2 Logarithmic model:

$$\gamma(r) = \log r \tag{6.13}$$

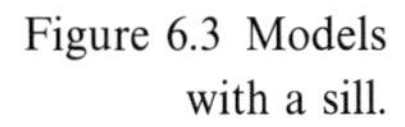

Figure 6.3 Models with a sill.

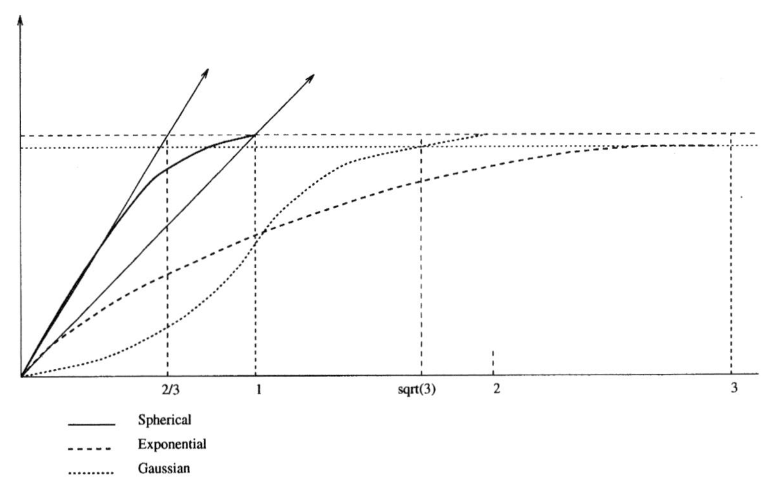

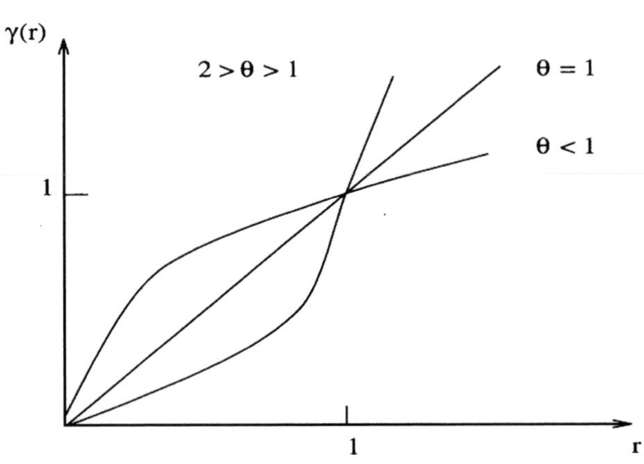

Figure 6.4 Models in r^θ.

Note that $\log(h) \to -\infty$ as $h \to 0$, so that the logarithmic function cannot be used to describe regionalizations on a strict point support. On the other hand, this model, once regularized on a non-zero support l, can be used as the model of a regularized variogram (Journel and Huijbregts, 1978; Matheron, 1970). The regularized variogram will have as expression:

$$\gamma_l(r) = \log\left(\frac{r}{l}\right) + \frac{3}{2} \qquad r \geq l \tag{6.14}$$

with $\gamma_l(0) = 0$.

The asymptotic behavior of these variograms is conditioned by

$$\lim_{r\to\infty} \frac{\gamma(r)}{r^2} = 0 \tag{6.15}$$

- Nugget effect:
  This is a discontinuity at the origin of the variogram. The limit of $\gamma(r)$ when r tends toward zero is a positive constant C_0 called *nugget constant*.

  This discontinuity can be linked to the presence of structures of lesser scale than those of the measure support, but can also be produced by measurement errors or by noise. When the discontinuity is total, we say that there is a pure nugget effect. The phenomenon is totally random and the variance is independent of the interval considered.

Every linear combination of these models is possible insofar as the coefficients are positive.

Experimental variogram. Let h be a vector of modulus $r = \mid h \mid$ and direction α. If there are N pairs of data separated by the vector h, then the experimental variogram in the direction α and for the distance h is expressed by Journel and Huijbregts (1978):

$$\gamma(r, \alpha) = \frac{1}{2N}\sum_{i=1}^{N}[F(x_i + h) - F(x_i)]^2 \tag{6.16}$$

Although these expressions are unique, the methods used in constructing variograms depend on the spatial configuration of the available data (Journel and Huijbregts, 1978).

We will place ourselves in the context of the hypothesis where most of the texture images present an isotropic characteristic (Ramstein, 1989). We will therefore suppose that the image variogram $\gamma(\vec{h})$ is independent of the orientation of the vector $\vec{h}$ and we will keep the notation $\gamma(h)$, h being the norm of the vector $\vec{h}$. The use of a directional variogram is justified in the case of anisotropic structures.

In practice, good numerical approximation of a two-dimensional image variogram is given by Ramstein (1989; Ramstein and Raffy, 1989):

$$\gamma(h) = \frac{1}{2}\left\{\frac{1}{N_c N_l}\sum_{x=1}^{N_c}\sum_{y=1}^{N_l}[I(x+h,y)-I(x,y)]^2 + \frac{1}{N_c N_l}\sum_{x=1}^{N_c}\sum_{y=1}^{N_l}[I(x,y+h)-I(x,y)]^2\right\} \quad (6.17)$$

where N_l and N_c are respectively the number of rows and columns of the image, and h is a distance expressed in pixels.

6.2.3 Kriging as an interpolator

At any given step of the matching procedure we have a set of matched points, which leads to the disparities ϵ_x, ϵ_y. Our aim is to fully map these functions, and for this purpose we will use kriging as an interpolation technique (Matheron, 1965).

Once the spatial structure of a regionalized variable has been demonstrated through computation of the variogram, the spatial structure can be used to estimate the value of the variable at unsampled locations.

Kriging is a means of weighted local averaging in which the weights $\lambda_i, i = 1, \cdots, n$ are chosen so as to give an unbiased estimate f^* at point x_0, while at the same time minimizing the estimation variance. Thus kriging can be thought of as a special kind of linear smoothing filter (Atkinson, 1991). Often, the weights are all the same and are therefore equal to $\frac{1}{n}$. When the property of interest is spatially dependent, however, a more precise estimate is obtained if the weights are chosen according to their influence on the point to be estimated. Kriging provides a means by which this is achieved.

Let $F(x)$ be a function whose values $F(x_1), F(x_2), \ldots, F(x_n)$ are known at $x_1, x_2, \ldots, x_n$, and for which we want to estimate the value at an unknown point x_0. We are then looking for the estimator $F^*(x_0)$, a linear combination of the random values $F(x_k)$, $k = 1, \ldots, n$ as:

$$F^*(x_0) = \sum_{k=1}^{n} \lambda_k F(x_k) \quad (6.18)$$

In the following, the summation will vary from 1 to n, n being the number of points used in the estimation.

$F^*(x_0)$ must meet the two following conditions:

$$E[\{F^*(x_0) - F(x_0)\}^2] \quad \text{minimum} \tag{6.19}$$

$$E[F^*(x_0) - F(x_0)] \quad = 0 \tag{6.20}$$

Equations (6.19) and (6.20) imply on the one hand that the estimator $F^*(x_0)$ is non-biased, and on the other hand that it is optimal (in the least mean square sense) as it minimizes the estimation error. By replacing (6.18) in (6.20) and by considering the intrinsic hypothesis (Journel and Huijbregts, 1978) we get:

$$E[F(x_0)]\{\sum_k \lambda_k - 1\} = 0 \tag{6.21}$$

which leads to the first kriging equation:

$$\boxed{\sum_k \lambda_k = 1} \tag{6.22}$$

Let us evaluate the two conditions that $F^*(x_0)$ must verify

$$\begin{array}{lcl} E[F^*(x_0) - F(x_0)^2] & & \text{minimum} \\ E[F^*(x_0) - F(x_0)] & = & 0 \end{array} \tag{6.23}$$

For this purpose, let us compute the two quantities $[F^*(x_0) - F(x_0)]$ and $[F^*(x_0) - F(x_0)]^2$. By definition and by eqn. (6.22) we have:

$$\begin{aligned} F^*(x_0) - F(x_0) &= \sum_k \lambda_k F(x_k) - F(x_0) \\ &= \sum_k \lambda_k [F(x_k) - F(x_0)] \end{aligned} \tag{6.24}$$

and

$$\begin{aligned} \{F^*(x_0) - F(x_0)\}^2 &= \left\{\sum_k \lambda_k F(x_k) - F(x_0)\right\}^2 \\ &= \sum_k \sum_l \lambda_k \lambda_l F(x_k) F(x_l) - 2 \sum_k \sum_l \lambda_k \lambda_l F(x_k) F(x_0) \\ &\quad + \sum_k \sum_l \lambda_k \lambda_l F^2(x_0) \end{aligned} \tag{6.25}$$

or

$$\gamma(h) = \frac{1}{2}E[\{F(x+h) - F(x)\}^2]$$
$$= E[F^2(x)] - E[F(x)F(x+h)] \tag{6.26}$$

We will have the expressions of $E[F(x_k)F(x_l)]$ and $E[F(x_k)F(x_0)]$ as a function of the variogram

$$E[F(x_k)F(x_l)] = E[F^2(x_l)] - \gamma(x_k - x_l)$$
$$E[F(x_k)F(x_0)] = E[F^2(x_0)] - \gamma(x_k - x_0) \tag{6.27}$$

which allows us to evaluate the value of $E[\{F^*(x_0) - F(x_0)\}^2]$ as a function of the variogram

$$\begin{aligned} E[\{F^*(x_0) - F(x_0)\}^2] &= 2\sum_k \sum_l \lambda_k \lambda_l \gamma(x_k - x_0) - \sum_k \sum_l \lambda_k \lambda_l \gamma(x_k - x_l) \\ &\quad + \sum_k \sum_l \lambda_k \lambda_l E[F^2(x_l)] - \sum_k \sum_l \lambda_k \lambda_l E[F^2(x_0)] \\ &= 2\sum_k \lambda_k \gamma(x_k - x_0) - \sum_k \sum_l \lambda_k \lambda_l \gamma(x_k - x_l) \end{aligned} \tag{6.28}$$

the variance being independent of x.

Let us minimize this expression with the constraint $\sum_k \lambda_k = 1$ using a Lagrange multiplier

$$\begin{cases} E[\{F^*(x_0) - F(x_0)\}^2] & \text{minimum} \\ \text{with the constraint} & \sum_k \lambda_k = 1 \end{cases}$$

which is equal to minimizing

$$2\sum_k \lambda_k \gamma(x_k - x_0) - \sum_k \sum_l \lambda_k \lambda_l \gamma(x_k - x_l) - \mu \sum_k \lambda_k \tag{6.29}$$

μ being the Lagrange multiplier, and $\sum_k \lambda_k$ the imposed constraint. We have then to solve the following system

$$\begin{cases} \dfrac{\partial}{\partial \lambda_1}\left[-\sum_k \sum_l \lambda_k \lambda_l \gamma(x_k - x_l) + 2\sum_k \lambda_k \gamma(x_k - x_0) - \mu \sum_k \lambda_k\right] = 0 \\ \vdots \qquad\qquad\qquad\qquad\qquad\qquad\qquad\qquad\qquad\qquad \vdots \;\; \vdots \\ \dfrac{\partial}{\partial \lambda_n}\left[-\sum_k \sum_l \lambda_k \lambda_l \gamma(x_k - x_l) + 2\sum_k \lambda_k \gamma(x_k - x_0) - \mu \sum_k \lambda_k\right] = 0 \end{cases} \tag{6.30}$$

which gives

$$\begin{cases} -\sum_l \lambda_l \gamma(x_1 - x_l) - \sum_k \lambda_k \gamma(x_k - x_1) + 2\gamma(x_1 - x_0) - \mu = 0 \\ \vdots \qquad\qquad\qquad\qquad \vdots \; \vdots \\ -\sum_l \lambda_l \gamma(x_n - x_l) - \sum_k \lambda_k \gamma(x_k - x_n) + 2\gamma(x_n - x_0) - \mu = 0 \end{cases} \tag{6.31}$$

The variogram is symmetric, and so $\sum_l \lambda_l \gamma(x_n - x_l) = \sum_k \lambda_k \gamma(x_k - x_n)$ and the system becomes

$$\begin{cases} -2\sum_k \lambda_k \gamma(x_1 - x_k) + 2\gamma(x_1 - x_0) - \mu &= 0 \\ \vdots & \vdots \;\; \vdots \\ -2\sum_k \lambda_k \gamma(x_n - x_k) + 2\gamma(x_n - x_0) - \mu &= 0 \end{cases} \tag{6.32}$$

This gives the kriging system:

$$\boxed{\begin{array}{l} \sum_i \lambda_i \gamma(x_i - x_j) + \mu = \gamma(x_j - x_0) \quad j = 1, \cdots, n \\ \sum_{i=1}^{n} \lambda_i = 1 \end{array}} \tag{6.33}$$

This can be written in matrix form after replacing $\gamma(0)$ by its value:

$$\begin{pmatrix} 0 & \gamma(x_2 - x_1) & \cdots & \gamma(x_n - x_1) & 1 \\ \gamma(x_1 - x_2) & 0 & \cdots & \gamma(x_n - x_2) & 1 \\ \vdots & \vdots & \vdots & \vdots & \vdots \\ \gamma(x_1 - x_n) & \gamma(x_2 - x_n) & \cdots & 0 & 1 \\ 1 & 1 & \cdots & 1 & 0 \end{pmatrix} \begin{pmatrix} \lambda_1 \\ \lambda_2 \\ \vdots \\ \lambda_n \\ \mu \end{pmatrix} = \begin{pmatrix} \gamma(x_1 - x_0) \\ \gamma(x_2 - x_0) \\ \vdots \\ \gamma(x_n - x_0) \\ 1 \end{pmatrix} \tag{6.34}$$

The kriging variance which is the estimation error is then given by:

$$E[\{f^*(x_0) - f(x_0)\}^2] = \sum_i \lambda_i \gamma(x_i - x_0) + \mu \tag{6.35}$$

In practice, a neighborhood of N points is defined outside which the observations carry so little weight that they can be ignored. We will call this neighborhood the kriging search window.

6.3 Disparity mapping with the wavelet transform

Our aim is to compute the disparity at each point of a pair of stereo images. For this purpose we will introduce an iterative process for the computation of the disparity values based on a multiresolution approach.

We begin by defining some of the terms used in the following. We will call:

- *real disparities*: the disparities computed on the discrete set of GCP,
- *disparity map*: the disparities estimated at each point of the image,
- *real variogram*: the variogram computed from the discrete set of GCP,
- *theoretical variogram model*: the theoretical model used in the kriging procedure which is based on a least squares fit of the real variogram values.

Let I_n, $n \in (1, N)$, $N = 2$, be the two stereoscopic images to be processed. Let us consider I_1 and I_2 as the reference and the working image respectively. Let M be the largest distance in the pixel space between two identical features. The matching must be first processed with the largest scale L, $2^{L-1} < M \leq 2^L$, in order to automatically match identical features without errors (Bijaoui and Guidicelli, 1991; Djamdji *et al.*, 1993a).

On each image I_n, we compute the wavelet transform with the à trous algorithm, up to the scale L. We then obtain N smoothed image $S_n(i, j)$ and $N \times L$ wavelet images $W_{nl}(i, j)$, $n \in (1, 2)$ and $l \in (1, L)$. The smoothed images are not used in the disparity computation procedure. The reference image will be for $n = 1$.

With L being the initial dyadic step, we perform on $W_{nl}(i, j)$ a detection procedure in order to detect the structures in the wavelet images and keep only those structures above a threshold of $(\theta \times \sigma_{n1})$, θ being a constant which increases when the resolution decreases, and σ_{n1} being the standard deviation of W_{n1} (Djamdji *et al.*, 1993a). We only retain from these structures their local maxima which will then act as GCPs. Our objective is to obtain the largest number possible of matched points in order to have a real disparity map which is as dense as possible.

Let (X, Y) be the coordinates of a maximum in the reference image and (x, y) the coordinates of the corresponding point in the working image. Let (x_1, y_1) be the coordinates (in the working frame) of the point corresponding to (X, Y) after applying the deformation model. If the model used describes correctly the geometrical deformation,

(x_1, y_1) must be very close or equal to (x, y). On the other hand, since the polynomial model does not model the deformations adequately due to the difference in the viewing angles, (x_1, y_1) is different from (x, y) and a correction term, the disparity, has to be introduced (Djamdji and Bijaoui, 1995a). We have:

$$\begin{cases} x_1 = f_l(X, Y) \\ y_1 = g_l(X, Y) \end{cases} \tag{6.36}$$

The disparity (ϵ_x, ϵ_y) is then computed at every point (X, Y) by:

$$\begin{cases} \epsilon_x(X, Y) = x - f_l(X, Y) = x - x_1 \\ \epsilon_y(X, Y) = y - g_l(X, Y) = y - y_1 \end{cases} \tag{6.37}$$

At step l, $l \neq L$, we carry out a new detection procedure over W_{il}. As previously, we detect the coordinates of the local maxima (X, Y) and (x, y) in the reference and the working image respectively. We then compute the variogram of the real disparities for step $l - 1$. The theoretical variogram model is then adjusted by least squares over these values. The coordinates (X, Y) of the GCP from step l are then transformed into the working frame using the deformation model (f_{l-1}, g_{l-1}) in order to get the new set of coordinates (x_1, y_1). The disparities (ϵ_x, ϵ_y) are estimated on each GCP (X, Y) between the points (x, y) and (x_1, y_1) (eqn. (6.37)) by kriging with the theoretical variogram model of step $l - 1$. These values are then used to correct the values (x_1, y_1) of step l, from the distortions due to the difference in the viewing angles. The corrected coordinates (x_2, y_2) (in the reference frame) are therefore obtained from:

$$\begin{cases} x_2 = x_1 + \epsilon_x(X, Y) \\ y_2 = y_1 + \epsilon_y(X, Y) \end{cases} \tag{6.38}$$

The points (x_2, y_2) are then matched with (x, y) and a new deformation model is computed between (X, Y) and (x, y). Next, the new real disparities associated with each GCP (X, Y) are computed and the process is reiterated until we reach the finest resolution (Djamdji and Bijaoui, 1995a,b).

At the lowest resolution, generally one, the real variogram is computed from the disparities at this resolution. The theoretical variogram is then adjusted over these values and the final disparity map is computed by kriging.

This approximation process can be seen as an inverse problem whose solution can be computed iteratively. The inverse problem is formulated as follows (Djamdji, 1993):

The two images being I_1 and I_2, by wavelet transform, thresholding and maxima detection, we get a list of maxima $L(I_1)$ and $L(I_2)$:

$$L(I_1) = (\text{Max} \circ \text{Thresh} \circ \text{WT})(I_1)$$
$$L(I_2^n) = (\text{Max} \circ \text{Thresh} \circ \text{WT})(I_2^n) \qquad (6.39)$$

with: WT wavelet transform operator,
Thresh thresholding operator,
Max maxima detection operator.

The goal is to obtain an image I_2^n in the same frame as I_1 and whose list $L(I_2^n)$ is identical to $L(I_1)$, i.e.:

$$\text{Distance}\{L(I_1)\,,\ L(I_2^n)\} \ \text{minimum} \qquad (6.40)$$

I_2^n is obtained by the application of an operator O^n over $I_2 = I_2^0$

$$I_2^n = O^n(I_2) \quad n \geq 1 \qquad (6.41)$$

O^n being the geometrical operator to be determined, and n being the iteration number.

The estimation of O^n must be refined until it converges toward a stable solution. The working image, at a given iteration n, is then registered using the deformation model associated with the final disparity map for that iteration. The entire process is then reiterated using the registered image as image I_2. The convergence is fast and attained after a few iterations, 3 in our case.

Once the iterative process has been carried out, the resulting disparity maps should be established for a given deformation model. We will use the deformation model (f_1, g_1) of the first iteration. For a procedure with only two iterations (Fig. 6.5), we have the following systems:

– First iteration:

$$\begin{cases} x_1 = f_1(x_1', y_1') \\ y_1 = g_1(x_1', y_1') \end{cases} \qquad (6.42)$$

$$\begin{cases} \epsilon_{x_1}(x_1', y_1') = x_0 - x_1 \\ \epsilon_{y_1}(x_1', y_1') = y_0 - y_1 \end{cases} \qquad (6.43)$$

– Second iteration:

$$\begin{cases} x_2 = f_2(x_2', y_2') \\ y_2 = g_2(x_2', y_2') \end{cases} \qquad (6.44)$$

$$\begin{cases} \epsilon_{x_2}(x_2', y_2') = x_1' - x_2 \\ \epsilon_{y_2}(x_2', y_2') = y_1' - y_2 \end{cases} \qquad (6.45)$$

Equations (6.42), (6.43), (6.44) and (6.45) allow us to establish the relation linking the coordinates (x'_2, y'_2) to (x_0, y_0) as a function of the disparities and the successive deformation models. We have the final expression:

$$\left\{\begin{array}{lcl} x_0 & = & f_1\{\, f_2(x'_2,y'_2) + \epsilon_{x_2}(x'_2,y'_2)\,,\; g_2(x'_2,y'_2) + \epsilon_{y_2}(x'_2,y'_2)\,\} \\ & & +\epsilon_{x_1}(x'_1,y'_1) \\ y_0 & = & g_1\{\, f_2(x'_2,y'_2) + \epsilon_{x_2}(x'_2,y'_2)\,,\; g_2(x'_2,y'_2) + \epsilon_{y_2}(x'_2,y'_2)\,\} \\ & & +\epsilon_{y_1}(x'_1,y'_1) \end{array}\right. \tag{6.46}$$

This model can be easily extended to any given number N of iterations.

We seek now to establish the disparity maps in X and Y linked to the deformation model (f_1, g_1),

$$\left\{\begin{array}{lcl} x'_0 & = & f_1(x'_2,y'_2) \\ y'_0 & = & g_1(x'_2,y'_2) \end{array}\right. \tag{6.47}$$

$$\left\{\begin{array}{lcl} \epsilon_x(x'_2,y'_2) & = & x_0 - x'_0 \\ \epsilon_y(x'_2,y'_2) & = & y_0 - y'_0 \end{array}\right. \tag{6.48}$$

The outcome is the final disparity map (ϵ_x, ϵ_y) associated with the deformation models (f_1, g_1). The flowchart of this algorithm is given in Fig. 6.6.

Pyramidal implementation. As in the case of geometric registration, this wavelet method is not well adapted for the processing of large

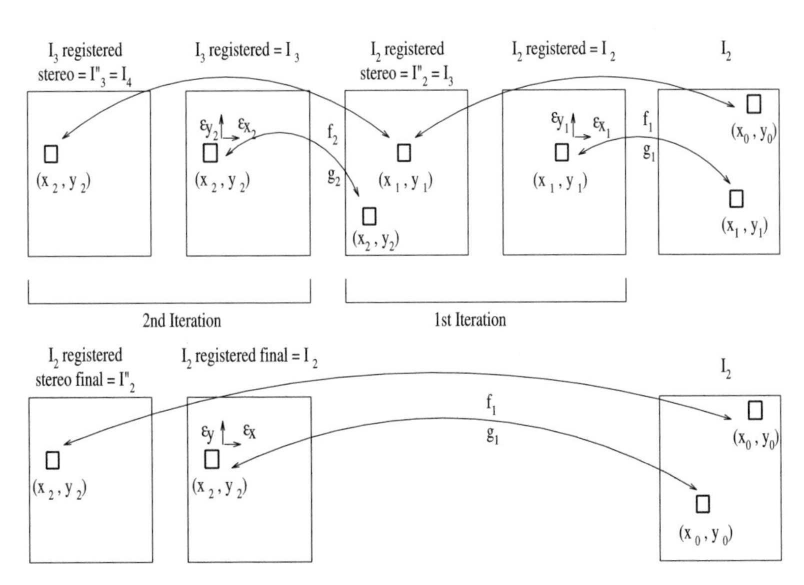

Figure 6.5 Iterative procedure: 2 iterations.

images mainly due to computation time which is very large in this case; and the very large disk space needed for the processing. One way to reduce these factors is to use a pyramidal implementation of the à trous algorithm. The process is globally quite similar to the one above. The image being reduced by a factor of 4 at each step, due to the decimation procedure, the matching must be done in the real coordinate space (Djamdji *et al.*, 1993b) and the disparity computation as well as the kriging must take into account this decimation. The flowchart of this algorithm is given in Fig. 6.7.

6.4 Image registration

Once the final disparity maps and the associated geometric deformation model (f_1, g_1) are computed, eqn. (6.4) can be rewritten:

$$\begin{aligned} x_i &= f_1(X_i, Y_i) + \epsilon_x(X_i, Y_i) = F_S(X_i, Y_i) \\ y_i &= g_1(X_i, Y_i) + \epsilon_y(X_i, Y_i) = G_S(X_i, Y_i) \end{aligned} \quad (6.49)$$

and the two stereoscopic images are registered as follows (Djamdji and Bijaoui, 1995c).

- For each output pixel location (i, j), we compute (k, l), $k = F_S(i, j)$ and $l = G_S(i, j)$, record the pixel value at the location (k, l) and assign it to the output pixel at (i, j). The process is iterated over the entire image and the output image is generated.
- The pixel locations (k, l) are generally not integers, so an interpolation must be carried out to compute the intensity value for the output pixel. Nearest-neighbors, bilinear, bicubic, spline interpolations are the most widely used.

6.5 Application to real images

This procedure was applied to the two following SPOT scenes:

- Scene number 148 – 319 dated 05 February 1991 taken at 07h_51mn_04s, composed of 3000 rows and 3000 columns, level 1a.
- Scene number 148–319 dated 02 April 1988 taken at 07h_34mn_40s, composed of 3003 rows and 3205 columns, level 1b.

These two scenes from the eastern region of Marib in the Republic of Yemen, were taken with different imaging directions. Level 1*a* (Spotimage, 1986) (Fig. 6.8) was taken with an incidence of 25.8

degrees left, while level 1*b* (Spotimage, 1986) (Fig. 6.9) was taken with an incidence of 6.3 degrees right. Two subscenes of 512×512 pixels were then extracted.

Image 1*b* will be considered as the reference image and image 1*a* as the working one. The noise level in the SPOT images being low, we reduced the threshold in the matching procedure in order to obtain the maximum number of GCPs. A four-iteration procedure is then applied to these two images with a kriging search window of ten points.

For each iteration, we compute the real disparity maps in X and Y by kriging. These maps, together with the associated deformation model, allow us to register the working image, by correcting the local distortions due to the viewing angles. The working corrected image is then used as the input working image for the next iteration, and the process is reiterated until convergence to a stable solution. A stability criterion can be determined from a statistical study of the real disparity maps for each iteration. The iterative process can be stopped if the standard deviation of the disparity maps (after kriging) reaches a certain threshold. The convergence is nevertheless quite fast and three iterations are sufficient. In this case, image *yemen1b* will be the reference frame and image *yemen1ai* the working one at iteration i. The working image at iteration i, $i \neq 1$, will be that of the $(i-1)$ iteration corrected for the distortions and registered.

The resulting final disparity maps in X and Y (Figs. 6.10, 6.11 and 6.12) are then built up from the disparity maps in X and Y for every iteration and from the associated deformation model, which is in this case the deformation model of the first iteration.

We used a six-order wavelet decomposition and a linear model for the theoretical variogram. The variogram was computed only over the last five resolutions, the number of GCPs in the first one (level 6) being insufficient for a variogram computation. The theoretical variogram model was fitted over the first 300 distance values h in pixels ($h \in [1, 300]$) of the real variogram $\gamma(h)$ for the three resolutions (levels 5,4,3) and on a distance of 100 pixels ($h \in [1, 100]$) for the last two resolutions (levels 2,1). In Table 6.1 we give the statistics on the final disparities for each iteration. Looking at this table, it can be easily seen that the fourth iteration was unnecessary and that three iterations would have sufficed.

The real and theoretical variograms in X and Y for the fourth iteration are given in Figs. 6.13 and 6.14. The real variogram is noisy, especially for the coarsest resolutions. This is due to the small number of points used for its estimation.

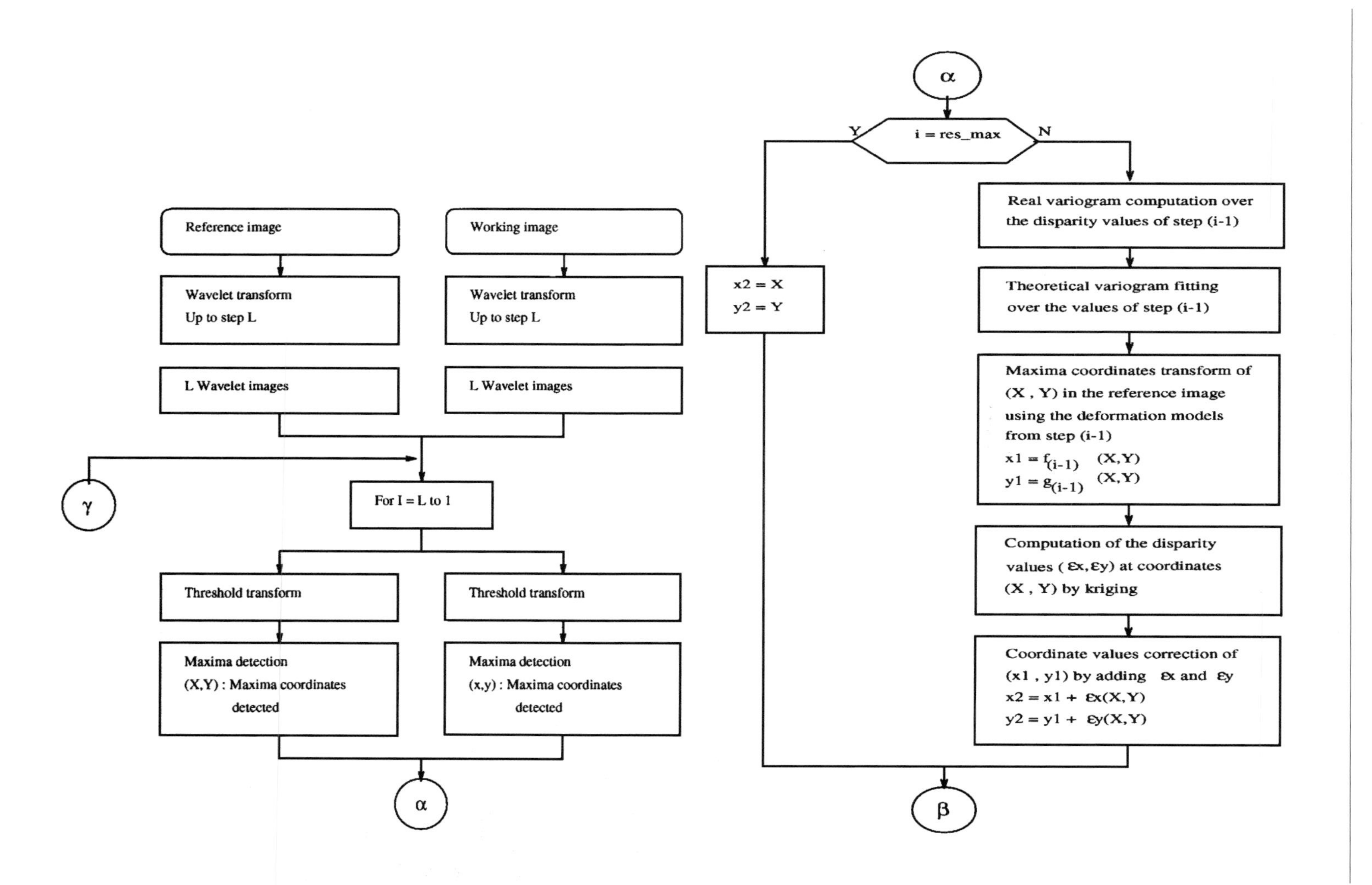
Reference image
Wavelet transform
Up to step L
L Wavelet images
Working image
Wavelet transform
Up to step L
L Wavelet images
γ
For I = L to 1
Threshold transform
Maxima detection
(X,Y) : Maxima coordinates detected
Threshold transform
Maxima detection
(x,y) : Maxima coordinates detected
α
α
i = res_max
Y
N
x2 = X
y2 = Y
Real variogram computation over the disparity values of step (i-1)
Theoretical variogram fitting over the values of step (i-1)
Maxima coordinates transform of (X , Y) in the reference image using the deformation models from step (i-1)
x1 = f(i-1) (X,Y)
y1 = g(i-1) (X,Y)
Computation of the disparity values (εx, εy) at coordinates (X , Y) by kriging
Coordinate values correction of (x1 , y1) by adding εx and εy
x2 = x1 + εx(X,Y)
y2 = y1 + εy(X,Y)
β

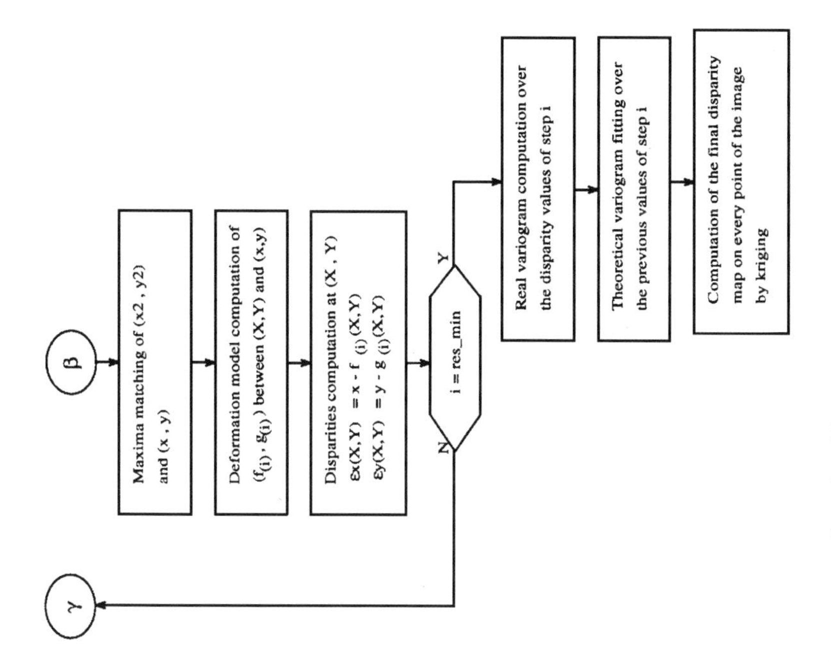

Figure 6.6 Flowchart of the disparity computation algorithm.

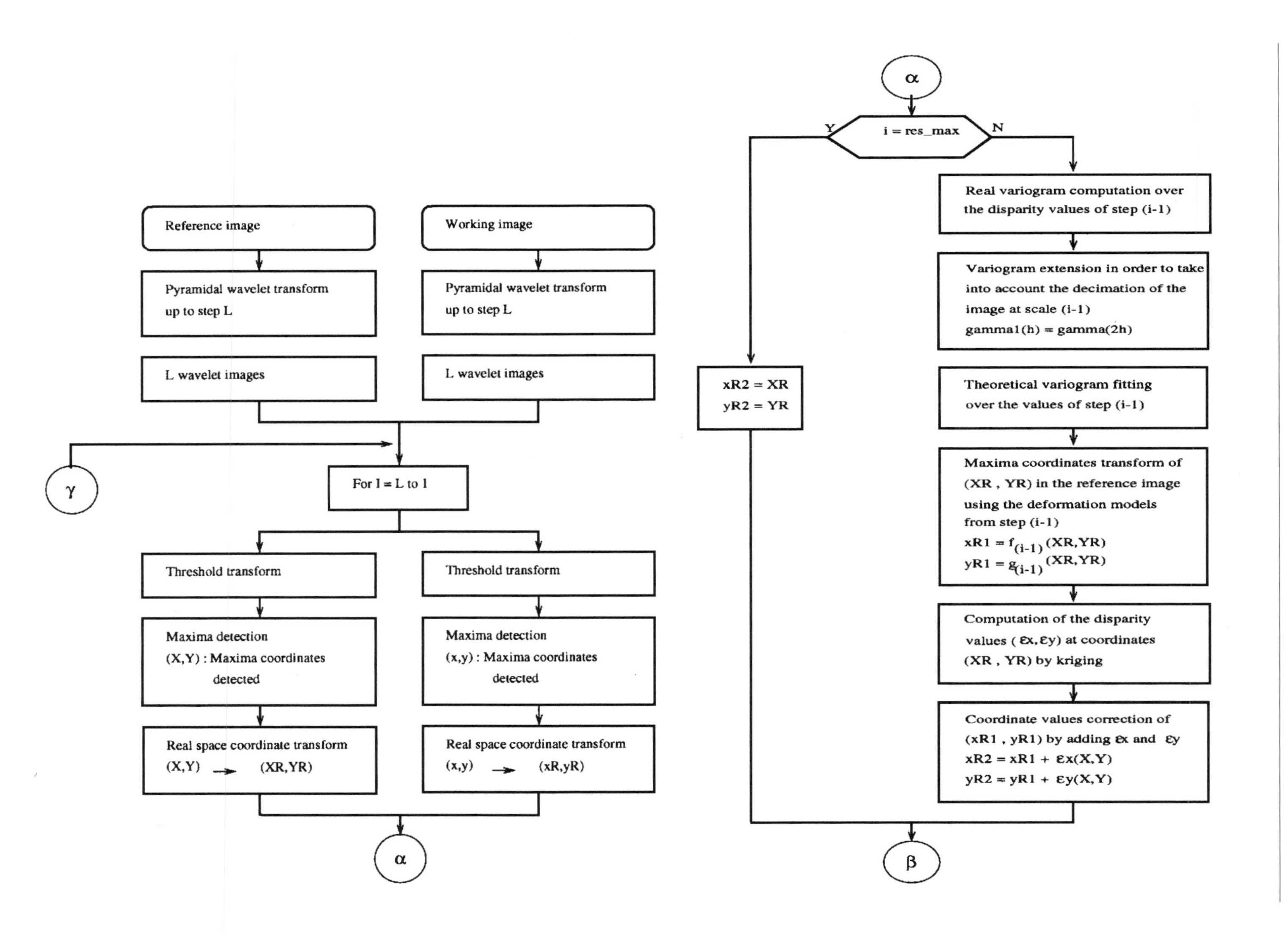
Reference image
Pyramidal wavelet transform up to step L
L wavelet images
Working image
Pyramidal wavelet transform up to step L
L wavelet images
γ
For l = L to 1
Threshold transform
Maxima detection (X,Y) : Maxima coordinates detected
Real space coordinate transform (X,Y) → (XR,YR)
Threshold transform
Maxima detection (x,y) : Maxima coordinates detected
Real space coordinate transform (x,y) → (xR,yR)
α
α
i = res_max
Y
N
xR2 = XR
yR2 = YR
Real variogram computation over the disparity values of step (i-1)
Variogram extension in order to take into account the decimation of the image at scale (i-1)
gamma1(h) = gamma(2h)
Theoretical variogram fitting over the values of step (i-1)
Maxima coordinates transform of (XR , YR) in the reference image using the deformation models from step (i-1)
xR1 = f(i-1) (XR,YR)
yR1 = g(i-1) (XR,YR)
Computation of the disparity values (εx, εy) at coordinates (XR , YR) by kriging
Coordinate values correction of (xR1 , yR1) by adding εx and εy
xR2 = xR1 + εx(X,Y)
yR2 = yR1 + εy(X,Y)
β

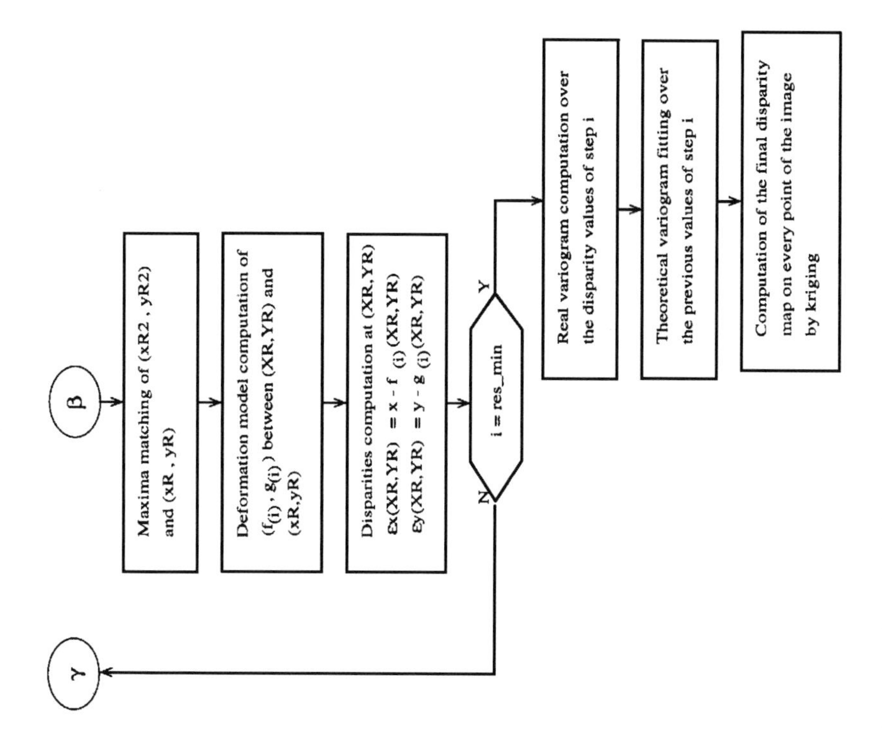

Figure 6.7 Flowchart of the pyramidal disparity computation algorithm.

The number of points detected in the working image decreases from the first iteration to the second. This is due to the interpolation introduced by the geometrical registration, which acts like a smoothing filter.

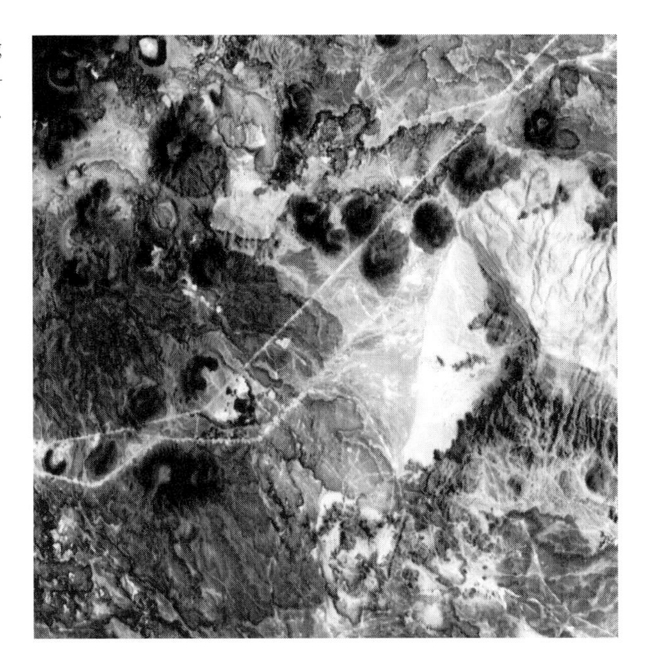

Figure 6.8 Working SPOT image – level 1a.

Figure 6.9 Reference SPOT image – level 1b.

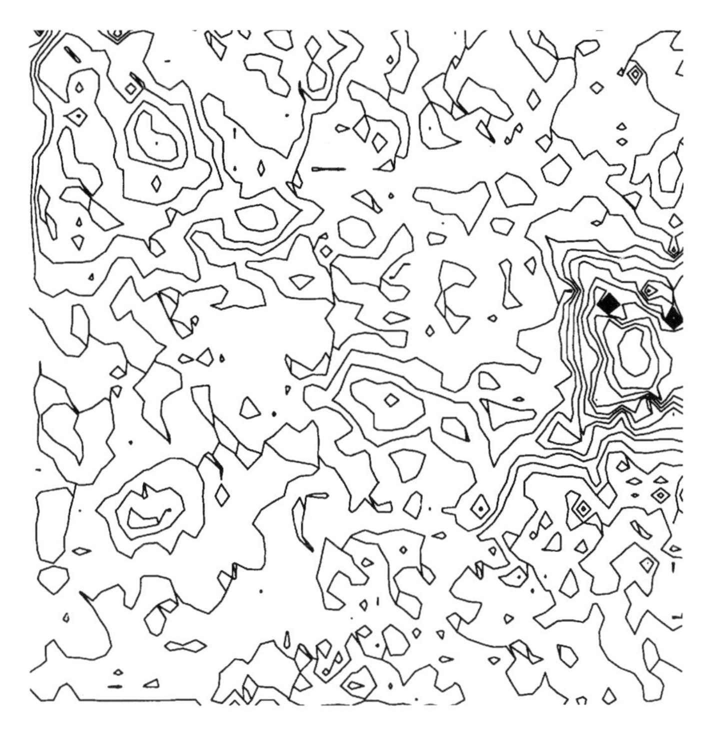

Figure 6.10 Isocontours of the final disparity map along the X axis.

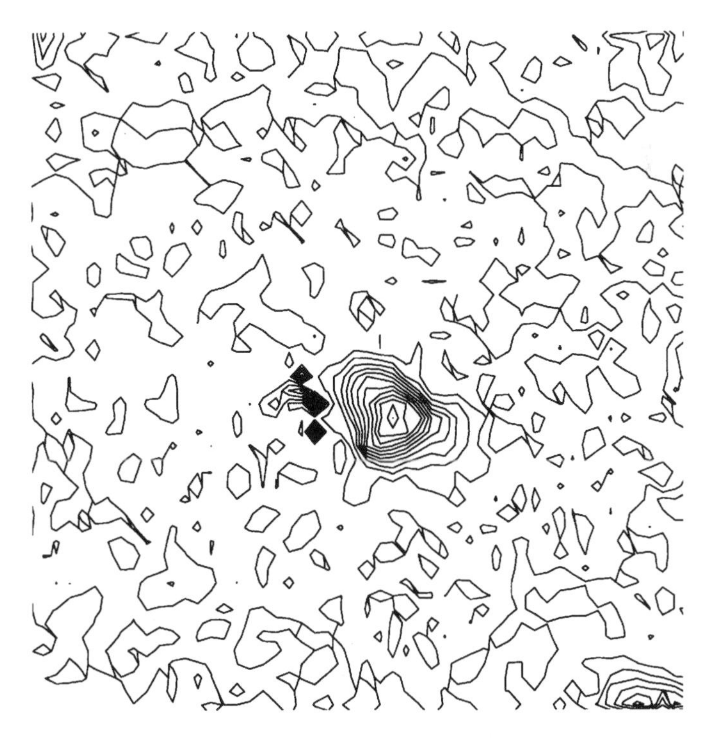

Figure 6.11 Isocontours of the final disparity map along the Y axis.

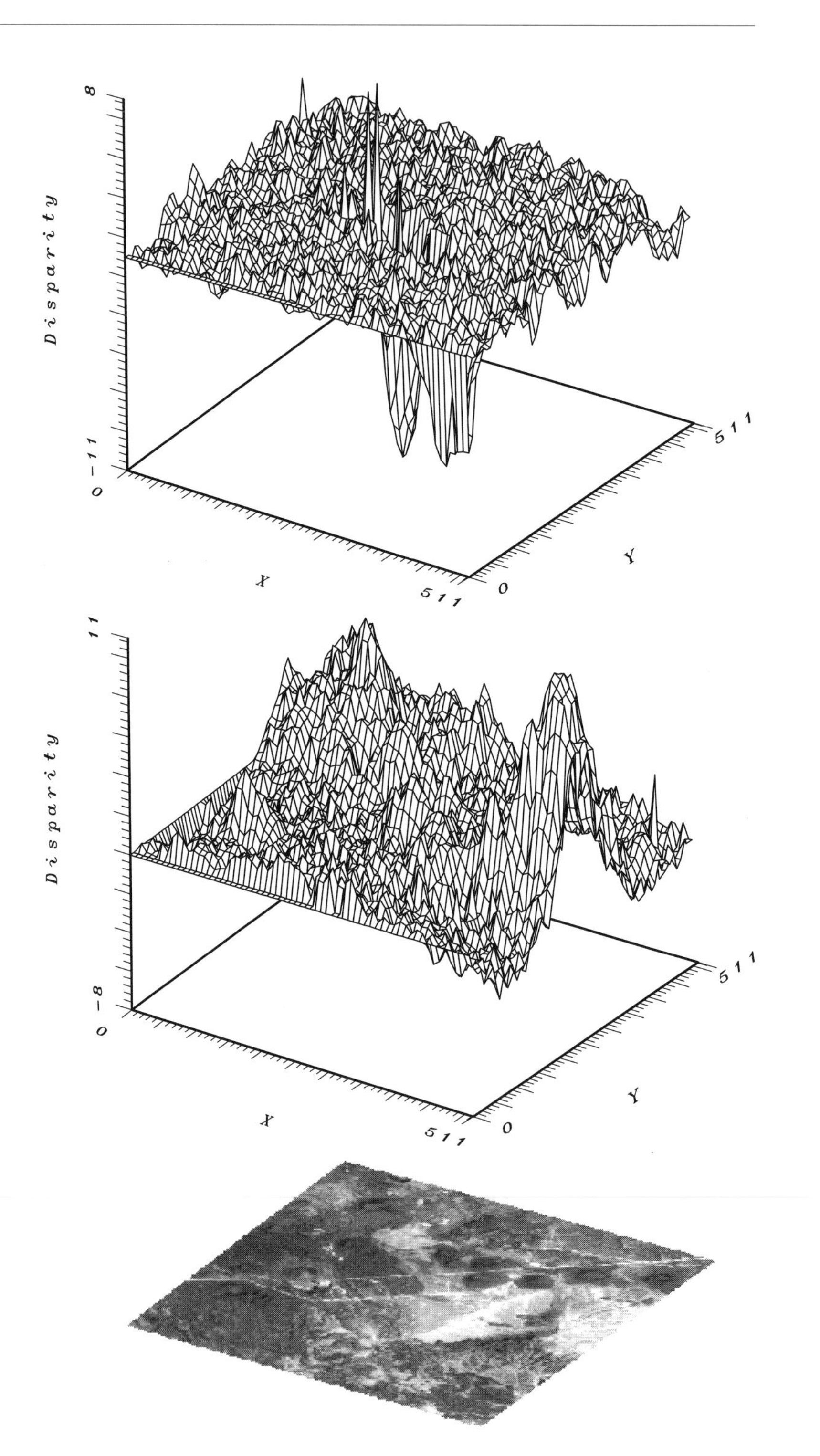

Figure 6.12 Perspective view of the final disparity maps along the Y and X axes, respectively, and plane view of the reference image.

The parameters of the second order polynomial deformation model are given by:

$$x' = a_x X^2 + b_x Y^2 + c_x XY + d_x X + e_x Y + f_x \qquad (6.50)$$

$$y' = a_y X^2 + b_y Y^2 + c_y XY + d_y X + e_y Y + f_y \qquad (6.51)$$

The accuracy of the final disparity map must be checked. In order to estimate the quality of the final disparity maps, we selected manually, on both the reference and the working image, 49 test ground control points (TGCP) uniformly distributed over the entire image (regions with high and low elevations). We then computed the disparities $(\epsilon\_\text{test}_x, \epsilon\_\text{test}_y)$ at the TGCP using the classical definition (eqn. (6.1)):

$$\epsilon\_\text{test}_{(x,i)} = x_{(\text{ref},i)} - x_{(\text{wrk},i)} \qquad (6.52)$$

$$\epsilon\_\text{test}_{(y,i)} = y_{(\text{ref},i)} - y_{(\text{wrk},i)} \qquad (6.53)$$

with $(x_{(\text{ref},i)}, y_{(\text{ref},i)})$ and $(x_{(\text{wrk},i)}, y_{(\text{wrk},i)})$ respectively the coordinates of the ith TGCP in the reference and working image. We then computed the kriged disparities $(\epsilon\_\text{krige}_x, \epsilon\_\text{krige}_y)$ at the TGCP, from the final disparity maps (I_{disp_x} and I_{disp_y}) and the associated deformation model (f_1, g_1) following the classical definition by:

$$\epsilon\_\text{krige}_{(x,i)} = x_{(\text{ref},i)} - \{f_1(x_{(\text{ref},i)}, y_{(\text{ref},i)}) + I_{\text{disp}\_x}(x_{(\text{ref},i)}, y_{(\text{ref},i)})\}$$

$$\epsilon\_\text{krige}_{(y,i)} = y_{(\text{ref},i)} - \{g_1(x_{(\text{ref},i)}, y_{(\text{ref},i)}) + I_{\text{disp}\_y}(x_{(\text{ref},i)}, y_{(\text{ref},i)})\}$$

and the residual (ρ_x, ρ_y) between the values $(\epsilon\_\text{test}_x, \epsilon\_\text{test}_y)$ and the

Table 6.1 *Statistics of the disparity images (after kriging) for different resolutions.*

Type of disparity	iteration	mean (pixel)	standard-deviation (pixel)	max	min
in X	1	−0.024	1.19	6.90	−10.54
in X	2	0.033	1.09	7.13	−3.37
in X	3	0.014	0.36	2.95	−1.35
in X	4	0.0062	0.29	1.81	−1.50
in Y	1	−0.047	1.01	1.82	−9.00
in Y	2	−0.0023	0.38	3.57	−1.63
in Y	3	−0.022	0.45	4.72	−2.23
in Y	4	−0.019	0.42	1.22	−3.39

corresponding values after kriging ($\epsilon\_krige_x, \epsilon\_krige_y$) for each TGCP:

$$\rho_{(x,i)} = \epsilon\_test_{(x,i)} - \epsilon\_krige_{(x,i)}$$
$$\rho_{(y,i)} = \epsilon\_test_{(y,i)} - \epsilon\_krige_{(y,i)} \quad (6.54)$$

On these residuals, we computed the minimum (Min), the Maximum (Max), the mean (E) and the standard deviation (σ). The results are given in Table 6.2. We can see from these results that the final disparity maps are well estimated, very few disparity pixels are inaccurate, and the precision of the TGCP selection is ± 1 pixel.

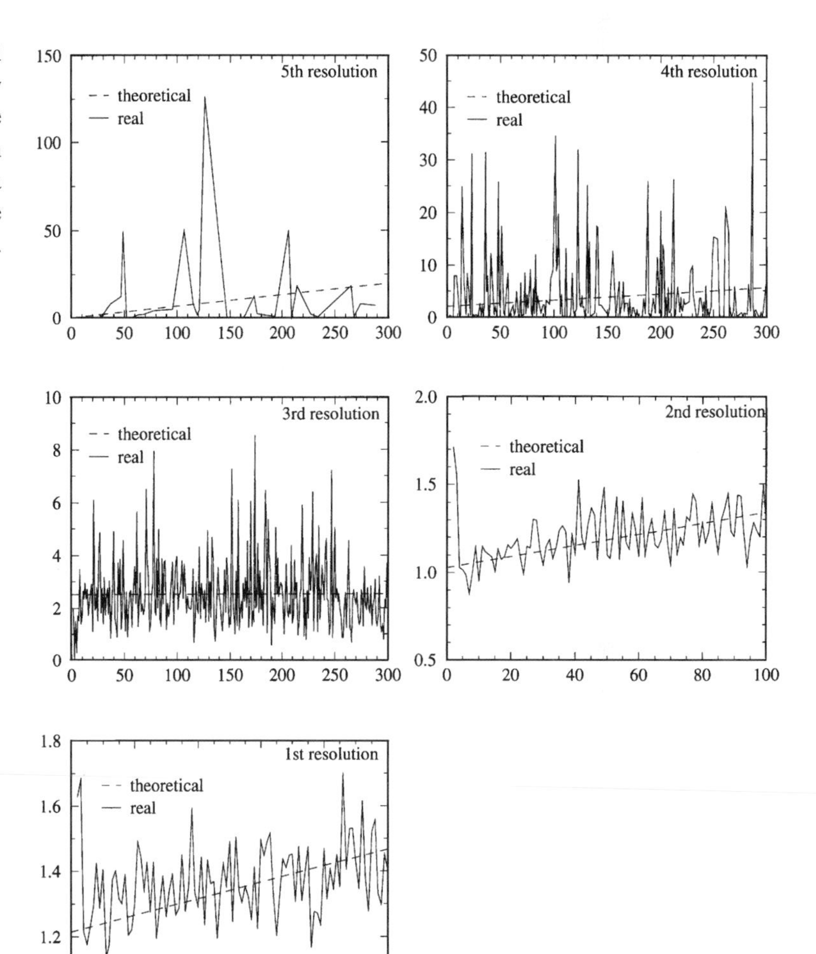

Figure 6.13 Real and theoretical disparity variogram along the X axis for the fourth iteration as a function of the distance in pixels.

Image registration. The resulting disparity maps are used to register the two original stereoscopic images. This is done by registering the working image with the deformation polynomials (f_1, g_1) and adding the disparity maps (eqn. (6.4)). The result is shown in Fig. 6.15. As in the case of the disparity evaluation, we selected manually, on both the reference and the working image, 40 test ground control points (TGCP), respectively $(x_{(\mathrm{ref},i)}, y_{(\mathrm{ref},i)})$ and $(x_{(\mathrm{wrk},i)}, y_{(\mathrm{wrk},i)})$, uniformly distributed over the entire image (regions with high and low elevations). Using the deformation model and the final disparity maps, we computed, for each TGCP in the reference frame $(x_{(\mathrm{ref},i)}, y_{(\mathrm{ref},i)})$, the corresponding point $(x^r_{(\mathrm{wrk},i)}, y^r_{(\mathrm{wrk},i)})$ in the working frame using eqn. (6.4).

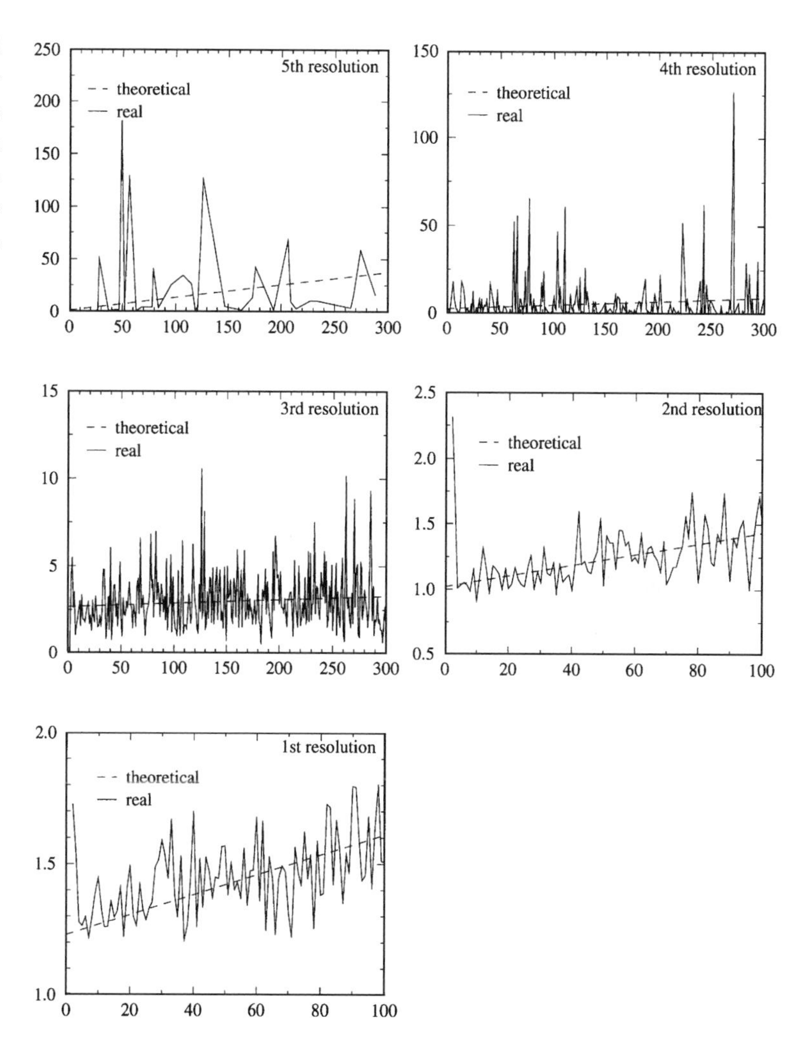

Figure 6.14 Real and theoretical disparity variogram along the Y axis for the fourth iteration as a function of the distance in pixels.

Then we computed the residual between the latter values and the TGCP of the working image $(x_{(\mathrm{wrk},i)}, y_{(\mathrm{wrk},i)})$ by:

$$\rho_i = \sqrt{((x_{(\mathrm{wrk},i)} - x^r_{(\mathrm{wrk},i)})^2 + (y_{(\mathrm{wrk},i)} - y^r_{(\mathrm{wrk},i)})^2)} \qquad (6.55)$$

On these residuals, we computed the minimum (Min), the maximum (Max), the mean (E) and the standard deviation (σ). The results are given in Table 6.3. We can see from these results that the geometrical registration is accurate, the precision of the TGCP selection being ± 1 pixel.

We also added the reference and the registered image in order to have a visual confirmation of the accuracy of the registration procedure (Fig. 6.16), which in our case is very good. We just recall that we were unable to have good registration of these images using a classical polynomial deformation model (Djamdji *et al.*, 1993a).

6.6 Conclusion on disparity analysis

We have presented a new method that allows the computation of the disparity maps along the X and Y axes, at each point of the image, between a pair of stereoscopic images with good accuracy, and without any knowledge of the position parameters of the satellite (ephemeris).

The procedure is fully automated and converges quite fast, three

Table 6.2 *Validation of the final disparity maps in X and Y after kriging (Nbh=Neighborhood).*

Disparity	Nbh	Number of TGCP N	Evaluation (pixel)			
			Min(ρ)	Max(ρ)	$E(\rho)$	$\sigma(\rho)$
x	10	49	−7.147	2.135	−0.199	1.419
y	10	49	−3.012	4.823	−0.073	1.128

Table 6.3 *Validation of geometrical registration of the stereoscopic images.*

Number of TGCP N	Evaluation (pixel)			
	Min(ρ)	Max(ρ)	$E(\rho)$	$\sigma(\rho)$
40	0.176	3.012	1.064	0.638

Figure 6.15 Registered working image with the help of the disparity maps.

Figure 6.16 The addition of the reference and the registered working image.

iterations being sufficient to achieve a good estimation of the disparity maps. The method has many other applications, one of which is the geometrical registration of images obtained under different viewing angles, a process which is achieved quite readily and with good accuracy. A pyramidal implementation (Djamdji and Bijaoui, 1995a) of this procedure reduces the processing time as well as the disk space needed, but can only be effective on large images due to the decimation introduced in the algorithm. Another important application is the computation of digital terrain models (DTM), which needs the ephemeris of the satellite.

7 Image compression

Image compression is required for preview functionality in large image databases (e.g. Hubble Space Telescope archive); with interactive sky atlases, linking image and catalog information (e.g. Aladin, Strasbourg Observatory); and for image data transmission, where more global views are communicated to the user, followed by more detail if desired.

Subject to an appropriate noise model, much of what is discussed in this chapter relates to faithful reproducibility of faint and sharp features in images from any field (astronomical, medical, etc.)

Textual compression (e.g. Lempel-Ziv, available in the Unix `compress` command) differs from image compression. In astronomy, the following methods and implementations have wide currency:

1. `hcompress` (White, Postman and Lattanzi, 1992). This method is most similar in spirit to the approach described in this chapter, and some comparisons are shown below. `hcompress` uses a Haar wavelet transform approach, whereas we argue below for a non-wavelet (multiscale) approach.
2. `FITSPRESS` (Press, 1992; Press *et al.*, 1992) is based on the non-isotropic Daubechies wavelet transform, and truncation of wavelet coefficients. The approach described below uses an isotropic multiresolution transform.
3. `COMPFITS` (Véran and Wright, 1994), relies on an image decomposition by bit-plane. Low-order, i.e. noisy, bit-planes may then be suppressed. Any effective lossless compression method can be used on the high-order bit-planes.
4. `JPEG` (Hung, 1993), although found to provide photometric and astrometric results of high quality (Dubaj, 1994), is not currently well-adapted for astronomical input images.

The basis of the approach described in the following (see Starck *et al.*, 1996) is astronomical image compression through noise removal. Noise is determined on the basis of the image's assumed stochastic properties. This is potentially a very powerful technique, since astronomical images are characterized by (i) the all-pervasive presence of noise, and (ii) knowledge of the detector's and image's noise properties, at least approximately. Rather than being open-ended in the amount of information which can be thrown away, the method described here has an inherent compressibility which is aimed at, – i.e. lossless compression of the noise-filtered image. The primary user parameter for controlling noise suppression is expressed in terms of the noise (e.g. a multiple of the noise variance).

We choose other multiresolution transforms than the wavelet transform for the reasons described in Chapter 1 (subsection 1.4.7).

7.1 Pyramidal median transform and image compression

7.1.1 Compression method

The principle of the method is to select the information we want to keep, by using the PMT (see Chapter 1, section 1.5), and to code this information without any loss. Thus the first phase searches for the minimum set of quantized multiresolution coefficients which produce an image of 'high quality'. The quality is evidently subjective, and we will define by this term an image such as the following:

- there is no visual artifact in the decompressed image.
- the residual (original image – decompressed image) does not contain any structure.

Lost information cannot be recovered, so if we do not accept any loss, we have to compress what we take as noise too, and the compression ratio will be low (3 or 4 only).

The method employed involves the following sequence of operations:

1. Determination of the multiresolution support (see Chapter 2).
2. Determination of the quantized multiresolution coefficients which gives the filtered image. (Details of the iterative implementation of this algorithm are dealt with below.)
3. Coding of each resolution level using the Huang-Bijaoui (1991) method. This consists of quadtree-coding each image, followed by Huffman-coding the quadtree representation. There is no information lost during this phase.

4 Compression of the noise if this is wished.
5 Decompression consists of reconstituting the noise-filtered image (+ the compressed noise if this was specified).

Note that we can reconstruct an image at a given resolution without having to decode the entire compressed file.

These four last phases will now be described.

7.1.2 Quantized multiresolution coefficients

We define the set $\mathcal{Q} = q_1, ..., q_n$ of quantized coefficients, q_j corresponding to the quantized multiresolution coefficients w_j. We have:

- $q_j(x, y) = 0$ if $M(j, x, y) = 0$
- $q_j(x, y) = \text{int}(w_j(x, y)/(k_{\text{signal}}\sigma_j))$ if $M(j, x, y) = 1$

Here, *int* denotes integer part. The image reconstructed from $\mathcal{Q}$ gives the decompressed image D. Good compression should produce D such that the image $R = I - D$ contains only noise. Due to the thresholding and to the quantization, this is not the case. So it can be useful to iterate if we want to compress the quantized coefficients which produce the best image. The final algorithm which allows us to compute both the quantized coefficients and the multiresolution support is:

1 Set $i = 0$, $R^i = I$
2 Set $M(j, x, y) = 0$ and $q_j^i(x, y) = 0 \quad \forall x, y, j$
3 Compute the PMT of R^i: we obtain w_j
4 If $i = 0$, estimate at each scale j the standard deviation of the noise σ_j.
5 New estimation of the multiresolution support:
for all j, x, y, if $\mid w_j(x, y) \mid > k\sigma_j$, $M(j, x, y) = 1$
6 New estimation of the set $\mathcal{Q}$:
for all j, x, y, if $\mid M(j, x, y) \mid = 1$,
$q_j(x, y) = q_j(x, y) + \text{int}(w_j(x, y)/(k_{\text{signal}}\sigma_j))$
7 Reconstruction of D^i from $\mathcal{Q}$
8 $i = i + 1$, $R^i = I - D^{i-1}$ and go to step 3

In step 6, $k_{\text{signal}} = 1.5$. (This value has been fixed experimentally and seems to be a good trade-off between quality and efficiency.) After a few iterations, the set $\mathcal{Q}$ contains the multiresolution coefficients. This allows a considerable compression ratio and the filtered image can be reconstructed without artifacts. The residual image R is our reconstruction error ($\text{rec}(\mathcal{Q}) + R = I$). The results without iterating are satisfying too, and are sufficient in most cases. If we are not limited by time computation during the compression, we can carry out a few iterations in order to have subsequently the best quality reconstruction.

7.1.3 Quadtree and Huffman encoding

We choose to code the multiresolution coefficients by using a quadtree (Samet, 1984) followed by a Huffman (Held and Marshall, 1987) encoding (fixed codes were used in our implementation).

The particular quadtree form we are using was described by Huang and Bijaoui (1991) for image compression.

- Divide the bitplane up into 4 quadrants. For each quadrant code as '1' if there are any 1-bits in the quadrant, else code as '0'.
- Subdivide each quadrant that is not all zero into 4 more sub-quadrants and code them similarly. Continue until one is down to the level of individual pixels.

7.1.4 Noise compression

If we want exact compression, the noise must be compressed too. There is no transform which allows better representation of the noise, and the noise compression ratio will be defined by the entropy. Generally, we do not need all the dynamic range of the noise, and the residual map R is not compressed but rather the image $R_q = \text{int}(R/(k_{\text{noise}}\sigma_R))$ with $k_{\text{noise}} = \frac{1}{2}$ in the applications below.

Lossless compression can be performed too, but this has sense only if the input data are integers, and furthermore the compression ratio will be very low.

7.1.5 Image decompression

The decompression is carried out scale-by-scale, starting from a low resolution, so it is not necessary to decompress the entire file if one is just interested in having a look at the image. Noise is decompressed and added at the end, if this is wanted. (The examples discussed below suppress the noise entirely.)

7.2 Examples and assessments

The examples cover digitized photographic (Figs. 7.1 and 7.4) and CCD cases (Fig. 7.7). Figures 7.1, 7.2 and 7.3 show comparative results of the approach described here with routines which we call `mr_comp` and mr_decomp, and hcompress (White *et al.*, 1992). Figures 7.4, 7.5 and 7.6 show another set of comparative results. Figures 7.7 and 7.8 show a result of compressing a CCD image.

Figure 7.1 shows a 256×256 section of the Coma cluster from a Space Telescope Science Institute POSS-I digitized plate. It was compressed using the approach described in this chapter to produce Fig. 7.2. Default options were used (which included: 6 scales in the multiresolution transform; thresholding at three times the noise standard deviation; signal quantization with a denominator of 1.5 times the standard deviation; and a Gaussian noise model). Figure 7.3 shows the decompressed image resulting from hcompress. Since the latter preserves textual image descriptors (our method deletes most of them; neither approach seeks to compress the textual data in the descriptors), we catered for this by having the hcompress decompressed image be appropriately larger. Figures 7.1, 7.2 and 7.3 show the images logarithmically transformed, and zoomed by a factor of two. The original 16-bit image had 131 072 bytes of image data. The mr_comp com-

Figure 7.1 The Coma cluster from a Space Telescope Science Institute POSS-I digitized plate.

pressed image had 4288 bytes, and the `hcompress` compressed image – by design, using a scaling factor of 2490 – had 4860 bytes. The compression factor was therefore 131 072/4288 = 30.6.

Visually, Fig. 7.2 (`mr_comp`) outperforms Fig. 7.3 (`hcompress`): one can notice all-pervasive box features in the latter at faint intensity levels. The RMS (root mean square) error between compressed and original images was 273.9 for `mr_comp` and 260.4 for `hcompress`. The SNR (signal-to-noise ratio, defined as 10 times the logarithm, base 10, of the variance of the input image minus the variance of the error image) was 14.1 dB for `mr_comp` and 14.5 dB for `hcompress`. Thus, in this case, these global quality measures favor `hcompress`.

In regard to such global quality measures, it may be noted that part of what is measured is how well the respective algorithms relate to the noisy (hence irrelevant aspect of the) input image. It may also encompass how well input noisy features differ from restored noise, following quantization. For the foregoing image, `hcompress` provided a better set of global quality measures, but we have found visual

Figure 7.2 Compressed/ decompressed using the PMT.

quality to often differ substantially from global quantitative values. Visual quality is of primary importance, since local and faint features are after all what one most often wishes to see respected. Nonetheless, we will make use of MSE (mean square error) and SNR quality measures here, given their indicative usefulness.

Figure 7.4 shows the central portion of a 900×900 subimage around NGC 5128 (Centaurus A), from the ESO Southern Sky Survey. This image and also Figs. 7.5 and 7.6 (respectively mr_comp and hcompress decompressed results for approximately equal compression rates) are all shown histogram-equalized. Compression to 104 Kbytes was attained in the case of mr_comp, with a slightly larger size for hcompress to compensate for preserving of descriptors. A scaling factor of 1480 was used by design by the latter. This gave therefore a compression rate for mr_comp of $1\,620\,000/103\,667 = 15.6$.

The RMS values relating to Figs. 7.5 and 7.6 are respectively 39.4 and 46.1. The SNR values were respectively 31.8 dB and 30.5 dB. In

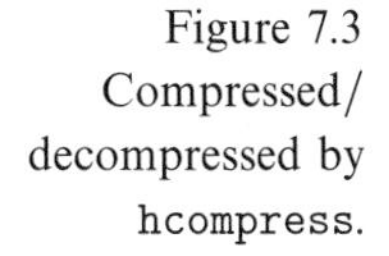

Figure 7.3 Compressed/ decompressed by hcompress.

this case, mr_comp performed better as regards these global measures. The faint box structures are visible in the hcompress output (Fig. 7.6).

Figure 7.7 shows an 800×800 Hubble Space Telescope WF/PC-2 (Wide Field/Planetary Camera 2) image of NGC 4321 (M100). The central part of this image, only, is shown in the figure. Figure 7.8 shows an mr_comp decompressed result. From 2 566 080 bytes of data in the input image, the compressed image contained 51 394 bytes, which implies a compression factor of $2\,566\,080/51\,394 = 49.9$. In Fig. 7.8, one can note that cosmic ray hits ('salt and pepper' noise) have not been treated in any way.

Many close variants on the implementation adopted could be worthy of consideration: a more finely tuned noise model; intervention to remove or doctor detector faults; different quantization schemes; dependence of compressibility on the image's content – e.g. large elliptical galaxy vs. star field. Many such were looked at in this work. The resulting implementation is one which we found to work convincingly and effectively.

Figure 7.4 NGC 5128 (Centaurus A), from the ESO Southern Sky Survey.

For a 900×900 image, mr_comp takes about 100 seconds, compared to hcompress which takes around 24 seconds. This mr_comp timing is with a median transform kernel of dimensions 3×3 (which can be varied, as a user parameter). The routine mr_comp aims at finding and separating out the noise. On the other hand, the scale factor used by hcompress in practice is set so that a certain compression rate is attained, and therefore this program may need to be run a few times to find the desired value. This difference in objectives between the two approaches should be noted.

7.3 Image transmission over networks

The decomposition of the image into a set of resolution scales, and furthermore the fact that they are available in a pyramidal data structure, can be used for effective transmission of image data (see Percival and White, 1993). Some current work on World-Wide Web progressive image transmission capability has used bit-plane decomposition

Figure 7.5 NGC 5128 compressed/decompressed using the PMT.

(Lalich-Petrich, Bhatia and Davis, 1995). Using resolution-based and pyramidal transfer and display with WWW-based information transfer is a further step in this direction.

We prototyped a number of approaches to image transmission, based on compressed images, and these will be briefly described. First, a few particular aspects of this issue will be noted.

1 There are a number of somewhat different image transmission scenarios. In the case of image delivery from a large image database (e.g. the Hubble Space Telescope archive) quick-look compressed images are available to guide the user in whether or not the images are really desired. In the case of storage management by an individual user, analogous to the use of a Unix command such as `compress` for text, efficient storage may be coupled with efficient on-the-fly uncompression and viewing. In the case of research collaborators sharing images (not on the same scale as the image database scenario above), network bandwidth may be a relatively

Figure 7.6 NGC 5128 compressed/decompressed using `hcompress`.

scarce resource. In the following, our thoughts are principally related to the last of these scenarios.

2 Astronomical images are noisy, and certainly if they are real-valued, then true lossless compression is highly unusual.

3 WWW browser support for in-lining of images does not currently extend to FITS. (FITS, Flexible Image Transport System, is an image format universally used in astronomy.) Therefore there are good grounds for providing GIF or JPEG versions of a FITS image, to facilitate viewing.

4 Actions can be defined for viewing (or performing other operations) at the client end of the transmission, based on the content type of the image file being transmitted. It is important to note that the client's local configuration usually does not override the WWW server's recommended content type.

Figure 7.7 NGC 4321 (Hubble Space Telescope WF/PC-2).

Three main options led to the following prototypes:

1 The web server which stores the multiresolution compressed images (we will use file extension .MRC, and take the decompression executable as `mr_decomp`), takes care of the uncompress process and sends back the requested image as a FITS file. This option only requires a FITS image viewer such as `SAOimage` (from the Smithsonian Astrophysical Observatory) on the client machine.

The main drawback of this option is the load on the network since a decompressed image is sent from the server to the client.

For this prototype, the server needs a CGI (the WWW Common Gateway Interface) script which calls the program to work on the MRC file of the requested image. And the client needs to configure the browser to recognize FITS images and locate the appropriate viewer. This configuration depends on the browser and the client machine. This is achieved by mapping the document MIME type, `image/x-fits`, to the `.fits` filename extension and to the FITS

Figure 7.8 NGC 4321 compressed/decompressed using the PMT.

viewer's application. (MIME is the Multipurpose Internet Mail Extension, a mechanisms for specifying and describing the format of Internet files.) On Unix, the `.mime.type` and `.mailcap` files are updated to do this.

2 The client decompresses the MRC file locally. The server sends back an MRC file to the client browser which calls `mr_decomp` to get a FITS image. Therefore `mr_decomp` must be installed on the client machine. This client machine must be powerful enough to run the decompression smoothly. This option saves the network's bandwidth: only compressed files are transferred.

 The decompression is made via the MIME scheme. For the Netscape browser, the X11 application default resource, `encodingFilters` is modified.

Another option has also been studied: an intermediate proxy server is used to decompress the MRC file and to send the FITS image to the browser. The proxy may also be used to cache the MRC file or the FITS image. This elegant option combines the main advantages of the previous options since it saves the wide area network bandwidth between the remote MRC file server and the local proxy server. The decompression program, `mr_decomp`, runs on a single machine, the proxy server (and so this saves the user any local computation requirement), and this process is transparent to a user who may access any MRC publishing server.

Another possible direction is to take advantage of the multiresolution process to send the image resolution-by-resolution (from the lowest to the highest). For example, one may request the lowest resolution of an image to have a 'quicklook' of this image and afterwards ask for the next resolutions until you are satisfied with the image's quality and noise level. The FITS image is reconstructed locally from the files that are sent one-by-one to the client. This is progressive image transmission.

Current directions which are being actively worked on include: (i) progressive transmission with user-control of the part(s) of the image to be transmitted with priority; and (ii) use of Java to send the decompressing code to the client in association with the compressed data.

7.4 Conclusion on image compression

We set out to disentangle signal from noise in astronomical imagery, and to use this to compress the signal part of the image (axiomatically, noise is incompressible). An innovative approach was used to do this, which has proven itself effective in practice. Some approximations and heuristics have been noted in this work, which point to possible directions of study.

For image preview systems, the issue of photometric fidelity to the true image is not of paramount importance. Instead, certain other issues are on the agenda, where choice is required. Among these are: ought cosmic ray hits be removed from the image, or should the user be presented with an image which shows 'warts and all'?; and should the noise image be stored separately for reconstitution of the original image in the preview system?

8 Object detection and point clustering

A pyramidal or other multiresolution representation of an image can be used to facilitate the extraction of information from an image. Imposing a certain 'syntax' on the image in this way may be of aid to the user in regard to the image's 'semantics': i.e. a structuring imposed on the image may help in interpretation of the image.

For large objects in an image, a well-resolved galaxy for example, or for superimposed objects, a multiscale approach is a very plausible one. Such an approach has been pursued in Bijaoui (1991a, 1993a,b), Bijaoui *et al.* (1994a), Bijaoui, Slezak and Mars (1989), Slezak *et al.* (1990, 1993).

Here we consider the case of images with sparsely located, small astronomical objects. For search and analysis of a particular class of objects, we use one privileged scale of a multiresolution transform. An object 'map' or boolean image may be derived from the multiresolution support. Such a boolean image may be further cleaned of detector faults, and unwanted objects, using mathematical morphology (or Minkowski) operators.

8.1 The problem and the data

Earlier work on our part aimed at finding faint edge-on galaxies in WF/PC images. For each object found, properties such as number of pixels in the object, peak-to-minimum intensity difference, a coefficient characterizing the azimuthal profile, and the principal axis ellipticity, were used to allow discrimination between potentially relevant objects, on the one hand, and faint stars or detector faults, on the other hand.

Here we are concerned with the study of globular cluster systems surrounding elliptical galaxies. NGC 4636 was discussed by Kissler *et al.* (1993), and characterized as a rich globular cluster system in a normal elliptical galaxy. Figure 8.1 shows an HST WF/PC image of NGC 4697, taken in May 1994 (hence pre-refurbishment). This image is of dimensions close to 1600×1600, where the optically inactive borders have been removed, and where one can see left-over features where the images produced by the four different CCD chips have been mosaiced together.

8.2 Pyramidal median transform and Minkowski operators

Pyramidal transforms based on the median transform were reviewed in Chapter 1. The pyramid data structure offers a storage-efficient structuring of the image. A given resolution level may be expanded up to the original image's dimensions, if this proves convenient for later processing. B-spline interpolation can be used to reconstruct the original dimensionality. We used the pyramidal median transform, with 4 resolution levels.

Figure 8.2 shows the level 3 resolution level (at full dimensionality) which has been booleanized on the basis of a 3σ threshold, and under the assumption that the image's noise characteristics were modeled

Figure 8.1 HST WF/PC image of NGC 4697. Globular clusters surrounding the galaxy are of interest.

correctly by additive Poisson and Gaussian read-out noise (the latter of zero mean, gain 7.5 e^-/DN, and variance 13 e^-/pixel). Cosmic ray hits were removed (at least to a first order) by consideration of a number of exactly similarly registered frames.

In Fig. 8.2, as a good approximation, the objects of interest are with high confidence among the contiguous boolean regions. Adaptivity for irregular background or objects superimposed on larger, diffuse objects, is built into our approach. Unlike traditional adaptive thresholding procedures for object detection in astronomy, the multiresolution transform used here takes much of the burden of defining background from the user. Detection thresholds are based on the image's noise properties, and thus on the detection of signal.

One also sees in Fig. 8.2 that this particular resolution level did not perform particularly well in revealing all of the large elliptical object, for which a different scale would be more appropriate. Detector faults (cold pixel areas, column overflow or bleeding) have been fairly well removed, but not all (see the very low-valued 'circle' of pixels in the upper right quadrant; or the just noticeable differences in the four quadrants in Fig. 8.2).

To 'fill in' parts of the large diffuse galaxy object, and thus to avoid

Figure 8.2 Resolution level 3, booleanized, version of previous figure.

later having to sift through parts of it which manifest themselves as small object-like, we used two dilations, using as structuring element:

$$s = \begin{pmatrix} 0 & 1 & 1 & 0 \\ 1 & 1 & 1 & 1 \\ 1 & 1 & 1 & 1 \\ 0 & 1 & 1 & 0 \end{pmatrix}$$

Such a structuring element is based on a priori knowledge of the object sought, viz. point symmetric, and of small size.

Following this, 5 openings were applied, i.e. 5 erosions to kill off smaller, spurious objects (left-overs of cosmic ray hits, thin detector faults) and 5 dilations to re-establish a sufficient hinterland around potentially relevant objects for later analysis. Figure 8.3 shows the resulting image.

The 'blobs' of Fig. 8.3 are then labeled; their corresponding original image pixel values are used to determine a range of parameters which are relevant for discrimination: size and magnitude information, and profile fits of Gaussians assessed by χ^2 discrepancy. In addition, we output for user convenience a plot of the object numeric labels at their center positions. Planned work will output astronomical coordinates

Figure 8.3 Following 2 dilations and 5 openings applied to previous figure.

for each object found in this way, to allow matching against relevant catalog information.

8.3 Conclusion on astronomical object detection

We have investigated this approach to finding objects with particular, clearly specified properties (faint globular clusters). For more massive, large-scale object trawling in image databases, we have found that varying assumptions about the desired types of objects can be partially met in this framework.

One aim of the broad-ranging application of this method is to characterize the information content of images, through carrying out a preliminary object detection and analysis. The image's content can be summarized through statistics related to number of objects present, their maximum and average sizes (numbers of pixels), and other other easily determined characterizations.

8.4 Object detection and clustering in point patterns

In Chapter 2, object detection theory for point pattern data was discussed. X-ray data typifies this problem. So also does point pattern clustering. A short review of the literature in this area is provided in Murtagh and Starck (1996). In all such cases, the data is sparse and in regimes characterized by low counts. Hence detection theory based on a low-count Poisson model is used.

Given a planar point pattern, a two-dimensional image is created by:

1 Considering a point at (x, y) as defining the value 1 at that point, yielding the tuple $(x, y, 1)$.
2 Projection onto a plane by (i) using a regular discrete grid (an image) and (ii) assigning the contribution of points to the image pixels by means of the interpolation function, ϕ, used by the chosen wavelet transform algorithm (in our case, the à trous algorithm with a B_3 spline).
3 The à trous algorithm is applied to the resulting image. Based on a noise model for the original image (i.e. tuples $(x, y, 1)$), significant structures are detected at each resolution level.

The cluster significance-testing procedure pursued here is similar to that used in Bijaoui *et al.* (1994a), Bury (1995), Slezak *et al.* (1988, 1990, 1993) in the study of large-scale cosmological structures.

8.4.1 Example 1: excellent recovery of Gaussian clusters

Figure 8.4 shows a point pattern set (a simulation for which the precise generation details are given below). Figure 8.5 shows the corresponding wavelet transform. Wavelet scales 1–6 are shown in sequence, left to right, starting at the upper right corner. The images shown in Fig. 8.5 may be summed pixel-wise to exactly reconstitute an *interpolated version* of Fig. 8.4, the interpolation being carried out, as mentioned above, by a B_3 spline. Two technical remarks regarding Fig. 8.5 are that (i) we rebinned each image to 128×128 from the input 256×256 to cut down on space, and (ii) this figure is shown histogram-equalized to more clearly indicate structure.

Figure 8.4 was generated with two Gaussian clusters designed with centers $(64, 64)$ and $(190, 190)$; and with standard deviations in x and y directions respectively $(10, 20)$ and $(18, 10)$. In the first (lower) of these clusters, there are 300 points, and there are 250 in the second. Background Poisson clutter was provided by 300 points. Figure 8.6 shows the 5th wavelet scale, following application of the significance threshold for positive values. The centroid values of the 'island' objects were found to be respectively $(62, 63)$ and $(190, 190)$ which are very good fits to the design values. The standard deviations in x and y were found to be respectively $(9.3, 14.4)$ and $(14.9, 8.9)$, again reasonable fits to the input data.

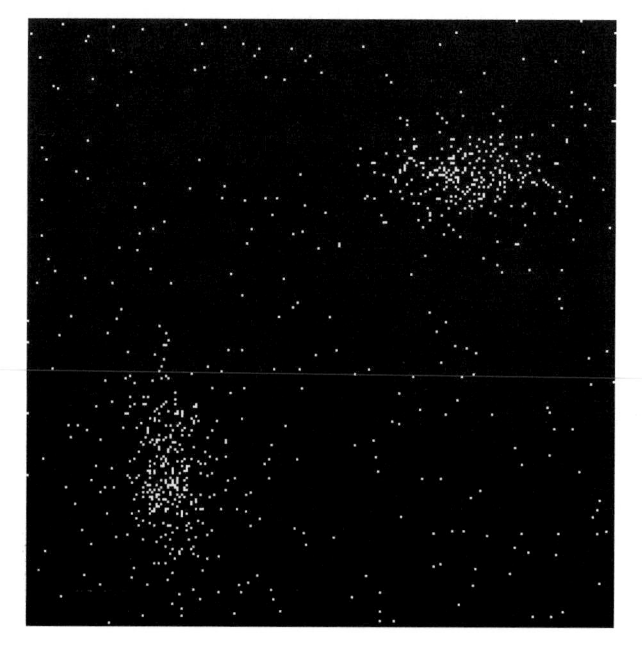

Figure 8.4 Simulated Gaussian clusters with 300 and 250 points; and background Poisson noise with 300 points.

The Poisson noise was successively increased, while keeping the same number of points in the Gaussian clusters. Table 8.1 shows the results obtained, even up to very substantial noise levels. These results show quite remarkable recovery of the Gaussian clusters' first two moments. The second, smaller cluster shows some 'eating into' the cluster by the noise as evidenced by the decreasing standard deviation. This assumes clearly that we know what shapes we are looking for, in this case, Gaussian-shaped clusters. To indicate the amount of noise

Figure 8.5 Wavelet transform (à trous method) of previous figure.

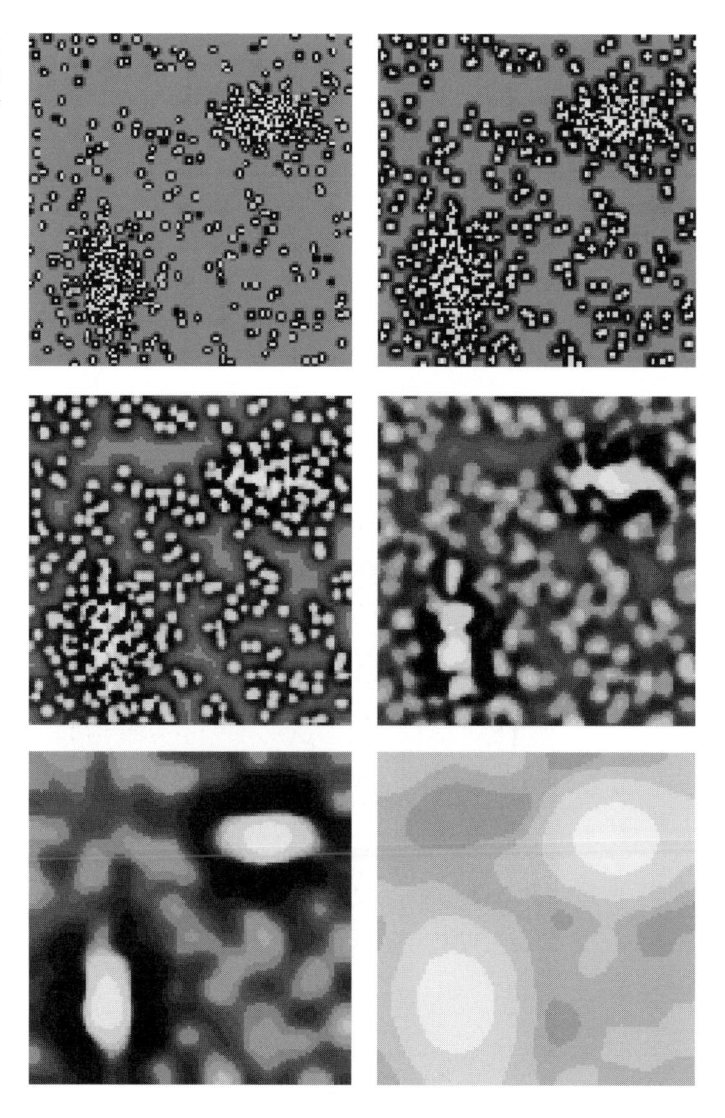

associated with 40 000 Poisson-distributed points, Fig. 8.7 shows this image.

A remark to make in regard to this result is that the wavelet method used prioritizes the finding of Gaussian-like shapes. The wavelet, as already mentioned, is associated with a B_3 spline. Its effect on cluster shapes which are less Guassian-like will be investigated next.

Table 8.1 *Recovery of two Gaussian-shaped clusters (centroids and standard deviations) with increasing amount of Poisson noise.*

Card noise	x_1	y_1	x_2	y_2	x_1-sd	y_1-sd	x_2-sd	y_2-sd
Input values	64	64	190	190	10	20	18	10
300	63	63	188	188	9.8	17.0	12.3	10.6
700	63	64	188	190	8.9	14.9	14.4	8.9
1100	65	63	188	188	8.5	19.1	17.4	9.3
1500	63	61	191	189	9.3	19.1	16.1	9.3
3500	62	62	190	188	9.3	12.7	14.0	9.3
7000	64	64	189	189	8.1	14.0	14.4	8.5
11000	63	63	184	189	7.6	14.9	13.6	7.2
15000	63	62	189	188	8.1	12.3	12.3	6.8
20000	65	63	174	186	5.1	8.5	8.1	6.4
30000	62	59	172	187	5.9	11.5	7.2	5.9
40000	61	70	187	186	6.4	8.1	7.2	6.4

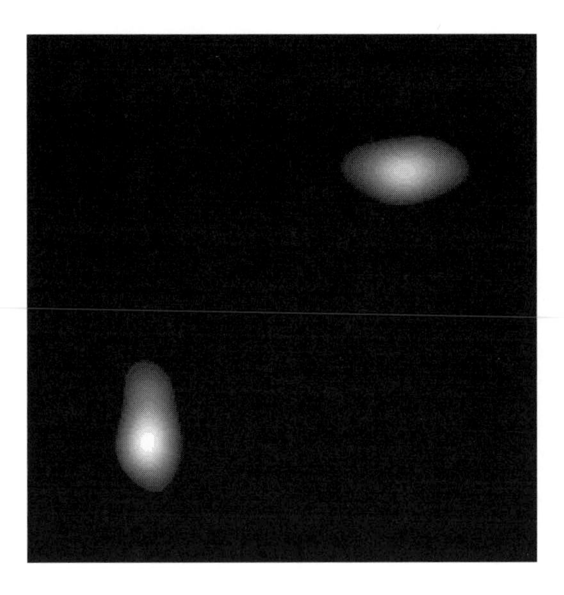

Figure 8.6 The support image at the 5th scale-level of the wavelet transform of Fig. 8.4 (multiplied by the wavelet resolution image at the same level), following significance-testing for positive image structure.

8.4.2 Example 2: diffuse rectangular cluster

Figure 8.8 shows a simulated minefield in coastal waters (the denser aligned set of points) with much clutter (Dasgupta and Raftery, 1995). For the Poisson clutter, 1500 points were used, and for the 'stochastic bar', 200 points were generated as Poisson-distributed with an additive Gaussian component. So we are seeking to find an approximate 200 somewhat more densely located points among the 1500 uninteresting points. Figure 8.9 shows the significant parts of this image, again taking the 5th wavelet scale. This result is visually quite good.

The input simulation used a bar in the x-interval (50, 200) and in the y-interval (95, 100). To this was added Gaussian noise of standard deviation 3. Recovery for one simulation of the case of 200 'signal' points and 1500 noise points yielded minimum and maximum x-values as (48, 196) and the corresponding y-values as (86, 109).

We then reduced the number of points to 50 'signal' points and 375 noise points. In this case, recovery (based on the 5th level of the multiscale transform) yielded an x-interval of (61, 188) and a y-interval of (88, 107). We can say that less information meant that our detection procedure allowed the cluster to be 'eaten into' to a slightly greater extent.

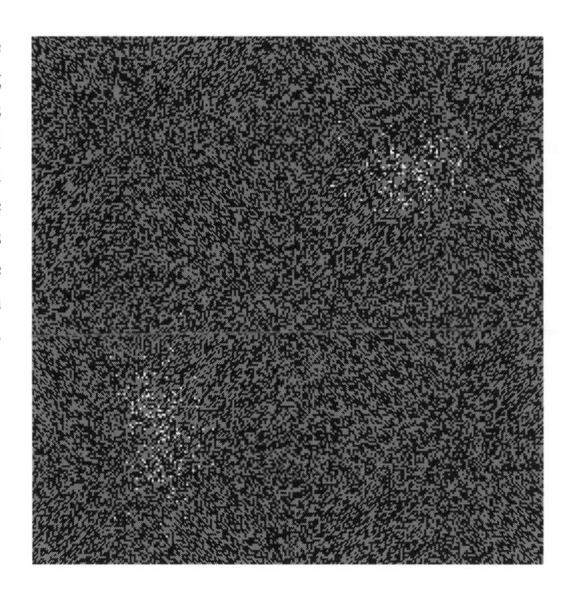

Figure 8.7 The image showing 550 'signal' points as two Gaussian-shaped clusters, with 40 000 Poisson noise points added. Details of recovery of the cluster properties can be seen in Table 8.1.

Figure 8.8 A point pattern with clutter – simulated coastal water minefield.

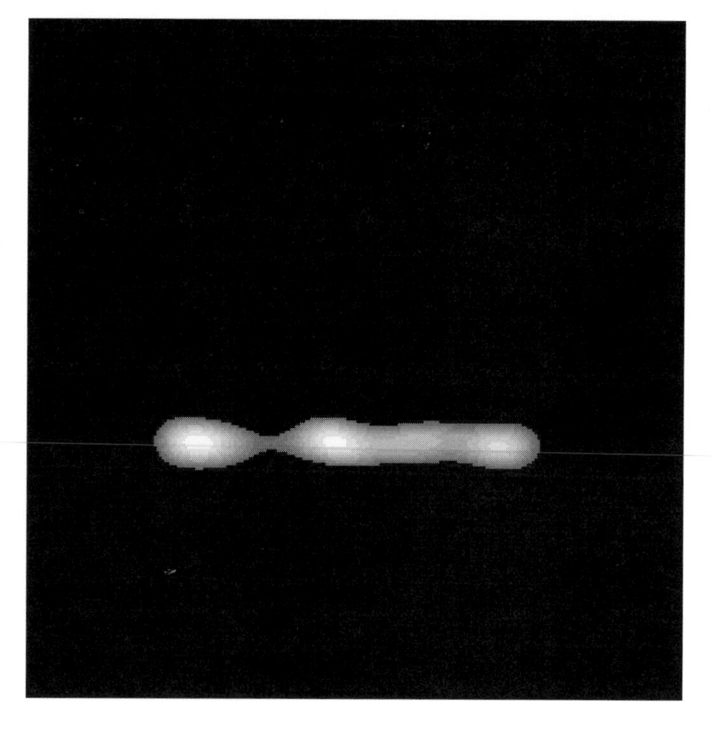

Figure 8.9 The support image (multiplied by the wavelet resolution image at the same level) at the 5th scale-level of the wavelet transform of the previous figure, following significance-testing for positive image structure.

8.4.3 Example 3: diffuse rectangle and faint Gaussian clusters

To study effects of scale, we used Fig. 8.10 (left): a diffuse rectangular bar containing 1300 points; less sharply defined, a Gaussian cluster of centroid (64, 64) and with marginal standard deviations in x and y of 10 and 30, respectively, which contained 600 points; and a Poisson noise background of 1000 points. The middle and right parts of Fig. 8.10 show the 4th and 5th level of the wavelet transform, restricted to support (i.e. significant) domains. The middle part of the figure shows the more high frequency information, with just a little of the Gaussian cluster. The right part of the figure shows more of the signal, whether relatively more or less diffuse.

8.5 Conclusion: cluster analysis in constant time

The overall method described here is very fast, taking less than about a minute on a Sparcstation 10. Note in fact the rather remarkable computational result that the method is $O(1)$, i.e. it is a *constant-time* algorithm. There is no dependence at all on the number of points constituting signal and noise. Thus the method described here shares this interesting property of the human visual system.

It may be appreciated of course that there is dependence on the resolution level, i.e. the image size which is used to represent the point pattern region. The computational requirements of the à trous method are linear as a function of the dimensionality of the image.

It is interesting to note that while our objective in this chapter has been to model noise accurately, thereby yielding the signal, an alternative strategy is to attempt directly to model the signal. An initial phase of noise modeling is commonplace in astronomy: traditionally carried out via local or global background modeling, perhaps using polynomial fitting; and more latterly carried out by multiscale noise-modeling. In this framework, fitting of idealized stellar shapes (close to Gaussian in form) to the objects found in an image is carried

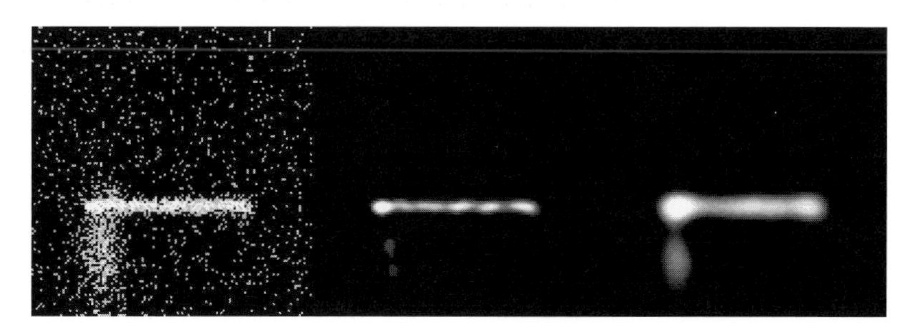

Figure 8.10 Left: input point pattern. Middle: support domain of wavelet coefficients at 4th scale. Right: same for 5th scale.

out following the estimation of the image 'background' or significant signal. This, then, is the starting point for discriminating between stars and extended objects such as galaxies.

As opposed to noise modeling, cluster modeling remains at a basic level in our method. We have noted how part of the 'microscope' which we are using to seek clusters is a B_3 spline, which is isotropic.

In more general multivariate data analysis, including the analysis of point patterns, Banfield and Raftery (1993) and Dasgupta and Raftery (1995) tackle the problem of finding clusters with prescribed shapes (e.g., highly elliptical multivariate Gaussian) in the presence of Poisson noise points. This is done using an agglomerative (or hierarchical) clustering algorithm, respecting the desired cluster shape. A stopping criterion, using a Bayesian argument, halts the agglomerations before too many noise points can be added. The precise definition of the cluster shape is set – based on the covariance structure or some other parametrization of the shape.

Our main aim is the filtering of noise or clutter in the image, in order to allow for accurate processing of the signal in a subsequent phase. Recovery of object (signal) parameters can be very accurate. As noted, our approach to noise modeling is computationally very efficient and has the intriguing human-like property of being of constant computational complexity.

9 Multiscale vision models

9.1 Artificial vision and astronomical images

Astronomical images contain typically a large set of point-like sources (the stars), some quasi point-like objects (faint galaxies, double stars) and some complex and diffuse structures (galaxies, nebulae, planetary stars, clusters, etc.). These objects are often hierarchically organized: star in a small nebula, itself embedded in a galaxy arm, itself included in a galaxy, and so on. We define a *vision model* as the sequence of operations required for automated image analysis. Taking into account the scientific purposes, the characteristics of the objects and the existence of hierarchical structures, astronomical images need specific vision models.

For robotic and industrial images, the objects to be detected and analyzed are solid bodies. They are seen by their surface. As a consequence, the classical vision model for these images is based on the detection of the surface edges. We first applied this concept to astronomical imagery (Bijaoui *et al.*, 1978). We chose the Laplacian of the intensity as the edge line. The results are independent of large-scale spatial variations, such as those due to the sky background, which is superimposed on the object images. The main disadvantage of the resulting model lies in the difficulty of getting a correct object classification: astronomical sources cannot be accurately recognized from their edges.

We encounter this vision problem of diffuse structures not only in astronomy, but also in many other fields, such as remote sensing, hydrodynamic flows or biological studies. Specific vision models were implemented for these kind of images. For reducing astronomical images, many procedures have been proposed using a model for which the image is the sum of a slowly variable background with superimposed small-scale objects (Slezak *et al.*, 1988; Stobie, 1986). A background mapping is first built (Bijaoui, 1980). For that purpose we need to introduce a scale: the background is defined in a given area. Each pixel with a value significantly greater than the background is considered to belong to a real object. The same label is given to each significant pixel belonging to the same connected field. For each field we determine the area, the position, the flux and some pattern parameters. Generally, this procedure leads to quite accurate measurements, with correct detection and recognition. The model works very well for content-poor fields. If this is not the case, a labeled field may correspond to many objects. The background map is constructed at a given scale. Larger objects are removed. This smoothing is only appropriate for star detection and not for larger objects.

We used this vision model on many sets of images. It failed to lead to a complete analysis because it is based on a single spatial scale for the adaptive smoothing and for the background mapping. A multiscale analysis allows us to get a background appropriate for a given object and to optimize the detection of different sized objects. The wavelet transform is expected to be the tool allowing us to build up an analysis, taking into account all the constraints.

9.2 Object definition in wavelet transform space

9.2.1 Choice of a wavelet transform algorithm

The 2D multiresolution analysis due to Mallat (1989) is generally performed separately by row and column. This does not lead to an isotropic perspective, three wavelet functions are used, and it is not easy to associate wavelet coefficients with a given pixel. Stars, and more generally astronomical sources, are quite isotropic sources, with no direction prioritized. Thus we choose an isotropic wavelet transform. We need also to make a connection between images at different scales. As the redundancy is not critical we prefer to avoid decimation. This leads us to the use of the à trous algorithm.

9.2.2 Bases of object definition

After applying the wavelet transform on the image, we have to detect, to extract, to measure and to recognize the significant structures. The wavelet space of a 2D direct space is a 3D one. An object has to be defined in this space. A general idea for object definition lies in the connectivity property. An object occupies a physical region, and in this region we can join any pixel to other ones. The connectivity in the direct space has to be transported to the wavelet transform space (WTS). In order to define the objects we have to identify the WTS pixels we can attribute to the objects.

At a given scale, it is clear that a given object leads to one connected field. A region labeling has to be made, scale-by-scale. A physical object can show features at different successive scales, and an interscale connectivity graph has then to be established. Connected trees are identified from the preceding graph. These correspond to a WTS region which can be associated with an object. This permits us to separate close components and to identify an object from its full components. The identification of WTS pixels related to a given object leads to the reconstruction of an image by partial restoration algorithms.

9.2.3 Significant wavelet coefficients

The statistical distribution of the wavelet coefficients depends on the noise process. Generally, we admit that we have stationary Gaussian noise for the image. As we have seen in Chapter 2, we can transform the pixel intensity by Anscombe's transform (1948) in the case of Poisson noise, and then process the data as Gaussian variables. This approach can be generalized to many kinds of noise. So, for a simple presentation of our vision model we further admit that the noise is Gaussian.

The means to compute the standard deviation $\sigma(i)$ of the wavelet coefficients at scale i due to the noise has been explained in Chapter 2. This has allowed us to extract the set of significant coefficients at each scale for a given decision level ϵ. The vision model is based only on these detected significant pixels.

9.2.4 Scale-by-scale field labeling

After the identification of the significant pixels we carry out an image segmentation scale-by-scale in WTS. In our present analysis we have examined only positive coefficients, which correspond to light sources. Significant negative pixels may be associated with absorbing regions, but they are generally associated with the wavelet bumps: around a peak we have always negative wavelet coefficients. The corresponding pixels do not belong to a real object, even if they are significant.

The region labeling is made by a classical growing technique. At each scale, we give a label to a pixel: 0 if the wavelet coefficient is smaller than the threshold, $n > 0$ for the contrary case. Neighboring significant pixels have the same label. We indicate by $L(i, k, l)$ the label corresponding to the pixel (k,l) at the scale i, and $D(i,n)$ a segmented field of label n at the same scale. We have:

$$D(i,n) = \{W(i,k,l) \text{ such that } L(i,k,l) = n\}. \quad (9.1)$$

An object could be defined from each labeled field, without taking into account the interscale neighborhood. We can restore an image of these objects from the known wavelet coefficients, but this restoration would not use all the information. A given physical object may lead to significant pixels at different scales. A correct restoration, however, needs all the information.

9.2.5 Interscale connection graph

An astronomical object is described as a hierarchical set of structures. So we have to link the labeled fields from one scale to the following one, in order to give the hierarchy. Let us consider the fields $D(i,n)$ at scale i and $D(i+1,m)$ at scale $i+1$. The pixel coordinates of the maximum coefficient $W(i,k_{i,n},l_{i,n})$ of $D(i,n)$ are $(k_{i,n},l_{i,n})$. The field $D(i,n)$ is said to be connected to $D(i+1,m)$ if the maximum position belongs to the field $D(i+1,m)$.

$$L(i+1,k_{i,n},l_{i,n}) = m \quad (9.2)$$

With this criterion of interscale neighborhood, a field of a given scale is linked to at most one field of the upper scale. Now we have a set of fields $D(i,n)$ and a relation $\mathcal{R}$:

$$D(i,n) \, \mathcal{R} \, D(i+1,m) \qquad \text{if} \qquad L(i+1,k_{i,n},l_{i,n}) = m \quad (9.3)$$

This relation leads to building the interscale connectivity graph, the vertices of which correspond to the labeled fields. Statistically, some significant structures can be due to the noise. They contain very few pixels and are generally isolated, i.e. connected to no field at upper and lower scales. So, to avoid false detection, the isolated fields are removed from the initial interscale connection graph.

9.2.6 An object as a tree

Let us consider an image without any noise, which contains one object (for instance a 2D Gaussian), and its wavelet transform. At each scale, one structure is detected by the thresholding and segmentation procedures. The evolution graph of the highest wavelet coefficient of the field of each scale has one maximum at a scale that increases with the object size. The coordinates of the highest coefficient do not vary with scale and correspond to the brightest pixel in the original image.

This wavelet property leads us to associate an object with each local maximum of the image wavelet transform. For each field $D(i, n_2)$ of the interscale connection graph, its highest coefficient $W(i, k_2, l_2)$ is compared with the corresponding coefficients of the connected fields of the upper scale, $W(i+1, k_+, l_+)$ and lower scale, $W(i-1, k_-, l_-)$. In the case of isolated simple objects, $W(i, k_2, l_2)$ only has to be compared with the coefficients of the same coordinates of the scale $i+1$ and $i-1$ in order to detect the object ($k_+ = k_- = k_2$ and $l_+ = l_- = l_2$) which is described by all the fields linked to $D(i, n_2)$. But generally, the object we want to identify is under the influence of the neighboring objects and may be contained in an object of larger size or itself include smaller objects. At the lowest scales, the division into small structures due to noise has to be taken into account too. At each scale, the position of the highest wavelet coefficient associated with the objects may change and several fields can be attributed to the object. Two problems have to be solved. First, the local maximum has to be detected correctly, i.e. $W(i+1, k_+, l_+)$ and $W(i-1, k_-, l_-)$ have to be localized precisely. Then the contribution of the object to each scale, i.e. the fields or part of fields that can be assigned to it, have to be identified. To determine $W(i+1, k_+, l_+)$ and $W(i-1, k_-, l_-)$, we have adopted the following rules.

- Given $D(i+1, n_3)$, the field of the scale $i+1$ connected to $D(i, n_2)$. If $D(i, n_2)$ is not linked to a field of the scale $i+1$, only the test with $W(i-1, k_-, l_-)$ will be processed. Let us note that $D(i, n_2)$ cannot be isolated, otherwise it has already been removed at the previous step: at least one field of the scale $i+1$ or $i-1$ connected to $D(i, n_2)$ exists.
- $D(i+1, n_3)$ has one antecedent $D(i, n_2)$, $W(i+1, k_+, l_+)$ corresponds to the highest coefficient of the field.
- $D(i+1, n_3)$ has several antecedents, so we have to isolate the sub-field of $D(i+1, n_3)$, $\widetilde{D}(i+1, n_2)$, the contribution of $D(i, n_2)$ to the upper scale. So $W(i+1, k_+, l_+)$ is the highest coefficient of this field. We have:

$$\widetilde{D}(i+1, n_2) = \{W(i+1, k, l) \in D(i+1, n_3) \quad \text{such that} \quad L(i, k, l) = n_2\} \tag{9.4}$$

and

$$W(i+1, k_+, l_+) = \text{Max}(\{W(i+1, k, l) \in \widetilde{D}(i+1, n_2)\}) \tag{9.5}$$

For $W(i-1, k_-, l_-)$, the same reasoning is applied:

- $D(i, n_2)$ has one antecedent, $D(i-1, n_1)$, $W(i-1, k_-, l_-)$ is equal to $W(i-1, k_1, l_1)$, the highest coefficient of $D(i-1, n_1)$.
- $D(i, n_2)$ is connected to several fields, and we select the field for which the position of its highest coefficient is nearest to (k_2, l_2). So $W(i-1, k_-, l_-)$ is this coefficient.

If $W(i-1, k_-, l_-) < W(i, k_2, l_2)$ and $W(i, k_2, l_2) > W(i, k_2, l_2)$, $D(i, n_2)$ corresponds to a local maximum of the wavelet coefficients. It defines an object. No other fields of the scale i are attributed to the object; $D(i, n_2)$ concentrates the main information which permits the object image to be reconstructed. Only the fields of the lower scales connected directly or indirectly to $D(i, n_2)$ are kept. So the object is extracted from larger objects that may contain it. On the other hand, some of these fields may define other objects. They are subobjects of the object. To get an accurate representation of the object cleaned of its components, the fields associated with the subobjects cannot be directly removed; as experiments show, their images will have to be restored and subtracted from the reconstructed global image of the object.

By construction, $D(i, n_2)$ is the root of a subgraph which mathematically defines a tree denoted $\mathscr{T}$ (Berge, 1967). The tree $\mathscr{T}$ expresses the hierarchical overlapping of the object structures and the 3D connectivity of the coefficients set that defined it in the WTS. Subtrees of $\mathscr{T}$ correspond to the subobjects:

$$\mathscr{T} = \{D(j, m) \quad \text{such that} \quad D(j, m)\, \mathscr{R} \cdots \mathscr{R}\, D(i, n_2)\} \tag{9.6}$$

9.2.7 Object identification

We can now summarize this method allowing us to identify all the objects in a given image:

1 We compute the wavelet transform with the à trous algorithm, which leads to a set $W(i, k, l)$, $i \leq N$;
2 We determine the standard deviation of $W(1, k, l)$ due to the noise;
3 We deduce the thresholds at each scale;
4 We threshold scale-by-scale and we do an image labeling;
5 We determine the interscale relations;
6 We identify all the wavelet coefficient maxima of the WTS;
7 We extract all the connected trees resulting from each WTS maximum;

Let us remark that this definition is very sensitive to the kind of wavelet transform used. We work with an isotropic wavelet, without decimation. With Mallat's algorithm this definition would have to be revised.

9.3 Partial reconstruction

9.3.1 The basic problem

Let us consider now an object $\mathcal{O}$ as previously defined. This corresponds to a volume $\mathcal{S}$ in WTS. This volume is associated with a set $\mathcal{V}$ of wavelet coefficients, such that:

$$\mathcal{O} \Longrightarrow \{\mathcal{V}(i,k,l), \text{ for } (i,k,l) \in \mathcal{S}\} \tag{9.7}$$

$\mathcal{F}$ is an image and $\mathcal{W}$ is its corresponding wavelet transform. $\mathcal{F}$ can be considered as a correct restored image of the object $\mathcal{O}$ if:

$$\mathcal{V}(i,k,l) = \mathcal{W}(i,k,l) \qquad \forall (i,k,l) \in \mathcal{S} \tag{9.8}$$

$P_{\mathcal{S}}$ denotes the projection operator in the subspace $\mathcal{S}$ and WT the operator associated with the wavelet transform, so that we can write:

$$\mathcal{V} = (P_{\mathcal{S}} \circ WT)(\mathcal{F}) \tag{9.9}$$

and we will term this transformation of $\mathcal{F}$, $A(\mathcal{F})$.

We have to solve the inverse problem which consists of determining $\mathcal{F}$ knowing A and $\mathcal{V}$. The solution of this problem depends on the regularity of A. In many papers and books authors have discussed the availability of a solution to this class of inverse problem (for example, see Demoment, 1989). The size of the restored image is arbitrary and it can be easily set greater than the number of known coefficients. It is certain that there exists at least one image $\mathcal{F}$ which gives exactly $\mathcal{V}$ in $\mathcal{S}$, i.e. the original one: the equation is consistent (Pratt, 1978). But generally we have an infinity of solutions, and we have to choose among them the one which is considered as correct. An image is always a positive function, which leads us to constrain the solution, but this is not sufficient to get a unique solution.

The choice of unique solution can be governed by a regularization condition. Many regularization conditions have been developed for restoration. Taking into account consistency, we first used a direct simple algorithm (Bijaoui and Rué, 1995), connected to Van Cittert's (1931) one for which the regularization is carried out by the limitation of the support. Then we applied the conjugate gradient algorithm (Lascaux and Théodor, 1994) which corresponds to minimization of the energy. This method is more efficient than the direct algorithm; the restoration quality and the convergence speed are improved.

9.3.2 Choice of the scale number

The set $\mathcal{V}$ associated with $\mathcal{O}$ is defined on N_m scales, N_m being the scale of the maximum of the wavelet coefficient maxima which has permitted the object to be identified. In order to have an accurate reconstruction, experiments show that $\mathcal{V}$ has to be extended to the upper scale $N_m + 1$, with $\tilde{D}(N_m + 1, n)$ the contribution to this scale of $D(N_m, n)$, the root of $\mathcal{T}$. If $D(N_m, n)$ is linked to no field of the upper scale, $\mathcal{V}$ cannot be extended and the image may not be well-restored (see below). Given $D(N_m + 1, p)$ the field linked to $D(N_m, n)$, two cases arise to determine this contribution:

- $D(N_m + 1, p)$ has one antecedent $D(N_m, n)$, $\tilde{D}(N_m + 1, n)$ is equal to $D(N_m + 1, p)$
- $D(N_m + 1, p)$ has several antecedents. $\tilde{D}(N_m + 1, n)$ contains the coefficients of $D(N_m+1, p)$ defined on the same support as $D(N_m, n)$:

$$\tilde{D}(N_m + 1, n) = \{W(N_m + 1, k, l) \in D(N_m + 1, p) \quad \text{such that} \quad L(N_m, k, l) = n\} \tag{9.10}$$

The number of scales used for the reconstruction is such that $N_r = N_m + 1$. The 3D support $\mathcal{S}$ of $\mathcal{V}$ becomes:

$$\begin{aligned} \mathcal{S} = &\{(i, k, l) \text{ such that } W(i, k, l) \in D(i, n) \text{ element of } \mathcal{T}\} \\ &\cup \{(N_m + 1, k, l) \text{ such that } W(N_m + 1, k, l) \in \tilde{D}(N_m + 1, n)\} \end{aligned} \tag{9.11}$$

An efficient restoration is ensured by having a set $\mathcal{V}$ that contains the maximum of the wavelet coefficients. But if $D(N_m, n)$ is linked to no field ($N_r = N_m$), for instance its local maximum is on the last scale of the wavelet analysis, the real maximum may belong to a higher scale. The reconstruction cannot be optimal in this case. There is some information missing and the discrepancy between the original and the restored image can be large.

9.3.3 Reconstruction algorithm

We use the least squares method to solve the relation $\mathcal{V} = A(F)$ which leads to seeking the image F which minimizes the distance $\|\mathcal{V} - A(F)\|$.

This distance is defined for a wavelet structure W by the relation:

$$\| W \| = \sqrt{\sum_{i=1}^{N_r} \sum_{k,l} W(i,k,l)^2} \tag{9.12}$$

$\| \mathcal{V} - A(F) \|$ is minimum if and only if F is a solution of the following equation:

$$\tilde{A}(\mathcal{V}) = (\tilde{A} \circ A)(F) \tag{9.13}$$

The initial eqn. (9.9) is thus modified with the introduction of $\tilde{A}$, the joint operator associated with A. $\tilde{A}$ is applied to a wavelet transform W and gives an image $\tilde{F}$:

$$\tilde{F} = \tilde{A}(W) = \sum_{i=1}^{N} (H(1) \cdots H(i-1)G(i))(W(i)) \tag{9.14}$$

$H(i)$ and $G(i)$ are respectively low- and high-pass filters used to compute the smoothed images and the wavelet levels. We have:

$$\begin{aligned} F(i,k,l) &= \sum_{n,m} h(n,m) F(i-1, k + 2^{i-1} n, l + 2^{i-1} m) \\ &= H(i)(F(i-1))(k,l) \end{aligned} \tag{9.15}$$

and

$$\begin{aligned} W(i,k,l) &= \sum_{n,m} g(n,m) F(i-1, k + 2^{i-1} n, l + 2^{i-1} m) \\ &= G(i)(F(i-1))(k,l) \end{aligned} \tag{9.16}$$

Initially, we implemented the reconstruction algorithm with this operator $\tilde{A}$, but several tests showed the existence of spurious rings around the objects in the restored images. This phenomenon is due to the positive bumps of the filtering of the negative components at each scale of the wavelet structures processed by $G(i)$. The artifacts are removed by suppressing this operator and we simply write:

$$\tilde{F} = \tilde{A}(W) = \sum_{i=1}^{N} (H(1) \cdots H(i-1))(W(i)) \tag{9.17}$$

Since we have a symmetric operator, $\tilde{A} \circ A$, the conjugate gradient method can be applied. Before describing the algorithm, some changes must be made:

$\mathcal{V}$ is not a real wavelet structure because it is only defined inside $\mathcal{S}$. To easily manipulate $\mathcal{V}$, we replace it by the wavelet structure $\mathcal{W}$

whose coefficients outside $\mathcal{S}$ are equal to zero.

$$\mathcal{W} = \begin{cases} = \mathcal{V}(i,k,l) & \text{if } (i,k,l) \in \mathcal{S} \\ = 0 & \text{if } (i,k,l) \notin \mathcal{S} \end{cases} \tag{9.18}$$

Now the inverse problem is to solve for F:

$$\tilde{A}(\mathcal{W}) = (\tilde{A} \circ A)(F) \tag{9.19}$$

The steps of the conjugate gradient algorithm we applied to this new system are now described. In these steps, $F^{(n)}$ is the estimation of the solution F at iteration n.

0. Initialization step: the estimated image $F^{(n)}$, the residual wavelet $W_r^{(n)}$ and image $F_r^{(n)}$ are initialized.

$$\begin{cases} F^{(0)} = R(\mathcal{W}) \\ W_r^{(0)} = \mathcal{W} - A(F^{(0)}) \\ F_r^{(0)} = \tilde{A}(W_r^{(0)}) \end{cases} \tag{9.20}$$

   R is the wavelet reconstruction operator. From a wavelet structure W, an image F is restored corresponding to the sum of the wavelet and the last smoothed images. W is not necessarily the wavelet transform of an image, so $WT(F)$ may not be equal to W.
1. Computation of the convergence parameter $\alpha^{(n)}$:

$$\alpha^{(n)} = \frac{\|\tilde{A}(W_r^{(n)})\|^2}{\|A(F_r^{(n)})\|^2} \tag{9.21}$$

2. An iterative correction is applied to $F^{(n)}$ to get the intermediate image $F_1^{(n+1)}$:

$$F_1^{(n+1)} = F^{(n)} + \alpha^{(n)} F_r^{(n)} \tag{9.22}$$

3. Finally, we obtain the restored image at iteration $n+1$, $F^{(n+1)}$, by thresholding $F_1^{(n+1)}$ in order to get a positive image:

$$F^{(n+1)}(k,l) = \mathrm{Max}(0, F_1^{(n+1)}(k,l)) \tag{9.23}$$

4. Residual wavelet computation where only the wavelet coefficients inside $\mathcal{S}$ are taken into account.

$$W_r^{(n+1)} = \mathcal{W} - A(F^{(n+1)}) \tag{9.24}$$

5. Test on the residual wavelet: if $\| W_r^{(n+1)} \|$ is less than a given threshold, the desired precision is reached and the procedure is stopped.
6. Computation of the convergence parameter $\beta^{(n+1)}$:

$$\beta^{(n+1)} = \frac{\|\tilde{A}(W_r^{(n+1)})\|^2}{\|\tilde{A}(W_r^{(n)})\|^2} \tag{9.25}$$

7. Residual image computation

$$F_r^{(n+1)} = \tilde{A}(W_r^{(n+1)}) + \beta^{(n+1)} F_r^{(n)} \tag{9.26}$$

8. Return to step 1

9.3.4 Numerical experiments

Restoration of a Gaussian pattern. In order to test the validity of the algorithms, we considered first a Gaussian pattern rather typical of the image of an astronomical object. We start without any noise in order to examine the quality of the inversion. The Gaussian pattern considered is shown in Fig. 9.1. On the right part of Fig. 9.1 we show the restored image. After about 20 iterations the convergence is correct. The restored image support is maximum at the first iteration and then decreases and converges to a compact field.

Restoration of noisy Gaussian patterns. We carried out a set of numerical experiments with noisy Gaussian patterns. In Fig. 9.2 we plotted the Gaussian pattern with SNR (SNR is defined as the ratio of the standard deviation of the signal to the standard deviation of the noise) equal to 10 and its corresponding restored image. The pattern is correctly restored, and the convergence is quite similar to the one without noise. In Fig. 9.3 SNR is equal to 1, some distortions exist but the accuracy is generally sufficient. In Fig. 9.4 SNR is equal to 0.1, the restoration is not too bad for a further analysis. These experiments show that it is possible to extract a significant pattern even for a small SNR. This would be quite impossible by the other vision models we described in the introduction to this chapter.

Figure 9.1 Gaussian pattern without noise (left) and its restored image (right).

Restoration of close objects. In the above experiments, we considered one isolated Gaussian, but generally astronomical objects may be close, and it is difficult to analyze them separately. We consider now two Gaussian patterns at a variable distance. In the right part of Fig. 9.5

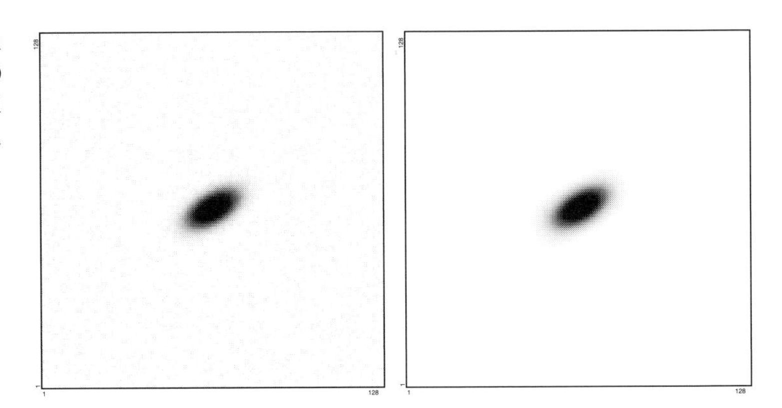

Figure 9.2 Gaussian pattern with SNR 10 (left) and its restored image (right).

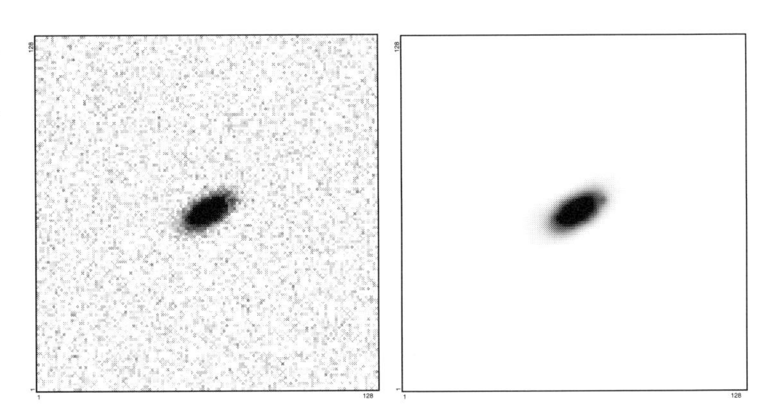

Figure 9.3 Gaussian pattern with SNR 1 (left) and its restored image (right).

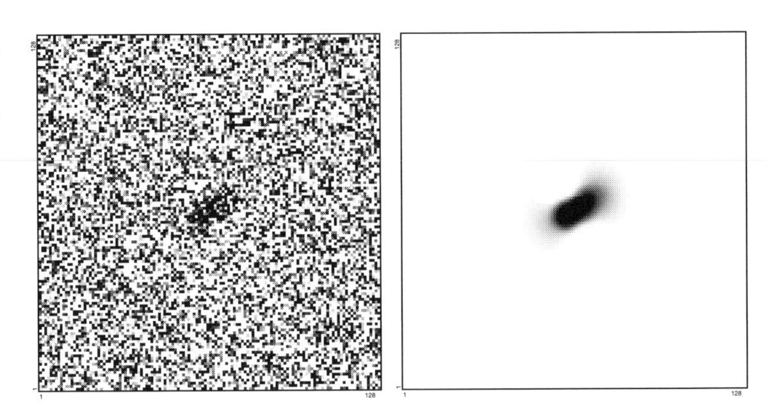

Figure 9.4 Gaussian pattern with SNR 0.1 (left) and its restored image (right).

we have plotted the Gaussian pattern considered while their sum at distance 4σ is plotted in the left part. If the distance between the two patterns is lower than 3σ, they cannot be separated. The restoration of the right component is plotted in the left part of Fig. 9.6. The reconstruction quality is quite bad because of the influence of the left pattern.

In the case of objects which are too close, the reconstruction can be improved by the following iteration. Let us consider an image $\mathscr{F}$, a sum of two objects with images $\mathscr{F}_1$ and $\mathscr{F}_2$. Their wavelet transforms are $\mathscr{W}_1$ and $\mathscr{W}_2$. Their significant wavelet structures are $\mathscr{V}_1$ (in $\mathscr{S}_1$) and $\mathscr{V}_2$ (in $\mathscr{S}_2$). If the objects are too close, $\mathscr{F}_2$ makes for a significant

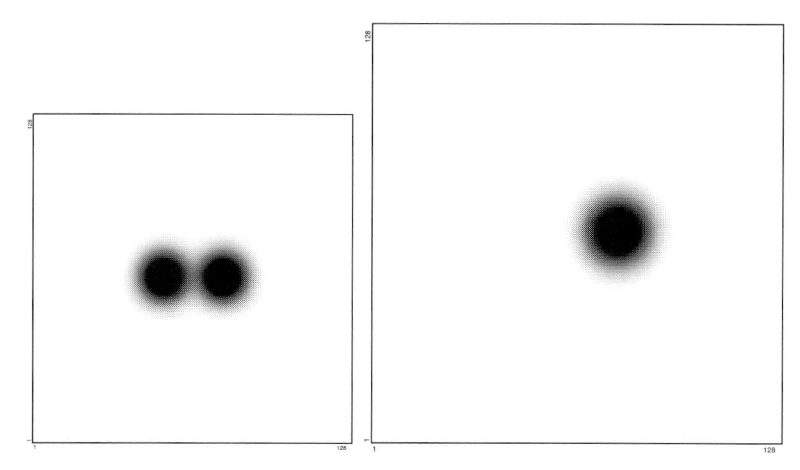

Figure 9.5 The sum of the Gaussian patterns at distance 4σ (left) and the original pattern (right).

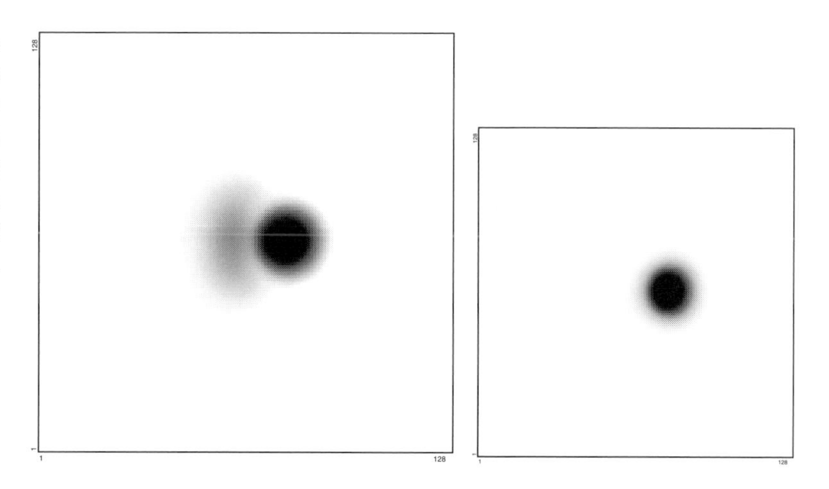

Figure 9.6 The reconstruction of the right Gaussian pattern without (left) and with (right) account taken of the influence of the left pattern.

contribution to $\mathcal{V}_1$, and $\mathcal{F}_1$ to $\mathcal{V}_2$:

$$\mathcal{V}_1 = \mathcal{V}_{11} + \mathcal{V}_{12} \tag{9.27}$$

where $\mathcal{V}_{11}$ is the $\mathcal{W}_1$ coefficients $\in \mathcal{S}_1$ and $\mathcal{V}_{12}$ the $\mathcal{W}_2$ coefficients $\in \mathcal{S}_1$.

We improve the $\mathcal{F}_1$ restoration if we reduce the component $\mathcal{V}_{12}$, i.e. the effect of $\mathcal{F}_2$. We get an approximate solution $\tilde{\mathcal{F}}_2$ of $\mathcal{F}_2$. We subtract $\tilde{\mathcal{F}}_2$ from the initial image $\mathcal{F}(k,l)$. The effect of $\mathcal{F}_2$ on $\mathcal{F}_1$ obviously decreases. We can do the same operation for $\mathcal{F}_1$. Then we iterate to convergence. This algorithm leads to a real improvement in the quality of the restoration, for intermediate distances. For patterns which are too close, the initial patterns are too far from the real ones, and the algorithm does not give a correct solution. In the right part of Fig. 9.6 we have plotted the effect on the right Gaussian component. The quality of the restoration is now acceptable, in spite of the interaction between the patterns. The algorithm could be applied to more patterns, but the complexity of the computations is a serious restriction on the number of objects to be restored.

Restoration of superimposed patterns. Finally we examine another important case, the case of superimposed patterns. In Fig. 9.7 we have plotted a central cut of the image of a narrow Gaussian function superimposed on a larger one. We have plotted in the same figure the original narrow pattern and the restored one. We remark that the influence of the larger background structure is negligible. The quality of the restoration depends on the separation between the patterns in WTS. We get no significant bias for a background pattern which can be considered as locally constant for the narrow Gaussian function.

9.4 Applications to a real image

We propose to compare the performance of the multiscale model with that of a well-known astronomical package for image analysis, INVENTORY (Kruszewski, 1989). For this, the two methods are tested on an image named L384-350 (see Fig. 9.8). This frame comes from a Schmidt plate (field number 384) of the ESO-R survey of the southern hemisphere. In its center, it contains the galaxy 384350 of the Lauberts catalog and other galactic field objects like the bright star juxtaposed to 384350. We limit the comparison to object detection and measurement.

9.4.1 Multiscale method

We performed a 7-scale wavelet transform of L384-350. 58 objects were detected. The restored image, made from the reconstructed images of each object, is given in Fig. 9.9 and the objects' mapping in Fig. 9.13.

The tree of the central galaxy object is plotted in Fig. 9.10. The corresponding restored image is plotted in Fig. 9.11. A subobject of the galaxy, which corresponds to a spiral arm, has been extracted; its image is shown in the same figure.

In the image center there is a large nebulous area that stands visually out against the background. The model does not permit this zone to be extracted. Indeed, its local maximum is hidden by the influence, up to high scales, of the local maxima of the two central bright objects, and cannot be detected. In order to isolate it, the vision model has only to be applied again on the image resulting from the difference between the initial image and the global restored images of all its objects. Figure 9.12 contains the restored image of the nebulous area whose local maximum belongs to the scale 6. Only the scales 5, 6 and 7

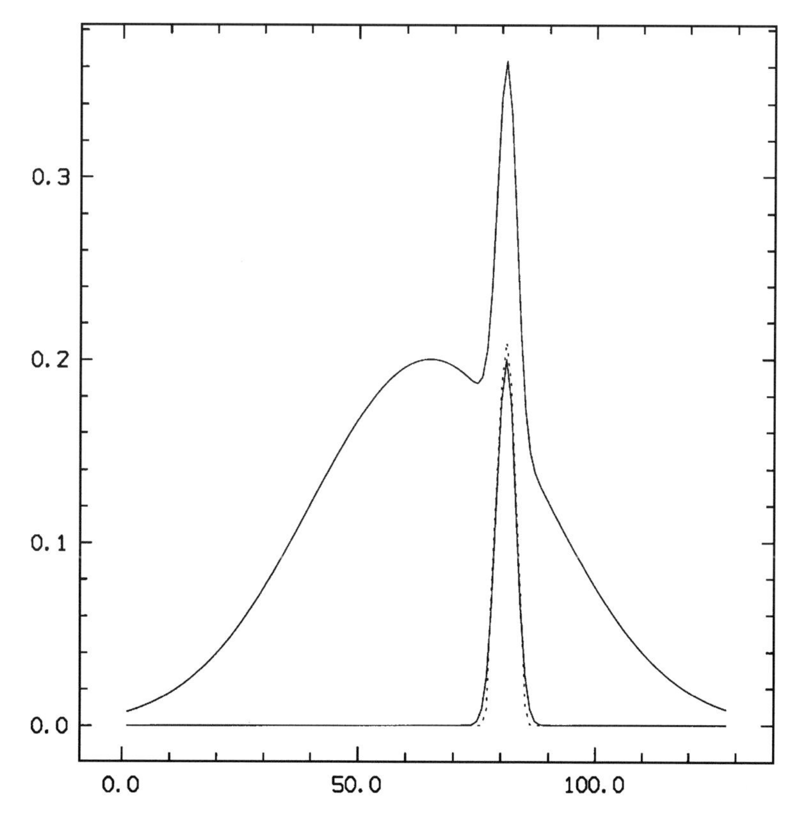

Figure 9.7 The central cut of the superimposed Gaussian patterns. The original narrow pattern is plotted with a solid line and the restored one with a dashed line.

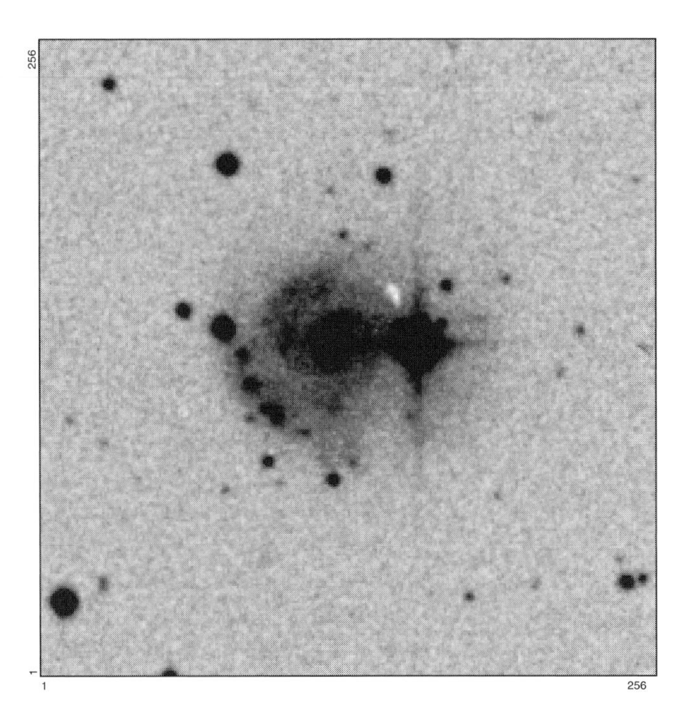

Figure 9.8 Image of L384-350.

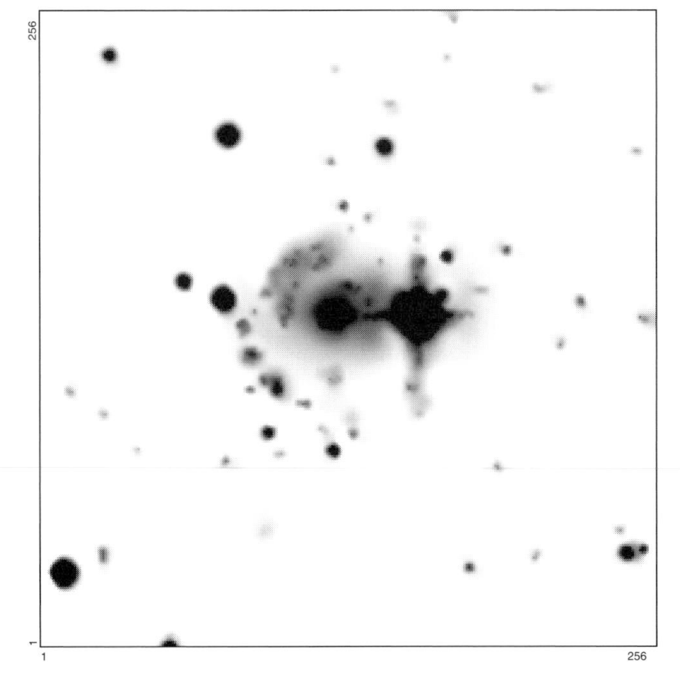

Figure 9.9 Restored image of L384-350.

have been used to reconstruct this image in order to reduce the effects of some artifacts of the difference image due to the imperfection of the restored object images. In spite of this precaution, the influence of the objects superimposed on the nebulous area cannot totally be removed.

9.4.2 INVENTORY method

INVENTORY is a searching, photometric, and classifying package included in the ESO-MIDAS image processing system. For L384-350, we take only an interest in the results concerning the searching and analysis of objects. The SEARCH routine of INVENTORY gives a preliminary list of objects. We obtained 59 objects using a detection threshold equal to 3 times the estimated standard deviation of the image noise. The ANALYSE routine verifies the previous detections and removes or adds some objects of the list: 60 objects remain for which many measures of magnitude and size are made. The corresponding mapping of objects is plotted in Fig. 9.13.

9.4.3 Comments

In the case of simple objects of small size, INVENTORY and the multiscale model give very similar results for the object localization and magnitude measurement as shown in Table 9.1 which gathers together results obtained with selected objects of the image (m_x and m_y are the coordinates of the brightest pixel of the object and *mag* is its magnitude in an arbitrary scale).

On the other hand, for an irregular and complex object such as the star or the galaxy of the image center, INVENTORY does not succeed

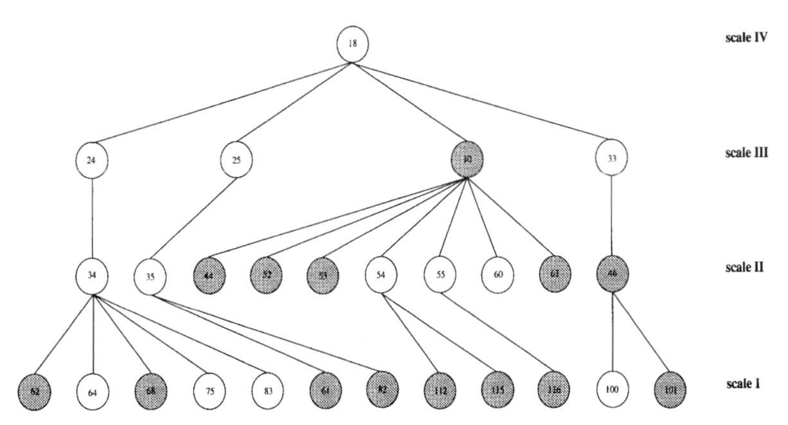

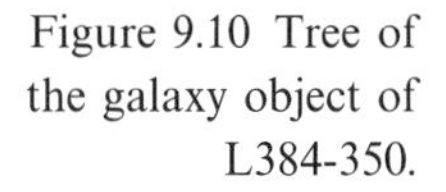
Figure 9.10 Tree of the galaxy object of L384-350.

in globally identifying them. There are multiple detections. Also, this method cannot bring out a structure hierarchy in the objects unlike the multiscale vision model that can decompose an object, complex as it may be, thanks to the notion of a subobject: the model identifies all the little objects it contains and points out their overlapping order.

9.5 Vision models and image classes

The vision model described above is based on the à trous algorithm. This algorithm is redundant, so the application to large images needs a lot of memory. The pyramidal algorithms provide smaller transforms, so a pyramidal vision model can easily be applied to these images.

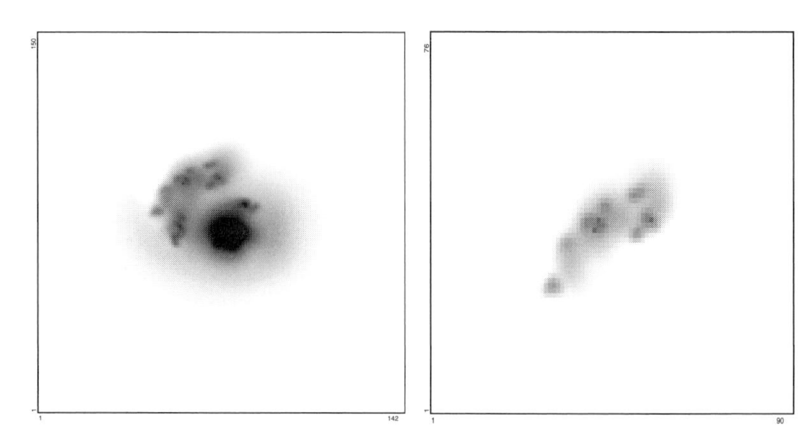

Figure 9.11 Restored images of the galaxy object and one of its subobjects.

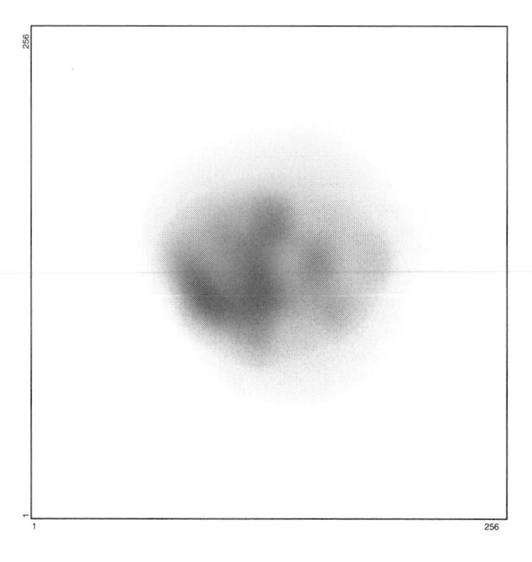

Figure 9.12 Restored image of the nebulous central area.

The vision model is based always on the significant structures. An interpolation is made in order to determine the interscale connectivity graph. The reconstruction algorithm is similar to the one developed for the à trous case, and the quality of the restoration is quite similar.

More generally, the vision model contains a set of operations which are independent of the multiscale transform which is applied: thresholding, labeling, interscale labeling, tree determination, extraction of

Table 9.1 *Results of the vision model and INVENTORY for some test objects.*

Object	Vision Model			INVENTORY		
	m_x	m_y	mag	m_x	m_y	mag
3-1	10.68	29.95	15.52	10.32	29.98	15.39
3-4	77.34	140.03	15.79	77.44	140.2	15.72
3-6	79.08	206.45	15.91	79.03	206.30	15.94
2-2	179.22	32.19	18.37	179.20	32.08	18.55
2-4	244.82	37.95	16.97	244.50	37.89	16.87
2-5	122.35	79.42	17.30	122.40	79.39	17.30
2-6	95.57	86.48	17.55	95.71	86.51	17.59
2-26	143.71	201.61	16.70	143.70	201.6	16.73
2-30	29.83	238.63	17.26	29.79	238.70	17.33
1-28	225.48	138.95	18.54	225.55	139.30	18.75

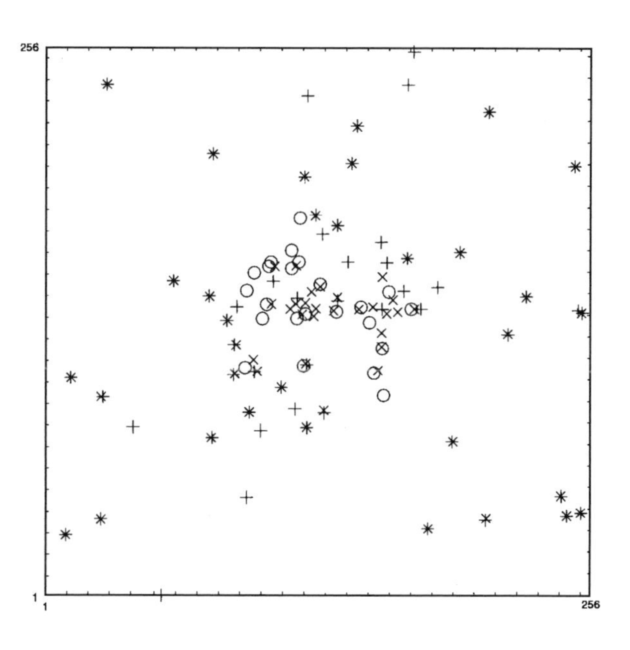

Figure 9.13 Mapping of detected objects (multiscale model: + (object) ∘ (sub-object) and INVENTORY: ×).

the coefficients corresponding to a given object. The specific transform is needed at first to compute the coefficients, and subsequently and finally to restore the objects. A class of multiscale vision models results for this strategy, by changing the transform.

In our comparison between the à trous and the pyramidal algorithms, we have established that the former furnishes more accurate geometric and photometric measurements. The quality was similar only if the data was previously oversampled for the pyramidal case. Clearly, in this algorithm information is lost, and it can be recovered by the oversampling.

We have not tested all the multiscale transforms, but it is a priori clear that the reconstructed objects will depend on the transform used. The choice of the best tool is probably associated with the image texture. We have applied the à trous algorithm on diffuse structures, which correspond generally to astronomical patterns. This transform works well for these objects, because the image is restored by cubic B-splines, which are also diffuse-like patterns. Now, if we consider the images of solid bodies, their representation by cubic B-splines is not optimal. Nonlinear multiscale transforms would be better adapted to this class of images.

A general framework is provided by the procedure described, but much effort has to be undertaken to determine the optimal multiscale vision model related to a given class of images.

APPENDIX A

Variance stabilization

A.1 Mean and variance expansions

Consider a variable x defined as $x = \alpha n + \gamma$ where α is the gain, γ is a Gaussian variable of mean g and standard deviation σ, and n is a Poisson variable of mean m_0.

A transformation of x is sought, such that the variance is constant irrespective of the value of x. The general form of this transformation is arrived at by the following argument. From the definition of x, its variance is given by $V(x) = \sigma^2 + \alpha^2 m_0$. Assuming the variability of x to be sufficiently small, the variance of a transformed x, $y(x)$, will be given by $V(y) = (\frac{dy}{dx})^2 V(x)$. Letting $V(y) = 1$, we get:

$$\frac{dy}{dx} = \frac{1}{\sqrt{\sigma^2 + \alpha^2 m_0}}$$

To a first order approximation, $\alpha m_0 = x - g$. Thus

$$\frac{dy}{dx} = \frac{1}{\sqrt{\sigma^2 - \alpha g + \alpha x}}$$

This leads to

$$y(x) = \frac{2}{\alpha}\sqrt{\alpha x + \sigma^2 - \alpha g}$$

This derivation of $y(x)$ has been based on a local approximation. More generally, we will now seek a transformation of this general form, i.e. $y = \sqrt{x + c}$. Series expansion about the mean will be used.

Define $E(x) = m$, and let $t = x - m$ and $m' = m + c$. An expansion of y as a function of t is then:

$$y = \sqrt{m' + t} = \sqrt{m'}\left[1 + \frac{1}{2}\frac{t}{m'} - \frac{1}{8}\frac{t^2}{m'^2} + \frac{1}{16}\frac{t^3}{m'^3} - \frac{5}{128}\frac{t^4}{m'^4} + \ldots\right] \tag{A.1}$$

Hence:

$$E(y) = \sqrt{m'}\left[1 - \frac{\mu_2}{8m'^2} + \frac{\mu_3}{16m'^3} - \frac{5\mu_4}{128m'^4} + \ldots\right] \tag{A.2}$$

where the μ_i are the centered moments of the variable t. We derive:

$$E^2(y) = m'\left[1 - \frac{\mu_2}{4m'^2} + \frac{\mu_3}{8m'^3} - \frac{5\mu_4}{64m'^4} + \frac{\mu_2^2}{64m'^4} + \ldots\right] \tag{A.3}$$

Hence the variance, $V(y)$, is given by:

$$V(y) = \frac{\mu_2}{4m'} - \frac{\mu_3}{8m'^2} - \frac{\mu_2^2 - 5\mu_4}{64m'^3} + \ldots \tag{A.4}$$

A.2 Determination of centered moments

The characteristic function of a sum of a Poisson variate and a Gaussian variate is given by:

$$\varphi(u) = \exp(-2\pi igu - 2\pi^2\sigma^2u^2 + m_0\exp(-2\pi i\alpha u) - 1) \tag{A.5}$$

Using the logarithm of $\varphi(u)$, taking its derivative, and considering the value of the derivative at the origin, yields:

$$\frac{\varphi'(u)}{\varphi(u)} = -4\pi^2\sigma^2u - 2\pi ig - \alpha m_0 2\pi ie^{-2\pi i\alpha u} \tag{A.6}$$

from which the first moment (mean) is:

$$m = g + \alpha m_0 \tag{A.7}$$

Using the second derivative yields:

$$\frac{\varphi''(u)}{\varphi(u)} - \frac{\varphi'^2(u)}{\varphi^2(u)} = -4\pi^2\sigma^2 - 4\pi^2\alpha^2m_0e^{-2\pi i\alpha u} \tag{A.8}$$

providing an expression for the second centered moment (variance):

$$\mu_2 = \sigma^2 + \alpha^2m_0 \tag{A.9}$$

The third derivative provides the third centered moment:

$$\mu_3 = m_3 - 3m_1m_2 + 2m_1^3 = \alpha^3m_0 \tag{A.10}$$

Finally, the fourth centered moment is:

$$\mu_4 = \alpha^4m_0 + 3(\sigma^2 + \alpha^2m_0)^2 \tag{A.11}$$

A.3 Study of the expansions for the variance

Given that $m' = m + c$, the following binomial expansions can be written:

$$\frac{1}{m'} = \frac{1}{m}\left[1 - \frac{c}{m} + \frac{c^2}{m^2} + \ldots\right] \tag{A.12}$$

$$\frac{1}{m'^2} = \frac{1}{m^2}\left[1 - 2\frac{c}{m} + 3\frac{c^2}{m^2} + \ldots\right] \tag{A.13}$$

$$\frac{1}{m'^3} = \frac{1}{m^3}\left[1 - 3\frac{c}{m} + \ldots\right] \tag{A.14}$$

We will consider the first two most significant terms of these. Consider eqn. (A.4) above, which is an equation for the variance which we seek to 'stabilize'. Expressions for μ_i for $i = 1, 2, 3, 4$ are provided by eqns. (A.7), (A.9)–(A.11). Finally expressions for $1/m', 1/m'^2$, and $1/m'^3$ are provided by eqns. (A.12)–(A.14). Substitution gives:

$$V = \frac{\alpha}{4} + \frac{\sigma^2 - \alpha g - c\alpha}{4m} - \frac{\alpha^2}{8m} + \frac{14\alpha^2}{64m} + \ldots \tag{A.15}$$

which further reduces to:

$$V = \frac{\alpha}{4} + \frac{16(\sigma^2 - \alpha g) - 16c\alpha + 6\alpha^2}{64m} \tag{A.16}$$

A.4 Variance-stabilizing transformation

To force the term in $1/m$ in eqn. (A.16) to disappear, we take:

$$c = \frac{3}{8}\alpha + \frac{\sigma^2 - \alpha g}{\alpha} \tag{A.17}$$

Using a normalizing term, $2/\sqrt{\alpha}$, will ensure that the variance will be stabilized to constant 1. The desired transformation is then:

$$t = \frac{2}{\alpha}\sqrt{\alpha x + \frac{3}{8}\alpha^2 + \sigma^2 - \alpha g} \tag{A.18}$$

Note how Anscombe's formula is obtained when $\alpha = 1$, and $\sigma = g = 0$.

APPENDIX B

Software information

A software package has been implemented by CEA (Saclay, France) and Nice Observatory. This software includes almost all applications presented in this book. Its goal is not to replace existing image processing packages, but to complement them, offering to the user a complete set of multiresolution tools. These tools are executable programs, which work on different platforms, independent of any image processing system, and allow the user to carry out various operations using multiresolution on his or her images such as a wavelet transform, filtering, deconvolution, etc.

The programs, written in C++, are built on three classes: the 'image' class, the 'multiresolution' class, and the 'noise-modeling class'. Figure B.1 illustrates this architecture. A multiresolution transform is applied to the input data, and noise modeling is performed. Hence the multiresolution support data structure can be derived, and the programs can use it in order to know at which scales, and at which positions, significant signal has been detected. Several multiresolution transforms are available (see Figure B.2), allowing much flexibility. Fig. 2.7 in Chapter 2 summarizes how the multiresolution support is derived from the data and our noise-modeling.

A set of IDL (Interactive Data Language) routines is included in the package which interfaces IDL and the C++ executables. The first release of the package will be in 1998 and more information will be available at http://ourworld.compuserve.com/homepages/ multires or from email address multires@compuserve.com.

Figure B.1 Software components diagram.

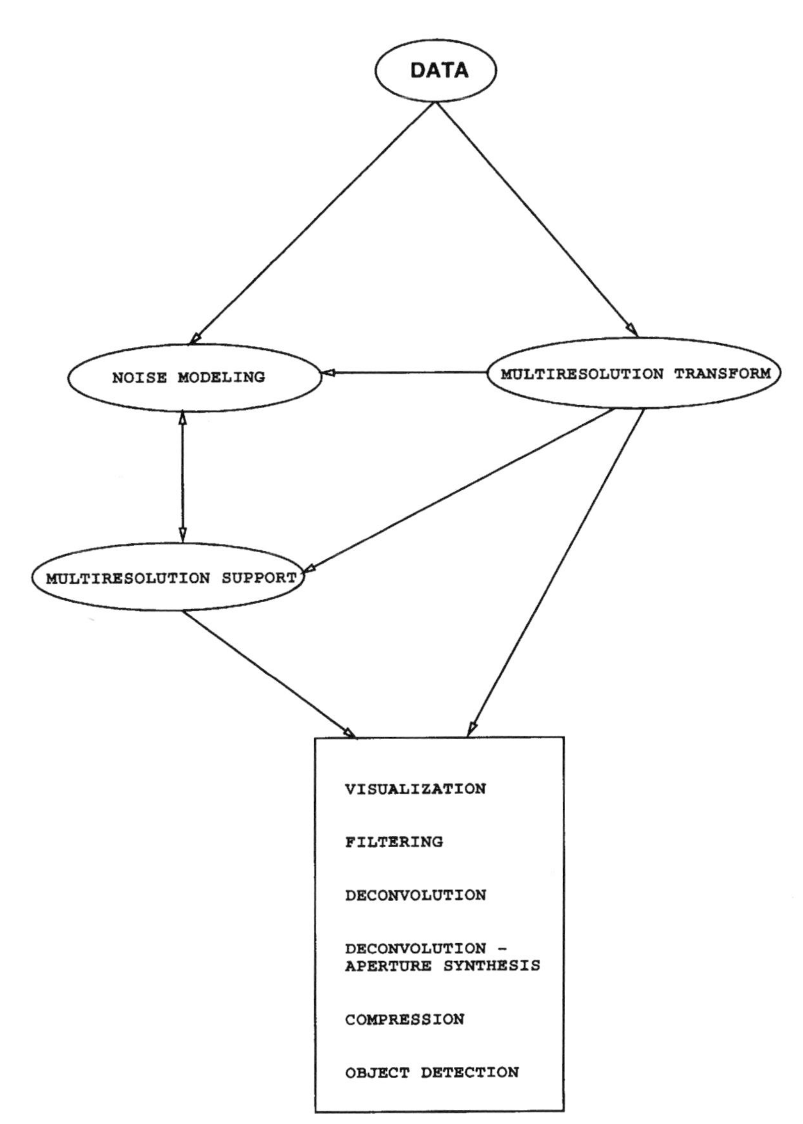

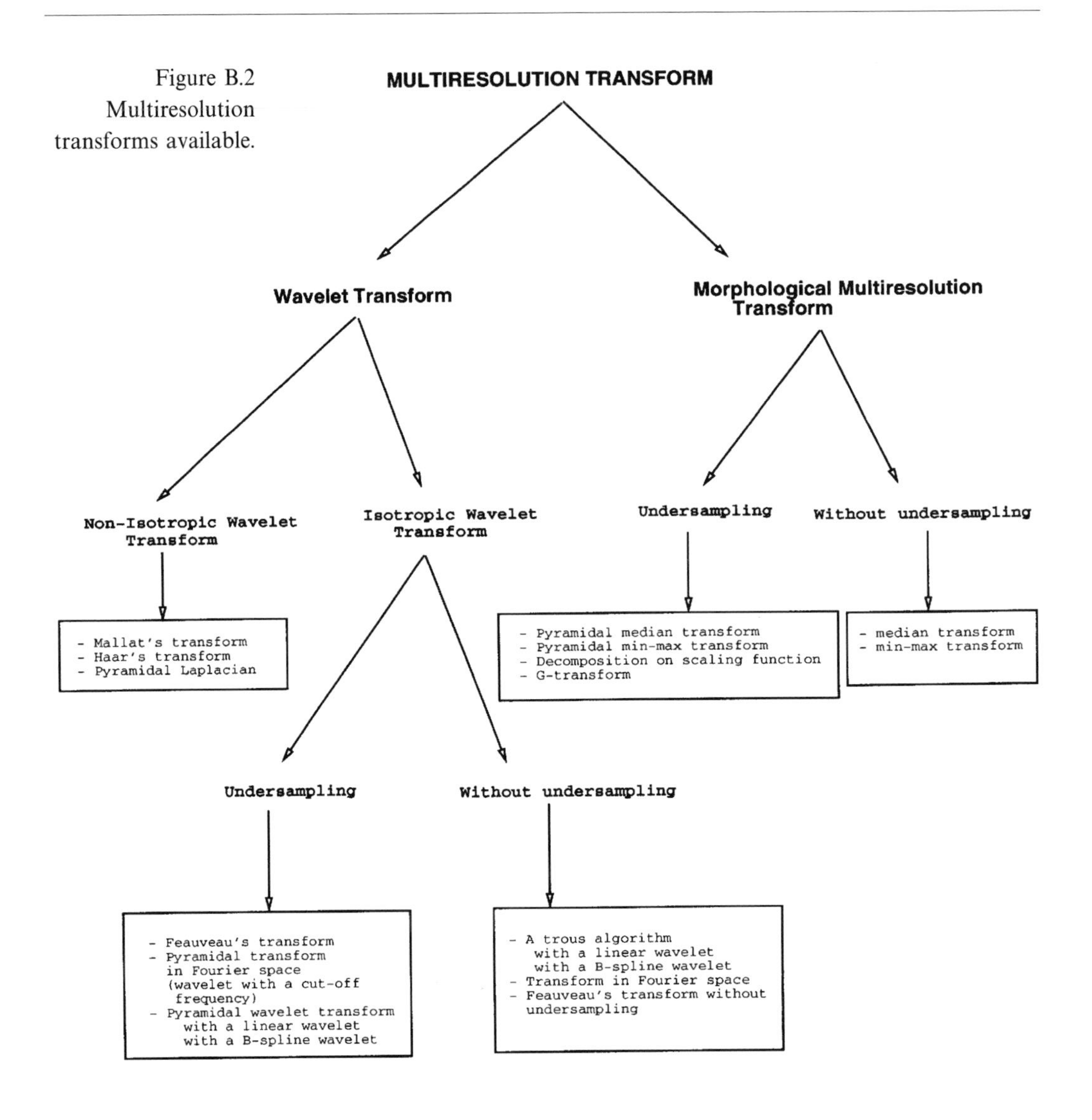

Figure B.2 Multiresolution transforms available.

APPENDIX C

Acronyms

1D, 2D, 3D	one-, two-, three-dimensional
AGB	Asymptotic Giant Branch
CCD	charge-coupled device
CGI	Common Gateway Interface
CWT	continuous wavelet transform
dB	decibel
DRNN	dynamic recurrent neural network
DWT	discrete wavelet transform
DTM	digital terrain model
ESO	European Southern Observatory
FFT	fast Fourier transform
FIR	finite input response
FOV	field of view
GCP	ground control point
HRV	High Resolution Visible
HST	Hubble Space Telescope
ICF	intrinsic correlation function
IRAS	Infrared Astronomical Satellite
ISO	Infrared Satellite Observatory
JPEG	Joint Photographic Expert Group
M	Messier
MAXENT	maximum entropy
ME	maximum entropy
MEM	maximum entropy method
MIDAS	Munich Image Data Analysis System
MIME	Multipurpose Internet Mail Extension
MLP	multilayer perceptron
MRC	multiresolution CLEAN
MSE	mean square error

MSS	Multispectral Scanner
MTF	modulated transfer function
NGC	New Galactic Catalog
PA	position angle
PCA	principal components analysis
PSF	point spread function
PMT	pyramidal median transform
POSS-I	Palomar Observatory Sky Survey I
RL	Richardson-Lucy
RMS	root mean square
RMSDE	root mean square distance error
SAO	Smithsonian Astrophysical Observatory
SNR	signal-to-noise ratio
SPOT	Satellite pour l'Observation de la Terre
STFT	short-time Fourier transform
TGCP	test ground control point
TIMMI	Thermal Infrared Multimode Instrument
TM	Thematic Mapper
WF/PC	Wide Field/Planetary Camera
WTS	wavelet transform space
XS	multispectral

Bibliography

J.G. Ables (1974), 'Maximum entropy spectral analysis', *Astronomy and Astrophysics Supplement Series*, 15, 383–393.

F. Achard, and F. Blasco (1990), 'Analysis of vegetation seasonal evolution and mapping of forest cover in West Africa with the use of NOAA AVHRR HRPT data', *Photogrammetric Engineering and Remote Sensing*, 56, 1359–1365.

E.H. Adelson, E. Simoncelli and R. Hingorani (1987), 'Orthogonal pyramid transforms for image coding', *SPIE Visual Communication and Image Processing II*, Vol. 845, 50–58.

H.-M. Adorf (1992), 'HST image restoration – recent developments', in P. Benvenuti and E. Schreier, eds., *Science with the Hubble Space Telescope*, European Southern Observatory, Garching, 227–238.

L.J. Allamandola, S.A. Sandford, A.G.G.M. Tielens, T.M. Herbst (1993), *Science*, 260, 64.

F.J. Anscombe (1948), 'The transformation of Poisson, binomial and negative-binomial data', *Biometrika*, 15, 246–254.

M. Antonini (1991), 'Transformée en ondelettes et compression numérique des images', PhD Thesis, Nice University.

P.N. Appleton, P.R. Siqueira and J.P. Basart (1993), 'A morphological filter for removing cirrus-like emission from far-infrared extragalactic IRAS fields', *The Astronomical Journal*, 106, 4, 1664–1678.

P.M. Atkinson (1991), 'Optimal ground-based sampling for remote sensing investigations: estimating the regional mean', *International Journal of Remote Sensing*, 12, 559–567.

A. Aussem (1995), 'Theory and Applications of Dynamical and Recurrent Neural Networks towards Prediction, Modeling and Adaptive Control of Dynamical Processes', Ph.D. Thesis (in French), Université René Descartes. Available at http://www.eso.org/meteo-seeing/aa-thesis/aa.html.

A. Aussem, F. Murtagh and M. Sarazin (1995), 'Dynamical recurrent neural networks – towards environmental time series prediction', *International Journal of Neural Systems*, 6, 145–170.

D.J. Axon, D.A. Allen, J. Bailey, J.H. Hough, M.J. Ward and R.F. Jameson (1982), 'The variable infrared source near HH100', *Monthly Notices of the Royal Astronomical Society*, 200, 239–245.

J.D. Banfield and A.E. Raftery (1993), 'Model-based Gaussian and non-Gaussian clustering, *Biometrics*, 49, 803–829.

S.T. Barnard and W.B. Thompson (1980), 'Disparity analysis of images', *IEEE Transactions on Pattern Analysis and Machine Intelligence*, PAMI-2, 333–340.

D.I. Barnea and H.F. Silverman (1972), 'A class of algorithms for fast digital image registration', *IEEE Transactions on Computers*, C-21, 179–186.

M.S. Bartlett (1936), *Journal of the Royal Statistical Society Supplement*, 3, 68.

J.M. Beckers (1991), 'Interferometric imaging with the Very Large Telescope', *Journal of Optics*, 22, 73–83.

P.M. Bentley and J.T.E. McDonnell (1994), 'Wavelet transforms: an introduction', *Electronics and Communication Engineering Journal*, 175–186.

C. Berge (1967), *Théorie des Graphes et ses Applications*, Dunod, Paris.

R. Bernstein (1975), 'Digital image processing', *Photogrammetric Engineering and Remote Sensing*, 41, 1465–1476.

R. Bernstein (1976), 'Digital image processing of Earth observation sensor data', *IBM Journal of Research and Development*, 20, 40–57.

M. Bhatia, W.C. Karl and A.S. Willsky (1996), 'A wavelet-based method for multiscale tomographic reconstruction', *IEEE Transactions on Medical Imaging*, 15, 92–101.

A. Bijaoui (1980), 'Skybackground estimation and application', *Astronomy and Astrophysics*, 84, 81–84.

A. Bijaoui (1991a), 'Wavelets and the analysis of astronomical objects', in J.D. Fournier and P.L. Sulem, eds., *Large-Scale Structures in Nonlinear Physics*, Springer-Verlag, Berlin, 340–347.

A. Bijaoui (1991b), 'Algorithmes de la transformation en ondelettes. Applications en astronomie', in P.J. Lions, ed., *Ondelettes et Paquets d'Ondes*, Cours CEA/EdF/INRIA.

A. Bijaoui (1993a), 'Wavelets and astronomical image analysis', in M. Farge, J.C.R. Hunt and J.C. Vassilicos, eds., *Wavelets, Fractals and Fourier Transforms: New Developments and New Applications*, Oxford University Press, Oxford, 195–212.

A. Bijaoui (1993b), 'Astronomical image inventory by the wavelet transform', in Y. Meyer and S. Roques, eds., *Wavelets and Applications*, Editions Frontières, Gif-sur-Yvette, 551–556.

A. Bijaoui (1994), 'Généralisation de la transformation d'Anscombe', technical report.

A. Bijaoui and M. Giudicelli (1991), 'Optimal image addition using the wavelet transform', *Experimental Astronomy*, 1, 347–363.

A. Bijaoui and F. Rué (1995), 'A multiscale vision model adapted to astronomical images', *Signal Processing*, 46, 345–362.

A. Bijaoui, G. Lago, J. Marchal and C. Ounnas (1978), 'Le traitement automatique des images en astronomie', in *Traitement des Images et Reconnaissance des Formes*, INRIA, 848–854.

A. Bijaoui, E. Slezak and G. Mars (1989), 'Détection des objets faibles dans des images célestes à l'aide de la transformation ondelette', *12ième Colloque du GRETSI*, 209–211.

A. Bijaoui, P. Bury and E. Slezak (1994a), 'Catalog analysis with multiresolution insights. I. Detection and characterization of significant structures', report.

A. Bijaoui, J.L. Starck and F. Murtagh (1994b), 'Restauration des images multi-échelles par l'algorithme à trous', *Traitement du Signal*, 11, 229–243.

Tj.R. Bontekoe, E. Koper and D.J.M. Kester (1994), 'Pyramid maximum entropy images of IRAS survey data', *Astronomy and Astrophysics*, 294, 1037–1053.

R. Bracho and A.C. Sanderson (1985), 'Segmentation of images based on intensity gradient information', in Proceedings of *CVPR-85 Conference on Computer Vision and Pattern Recognition*, San Francisco, 341–347.

L. Breiman, J.H. Friedman, R. Olshen and C.J. Stone (1984), *Classification and Regression Trees*, Wadsworth, Belmont.

A. Bruce and H.-Y. Gao (1994), *S+Wavelets User's Manual*, Version 1.0, StatSci Division, MathSoft Inc., Seattle, WA.

J.P. Burg (1978), *Annual Meeting International Society Exploratory Geophysics*, 1967. Reprinted in *Modern Spectral Analysis*, 1978, D.G. Childers, ed., IEEE Press, New York, 34–41.

P.J. Burt and A.E. Adelson (1983), 'The Laplacian pyramid as a compact image code', *IEEE Transactions on Communications*, 31, 532–540.

P. Bury (1995), 'De la Distribution de Matière à grande Echelle à Partir des Amas D'Abell', PhD Thesis, Nice University.

J.R. Carr and D.E. Myers (1984), 'Application of the theory of regionalized variables to the spatial analysis of Landsat data', *IEEE 1984 Proceedings of Pecora 9 – Spatial Information Technologies For Remote Sensing Today and Tomorrow*, IEEE Computer Society Press, New York, 55–61.

K.R. Castleman (1979), *Digital Image Processing*, Prentice-Hall, Englewood Cliffs.

A. Caulet and W. Freudling (1993), 'Distant galaxy cluster simulations – HST and ground-based', *ST-ECF Newsletter*, No. 20, 5–7.

M.K. Charter (199), 'Drug absorption in man, and its measurement by MaxEnt', in P.F. Fougère, ed., *Maximum Entropy and Bayesian Methods*, Kluwer, Dordrecht, 325–339.

W.P. Chen and J.A. Graham (1993), 'Ice grains in the Corona Australis molecular cloud', *Astrophysical Journal*, 409, 319–326.

J.E. Chiar, A.J. Adamson, T.H. Kerr and D.C.B. Whittet (1995), 'High-resolution studies of solid CO in the Taurus dark cloud: characterizing the ices in quiescent clouds', *Astrophysical Journal*, 455, 234–243.

J.P. Chiles and A. Guillen (1984), 'Variogrammes et krigeages pour la gravimétrie et le magnétisme', in J.J. Royer, ed., *Computers in Earth Sciences for Natural Resources Characterization, Annales de l'École Nationale Supérieure de Géologie Appliquée et de Prospection Minière*, Université de Nancy, 455–468.

C.H. Chui (1992), *Wavelet Analysis and Its Applications*, Academic Press, New York.

A. Cohen, I. Daubechies and J.C. Feauveau (1992), 'Biorthogonal bases of compactly supported wavelets', *Communications in Pure and Applied Mathematics*, 45, 485–560.

J. Cohen (1991), 'Tests of the photometric accuracy of image restoration using the maximum entropy algorithm', *Astrophysical Journal*, 101, 734–737.

T.J. Cornwell (1989), 'Image restoration', in D.M. Alloin and J.M. Mariotti, eds., *Diffraction-Limited Imaging with Very Large Telescopes*, Kluwer, Dordrecht, 273–292.

T.J. Cornwell (1992), 'Deconvolution for real and synthetic apertures', in D.M. Worrall, C. Biemesderfer and J. Barnes, eds., *Astronomical Data Analysis Software and Systems I*, 163–169.

G. Coupinot, J. Hecquet, M. Aurière and R. Futaully (1992), 'Photometric analysis of astronomical images by the wavelet transform', *Astronomy and Astrophysics*, 259, 701–710.

A. Dasgupta, and A.E. Raftery (1995), 'Detecting features in spatial point processes with clutter via model-based clustering', Technical Report 295, Department of Statistics, University of Washington (1995). Available at http://www.stat.washington.edu/tech.reports/. *Journal of the American Statistical Association*, in press.

I. Daubechies (1988), 'Orthogonal bases of compactly supported wavelets', *Communications on Pure and Applied Mathematics*, 41, 909–996.

I. Daubechies (1992), *Ten Lectures on Wavelets*, Society for Industrial and Applied Mathematics (SIAM), Philadelphia.

G. Demoment (1989), 'Image reconstruction and restoration: overview of common estimation structures and problems', *IEEE Transactions on ASSP*, 37, 2024–2036.

J.P. Djamdji (1993), 'Analyse en Ondelettes et Mise en Correspondance en Télédétection', PhD Thesis, Université de Nice Sophia Antipolis.

J.P. Djamdji and A. Bijaoui (1995a), 'Disparity analysis: a wavelet transform approach', *IEEE Transactions on Geoscience and Remote Sensing*, 33, 67–76.

J.P. Djamdji and A. Bijaoui (1995b), 'Disparity analysis in remotely sensed images using the wavelet transform', in *SPIE International Symposium on Aerospace Defense Sensing and Dual-Use Photonics, Wavelet Applications*, Orlando.

J.P. Djamdji and A. Bijaoui (1995c), 'Earth science and remote sensing disparity analysis and image registration of stereoscopic images using the wavelet transform', in *European Symposium on Satellite Remote Sensing II – Image and Signal Processing for Remote Sensing*, Paris, Vol. 2579, 11–21.

J.P. Djamdji, A. Bijaoui and R. Manière (1993a), 'Geometrical registration of images: the multiresolution approach', *Photogrammetric Engineering and Remote Sensing*, 59, 645–653.

J.P. Djamdji, A. Bijaoui and R. Manière (1993b), 'Geometrical registration of remotely sensed images with the use of the wavelet transform', *SPIE International Symposium on Optical Engineering and Photonics*, Vol. 1938, Orlando, 412–422.

J.P. Djamdji, J.L. Starck, and A. Claret (1996), 'ISOCAM image registration', CEA-SAP technical report.

D.L. Donoho (1993), 'Nonlinear wavelet methods for recovery of signals, densities, and spectra from indirect and noisy data', Proceedings of *Symposia in Applied Mathematics*, 47.

D.L. Donoho, and I.M. Johnstone (1993), 'Ideal spatial adaptation by wavelet shrinkage', Stanford University, Technical Report 400, (available by anonymous ftp from playfair.stanford.edu:/pub/donoho).

D. Dubaj (1994), 'Evaluation de méthodes de compression d'images pour le stockage et la distribution d'images astronomiques', DEA dissertation (under the direction of F. Bonnarel and M. Louys), Strasbourg Observatory.

R.O. Duda and P.E. Hart (1973), *Pattern Recognition and Scene Analysis*, Wiley, New York.

J.C. Feauveau (1990), 'Analyse multirésolution par ondelettes non-orthogonales et bancs de filtres numériques', PhD Thesis, Université Paris Sud.

G.E. Ford and C.I. Zanelli (1985), 'Analysis and quantification of errors in the geometric correction of satellite images', *Photogrammetric Engineering and Remote Sensing*, 51, 1725–1734.

W. Freudling and A. Caulet (1993), 'Simulated HST observations of distant clusters of galaxies', in P. Grosbøl, ed., Proceedings of *5th ESO/ST-ECF Data Analysis Workshop*, European Southern Observatory, Garching, 63–68.

M.O. Freeman (1993), 'Wavelets: signal representations with important advantages', *Optics and Photonics News*, 8–14.

D.L. Fried (1992), 'Superresolution', *Signal Recovery and Synthesis IV Conference*, Optical Society of America, New Orleans.

B.R. Frieden (1975), 'Image enhancement and restoration', *Topics in Applied Physics*, 6, Springer-Verlag, Berlin, 177–249.

D. Gabor (1946), 'Theory of communication', *Journal of the IEE*, 93, 429–441.

N.P. Galatsanos and A.K. Katsaggelos (1992), 'Methods for choosing the regularization parameter and estimating the noise variance in image restoration and their relation', *IEEE Transactions on Image Processing*, 1, 322–336.

I.F. Glass (1974), *Monthly Notices of the Royal Astronomical Society*, 33, 53.

C. Glass, J. Carr, H.M. Yang and D. Myers (1987), 'Application of spatial statistics to analysing multiple remote sensing data set', in A.I. Johnson and C. B. Petterson, eds., *Geotechnical Applications of Remote Sensing and Remote Data Transmission*, American Society for Testing and Materials, Philadelphia, 139–150.

G.H. Golub, M. Heath, and G. Wahba (1979), 'Generalized cross-validation as a method for choosing a good ridge parameter', *Technometrics*, 21, 215–223.

P. Goupillaud, A. Grossmann and J. Morlet (1985), 'Cycle-octave and related transforms in seismic signal analysis', *Geoexploration*, 23, 85–102.

A. Graps (1995), 'An introduction to wavelets', *IEEE Computational Science and Engineering*, 2. Paper, with other documentation, available at http://www.best.com/~agraps/current/wavelet.html

R. Gredel (94), 'Near-infrared spectroscopy and imaging of Herbig-Haro objects', *Astronomy and Astrophysics*, 292, 580–592.

A. Grossmann and J. Morlet (1984), 'Decomposition of Hardy functions into sqare integrable wavelets of constant shape', *SIAM Journal of Mathematical Analysis*, 15, 723–736.

S.F. Gull and J. Skilling (1991), MEMSYS5 User's Manual.

R. Hanisch, Ed. (1993), 'Restoration – Newsletter of STScI's Image Restoration Project', Space Telescope Science Institute, Baltimore, 1993.

R.M. Haralick, S.R. Sternberg and Xinhua Zhuang (1987), 'Image analysis using mathematical morphology', *IEEE Transactions on Pattern Analysis and Machine Intelligence*, PAMI-9, 532–550, 1987.

J.L. Harris (1964), 'Diffraction and resolving power', *Journal of the Optical Society of America*, 54, 931–936.

G. Held and T.R. Marshall (1987), *Data Compression*, Wiley, New York.

J. Herman, B. Baud, H. J. Haking and A. Wimberg (1985), 'VLA line observations of OH/IR stars', *Astronomy and Astrophysics*, 143, 122–135, 1985.

P.H.Y. Hiernaux and C.O. Justice (1986), 'Monitoring the grasslands of the Sahel using NOAA AVHRR data: Niger 1983', *International Journal of Remote Sensing*, 11, 1475–1495.

J.A. Högbom (1974), 'Aperture synthesis with a non-regular distribution of interferometer baselines', *Astronomy and Astrophysics Supplement Series*, 15, 417–426.

M. Holschneider, R. Kronland-Martinet, J. Morlet and Ph. Tchamitchian (1989), 'A real-time algorithm for signal analysis with the help of the wavelet transform', in J.M. Combes, A. Grossman and P. Tchmitchian, eds., *Wavelets: Time-Frequency Methods and Phase Space*, Springer-Verlag, Berlin, 286–297.

L. Huang and A. Bijaoui (1991), *Experimental Astronomy*, 1, 311–327.

A.C. Hung (1993), PVRG-JPEG CODEC 1.0, Portable Video Research Group, Stanford University (anonymous ftp to: havefun.stanford.edu:/pub/jpeg).

P.T. Jackway and M. Deriche (1996), 'Scale-space properties of the multiscale morphological dilation-erosion', *IEEE Transactions on Pattern Analysis and Machine Intelligence*, 18, 38–51.

C.E. Jacobs, A. Finkelstein and D.H. Salesin (1995), 'Fast multiresolution image querying', *Computer Graphics Proceedings, Annual Conference Series, 1995*, SIGGRAPH 95, Los Angeles, 277–286.

S. Jaffard (1989), 'Algorithmes de transformation en ondelettes', *Annales des Ponts et Chaussées.*

E.T. Jaynes (1957), *Physical Review*, 106, 620–630.

R. Jeansoulin (1982), *Les images multi-sources en télédétection: Mise en correspondance numérique et analyse de la structure géométrique.* PhD thesis, Université Paul Sabatier de Toulouse.

A.G. Journel and Ch.J. Huijbregts (1978), *Mining Geostatistics*, Academic Press, Chicago.

C.O. Justice and P.H.Y. Hiernaux (1986), 'Suivi du développement végétal au cours de l'été 1984 dans le sahel malien'. *International Journal of Remote Sensing*, 11, 1515–1531.

A.K. Katsaggelos (1991), ed., *Digital Image Restoration*, Springer-Verlag, Berlin.

R.G. Keys (1981), 'Cubic convolution interpolation for digital image processing', *IEEE Transactions on Acoustics, Speech and Signal Processing*, ASSP-29, 1153–1160.

M. Kissler, T. Richtler, E.V. Held, E.K. Grebel, S. Wagner and M. Cappaccioli (1993), 'NGC 4636 – a rich globular cluster system in a normal elliptical galaxy', *ESO Messenger*, No. 73, 32–34.

J.B. Kruskal (1964), 'Multidimensional scaling by optimizing goodness of fit to a nonmetric hypothesis', *Psychometrika*, 29, 1–27.

A. Kruszewski (1989), 'INVENTORY – Searching, photometric and classifying package', in P.J. Grosbøl, F. Murtagh and R.H. Warmels, eds., *1st ESO/ST-ECF Data Analysis Workshop*, European Southern Observatory, Garching, 29–33.

A. Labeyrie (1978), 'Stellar interferometry methods', *Annual Review of Astronomy and Astrophysics*, 16, 77–102.

P.O. Lagage and E. Pantin (1994), 'Dust depletion in the inner disk of β Pictoris as a possible indicator of planets', *Nature*, 369, 628–630.

V. Lalich-Petrich, G. Bhatia, and L. Davis (1995), in R. Holzapfel, ed., Poster Proceedings, *Third International WWW Conference*, Darmstadt '95, Frauenhofer Institute for Computer Graphics, Darmstadt, 159.

L. Landweber (1951), 'An iteration formula for Fredholm integral equations of the first kind', *American Journal of Mathematics*, 73, 615–624.

A. Lannes, S. Roques and M.J. Casanove (1987), 'Resolution and robustness in image processing: a new regularization principle', *Journal of the Optical Society of America Series A*, 4, 189–199.

P. Lascaux and R. Théodor (1994), *Analyse Numérique Matricielle Appliquée à l'Art de l'Ingénieur*, Vol. 2, chap. 8, pp. 405–458.

S.M. Lea, L.A. Keller (1989), 'An algorithm to smooth and find objects in astronomical images', *Astronomical Journal*, 97, 1238–1246.

T. Le Bertre (1984), PhD Thesis, University of Paris 7.

T. Le Bertre, N. Epchtein, C. Gouiffes, M. Heydari-Malayri and C. Perrier (1989), 'Optical and infrared observations of four suspected photo-planetary objects', *Astronomy and Astrophysics*, 225, 417–431.

J.S. Lee (1981), 'Refined filtering of image noise using local statistics', *Computer Vision, Graphics and Image Processing*, 15, 380–389.

J.S. Lee (1983), 'Digital image smoothing and the sigma filter', *Computer Vision, Graphics and Image Processing*, 24, 255–269, 1983.

J.S. Lee and K. Hoppel (1989), 'Noise modelling and estimation of remotely-sensed images', in Proceedings of *1989 International Geoscience and Remote Sensing*, Vancouver, Vol. 3, 1005–1008.

E.-H. Liang and E.K. Wong (1993), 'Hierarchical algorithms for morphological image processing', *Pattern Recognition*, 26, 511–529.

T. Lillesand and R. Kiefer (1987), *Remote Sensing and Image Interpretation*, Wiley, New York, 2nd edn.

T. Lindeberg (1994), *Scale-Space Theory in Computer Vision*, Kluwer, Dordrecht.

J. Littlewood and R. Paley (1931), *Journal of the London Mathematical Society*, 6, 20.

J. Llacer and J. Núñez (1990), 'Iterative maximum likelihood estimator and Bayesian algorithms for image reconstruction in astronomy', in R.L. White and R.J. Allen, eds., *The Restoration of HST Images and Spectra*, Space Telescope Science Institute, Baltimore, 62–70.

H. Lorenz, G.M. Richter, 'Adaptive filtering of HST images: preprocessing for deconvolution', in P. Benvenuti and E. Schreier, eds., *Science with the Hubble Space Telescope*, European Southern Observatory, Garching, 203–206.

L.B. Lucy (1974), 'An iteration technique for the rectification of observed distributions', *Astronomical Journal*, 79, 745–754.

S. Mallat (1989), 'A theory for multiresolution signal decomposition: the wavelet representation', *IEEE Transactions on Pattern Analysis and Machine Intelligence*, 11, 674–693.

P. Maragos and R.W. Schaffer (1990), 'Morphological systems for multidimensional signal processing', *Proceedings of the IEEE*, 78, 690–710.

G.A. Mastin (1985), 'Adaptive filters for digital noise smoothing, an evaluation', *Comp. Vision Graphics Image Process*, 31, 103–121.

G. Matheron (1965), *Les Variables Régionalisées et leur Estimation*, Masson, Paris.

G. Matheron (1970), 'La théorie des variables régionalisées et ses applications', *Cahier du Centre de Morphologie Mathématique de Fontainebleau*. École des Mines, Paris.

A.B. McBratney, R. Webster and T.M. Burgess (1981a), 'The design of optimal sampling schemes for local estimation and mapping of regionalized variables. I: Theory and method', *Computers and Geosciences*, 7, 331–334.

A.B. McBratney, R. Webster and T.M. Burgess (1981b), 'The design of optimal sampling schemes for local estimation and mapping of regionalized variables. II: Program and examples', *Computers and Geosciences*, 7, 335–365.

P. Meer, J. Jolian and A. Rosenfeld (1990), 'A fast parallel algorithm for blind estimation of noise variance', *IEEE Transactions on Pattern Analysis and Machince Intelligence*, 12, 216–223.

C. Meneveau (1991), 'Analysis of turbulence in the orthonormal wavelet representation', *Journal of Fluid Mechanics*, 232, 469–520, 1991.

Y. Meyer (1991), *Proc. Ondelettes et Paquets d'Ondes*, INRIA, Rocquencourt.

Y. Meyer (1990, 1993), *Ondelettes*, Hermann, Paris. *Wavelets: Algorithms and Applications*, SIAM, Philadelphia.

Y. Meyer, S. Jaffard and O. Rioul (1987), 'L'analyse par ondelettes', *Pour la Science*, 28–37.

J.P.D. Mittaz, M.V. Penston, and M.A.J. Snijders (1990), 'Ultraviolet variability of NGC 4151: a study using principal components analysis', *Monthly Notices of the Royal Astronomical Society*, 242, 370–378.

R. Molina (1994), 'On the hierarchical Bayesian approach to image restoration. Applications to astronomical images', *IEEE Transactions on Pattern Analysis and Machine Intelligence*, 16, 1122–1128.

R. Murenzi (1988), 'Wavelet transforms associated with the *n*-dimensional Euclidean group with dilations: signal in more than one dimension', in J.M. Combes, A. Grossman and P. Tchamitchian, eds., *Wavelets: Time-Frequency Methods and Phase Space*, Springer-Verlag, Berlin, 239–246.

F. Murtagh (1985), *Multidimensional Clustering Algorithms*, Physica-Verlag, Würzburg.

F. Murtagh and M. Hernández-Pajares (1995), 'The Kohonen self-organizing feature map method: an assessment', *Journal of Classification*, 12, 165–190.

F. Murtagh and J.L. Starck (1994), 'Multiresolution image analysis using wavelets', *ST-ECF Newsletter*, No. 21.

F. Murtagh and J.-L. Starck (1996), 'Pattern clustering based on noise modeling in wavelet space', *Pattern Recognition*, submitted, in press.

F. Murtagh, J.L. Starck and A. Bijaoui (1995), 'Image restoration with noise suppression using a multiresolution support', *Astronomy and Astrophysics Supplement Series*, 112, 179–189.

D.E. Myers (1987), 'Interpolation with positive definite functions', In *Etudes Géostatistiques V – Séminaire C.F.S.G. sur la Géostatistique*, 251–265. Sciences de la Terre, Série Inf – Université de Nancy.

D.E. Myers (1991), 'Interpolation and estimation with spatially located data', *Chemometrics and Intelligent Laboratory Systems*, 11, 209–228.

R. Narayan and R. Nityananda (1986), 'Maximum entropy image restoration in astronomy', *Annual Reviews of Astronomy and Astrophysics*, 24, 127–170.

G.P. Nason and B.W. Silverman (1994), 'The discrete wavelet transform in S', *Journal of Computational and Graphical Statistics*, 3, 163–191. Software documentation, available at Statlib repository, http://lib.stat.cmu.edu/

W. Niblack (1986), *An Introduction to Digital Image Processing*, Prentice-Hall, Englewood Cliffs.

S.I. Olsen (1993), 'Estimation of noise in images: an evaluation', *CVGIP: Graphical Models and Image Processing*, 55, 319–323.

E. Pantin and J.L. Starck (1996), 'Deconvolution of astronomical images using the multiresolution maximum entropy method', *Astronomy and Astrophysics Supplement Series*, 118, 575–585.

S. Park and R. Showengerdt (1983), 'Image reconstruction by parametric cubic convolution', *Computer Vision, Graphics and Image Processing*, 23, 258–272.

J.W. Percival and R.L. White (1993), 'Efficient transfer of images over networks', in R.J. Hanisch, R.J.V. Brissenden and J. Barnes, eds., *Astronomical Data Analysis Software and Systems II*, Astronomical Society of the Pacific, San Francisco, 321-325.

C. Perrier (1982), Thesis, University of Paris 7.

C. Perrier (1986), 'ESO infrared specklegraph', *The Messenger*, No. 45, 29–32.

C. Perrier (1989), 'Amplitude estimation from speckly interferometry', in D.M. Alloin and J.M. Mariotti, eds., *Diffraction-Limited Imaging with Very Large Telescopes*, 99–119, Kluwer, Dordrecht, 1989.

I. Pitas and A.N. Venetsanopoulos (1990), *Nonlinear Digital Filters*, Kluwer, Dordrecht.

W.K. Pratt (1978), *Digital Image Processing*, Wiley, New York.

F.P. Preparata and M.I. Shamos (1985), *Computational Geometry*, Springer-Verlag, New York.

W.H. Press (1992), 'Wavelet-based compression software for FITS images', in D.M. Worrall, C. Biemesderfer and J. Barnes, eds., *Astronomical Data Analysis Software and Systems I*, ASP, San Francisco, 3–16.

W.H. Press, S.A. Teukolsky, W.T. Vetterling and B.P. Flannery (1992), *Numerical Recipes*, 2nd ed., Chapter 13, Cambridge University Press, New York.

M.B. Priestley (1981), *Spectral Analysis and Time Series*, Academic Press, New York.

S.D. Prince, C.J. Tucker and C.O. Justice (1986), 'Monitoring the grasslands of the Sahel: 1984-1985', *International Journal of Remote Sensing*, 11, 1571–1582.

G. Ramstein (1989), 'Structures spatiales irrégulière dans les images de télédétection', PhD Thesis, Université Louis Pasteur, Strasbourg.

G. Ramstein and M. Raffy (1989), 'Analysis of the structure of radiometric remotely-sensed images', *International Journal of Remote Sensing*, 10, 1049–1073.

W.H. Richardson (1972), 'Bayesian-based iterative method of image restoration', *Journal of the Optical Society of America*, 62, 55–59.

S.S. Rifman (1973), 'Digital rectification of ERTS multispectral imagery', in *Proc. Symposium on Significant Results Obtained from ERTS-1*, Vol. I, Section B, NASA SP-327, 1131–1142.

S.S. Rifman and D.M. McKinnon (1974), 'Evaluation of digital correction techniques for ERTS images', TRW Report 20634-6003-TV-02.

P.J. Rousseeuw and A.M. Leroy (1987), *Robust Regression and Outlier Detection*, Wiley, New York.

M.M. Ruskai, G. Beylkin, R. Coifman, I. Daubechies, S. Mallat, Y. Meyer and L. Raphael (1992), *Wavelets and Their Applications*, Jones and Barlett.

H. Samet (1984), 'The quadtree and related hierarchical data structures', *ACM Computing Surveys*, 16, 187–260.

J.W. Sammon (1969), 'A nonlinear mapping for data structure analysis', *IEEE Transactions on Computers*, C-18, 401–409.

J.D. Scargle (1997), 'Wavelet methods in astronomical time series analysis', in T. Subba Rao, M.B. Priestley and O. Lessi, eds., *Applications of Time Series Analysis in Astronomy and Meteorology*, University of Padua, 226–248.

R.J. Schalkoff (1989), *Digital Image Processing and Computer Vision*, Wiley, New York.

W.A. Schutte (1995), 'The role of dust in the formation of stars', ESO Workshop and Conference Proceedings, eds. Käufl & Siebenmorgen, Springer-Verlag.

W.A Schutte, P.A. Gerakines, T.R. Geballe, E.F. Van Dishoeck and J.M. Greenberg (1996), 'Discovery of solid formaldehyde toward the protostar GL 2136: observations and laboratory simulation', *Astronomy and Astrophysics*, 309, 633–647.

U.J. Schwarz (1978), 'Mathematical-statistical description of the iterative beam removing technique (method CLEAN)', *Astronomy and Astrophysics*, 65, 345–356.

J. Serra (1982), *Image Analysis and Mathematical Morphology*, Academic Press, New York.

C.E. Shannon (1948), *Bell Systems Technical Journal*. 27, 379 and 623.

M.J. Shensa (1992), 'Discrete wavelet transforms: wedding the à trous and Mallat algorithms", *IEEE Transactions on Signal Processing*, 40, 2464–2482.

L.A. Shepp and Y. Vardi (1982), 'Maximum likelihood reconstruction for emission tomography', *IEEE Transactions on Medical Imaging*, MI-1, 113–122.

S. Shlien (1991), 'Geometric correction, registration and resampling of Landsat Imagery', *IEEE Transactions on Geoscience and Remote Sensing*, 29, 292–299.

R. Showengerdt (1983), *Techniques for Image Processing and Classification in Remote Sensing*, Academic Press, New York.

H. Silverman and R. Bernstein (1971), 'Digital techniques for Earth resource image data processing', in *Proceedings of the American Institute of Aeronautics and Astronautics, 8th Annual Meeting*, Vol C-21, AIAA paper No. 71-978455–468.

J. Skilling (1989), 'Classic maximum entropy', in: Skilling. J, ed., *Maximum Entropy and Bayesian Methods*, Kluwer, Dordrecht, 45–52.

J. Skilling and R.K. Bryan (1984), 'Maximum entropy image reconstruction: general algorithm', *Monthly Notices of the Royal Astronomical Society*, 211, 111–124.

J. Skilling and S.F. Gull (1984), *Proceedings of the American Mathematical Society*, SIAM, 14, 167.

E. Slezak, G. Mars, A. Bijaoui, C. Balkowski and P. Fonatanelli (1988), 'Galaxy counts in the Coma supercluster field: automatic image detection and classification', *Astronomy and Astrophysics Supplement Series*, 74, 83–106.

E. Slezak, A. Bijaoui and G. Mars (1990), 'Structures identification from galaxy counts: use of the wavelet transform', *Astronomy and Astrophysics*, 227, 301–316.

E. Slezak, V. de Lapparent and A. Bijaoui (1993), 'Objective detection of voids and high density structures in the first CfA redshift survey slice', *Astrophysical Journal*, 409, 517–529.

M.J.T. Smith and T.P. Barnwell (1988), 'Exact reconstruction technique for tree structured subband coders', *IEEE Transactions on ASSP*, 34, 434–441.

H. Späth (1985), *Cluster Dissection and Analysis*, Ellis Horwood, Chichester.

Spotimage (1986), *Guide des Utilisateurs de Données SPOT*, Vol. 2, CNES and SPOTIMAGE.

J.L. Starck (1992), 'Analyse en ondelettes et imagerie à haute résolution angulaire', PhD Thesis, Nice University.

J.L. Starck (1993), in MIDAS Manual, Release 93NOV, Image Processing Group, European Southern Observatory, Garching.

J.L. Starck and A. Bijaoui (1991), 'Wavelets and multiresolution Clean', in *High Resolution Imaging by Interferometry II*, European Southern Observatory, Garching.

J.L. Starck and A. Bijaoui (1992), 'Deconvolution from the pyramidal structure', in *International Conference on Wavelets and Applications*, Toulouse.

J.L. Starck and A. Bijaoui (1994a), 'Filtering and deconvolution by the wavelet transform', *Signal Processing*, 35, 195–211.

J.L. Starck and A. Bijaoui (1994b), 'Multiresolution deconvolution', *Journal of the Optical Society of America*, 11, 1580–1588.

J.L. Starck and F. Murtagh (1994), 'Image restoration with noise suppression using the wavelet transform', *Astronomy and Astrophysics*, 288, 343–348.

J.L. Starck and E. Pantin (1996), 'Multiscale maximum entropy image restoration', *Vistas in Astronomy*, 40, 4, 1–6.

J.L. Starck, F. Murtagh and A. Bijaoui (1995), 'Multiresolution support applied to image filtering and deconvolution', *Graphical Models and Image Processing*, 57, 420–431.

J.L. Starck, A. Bijaoui, B. Lopez and C. Perrier (1994), 'Image reconstruction by the wavelet transform applied to aperture synthesis', *Astronomy and Astrophysics*, 283, 349–360.

J.L. Starck, F. Murtagh, B. Pirenne and M. Albrecht (1996), 'Astronomical image compression based on noise suppression', *Publications of the Astronomical Society of the Pacific*, 108, 446-455.

R.S. Stobie (1986), 'The COSMOS image analyzer', *Pattern Recognition Letters*, 4, 317–324.

E.J. Stollnitz, T.D. DeRose and D.H. Salesin (1995), 'Wavelets for computer graphics: a primer, Part 1', *IEEE Computer Graphics and Applications*, May, 76–84. 'Part 2', July, 75–85.

G. Strang (1989), 'Wavelets and dilation equations: a brief introduction', *SIAM Review*, 31, 614–627.

G. Strang and T. Nguyen (1996), *Wavelets and Filter Banks*, Wellesley-Cambridge Press, Wellesley, MA.

K.M. Strom, S.E. Strom and G.L. Grasdalen (1974), 'An infrared source associated with a Harbig-Haro object', *Astrophysical Journal*, 187, 83–86.

A. Stuart and M.G. Kendall (1973), *The Advanced Theory of Statistics*, Vol. 2. Charles Griffin, London, 3rd edn.

R. Sucher (1995), 'An adaptive nonlinear filter for detection and removal of impulses', in I. Pitas, ed., *1995 IEEE Workshop on Nonlinear Signal and Image Processing*, Neos Marmaras Halkidiki, Greece.

M. Tanaka, T. Nagata, S. Sato and T. Yamamoto (1994), 'The nature of CO and H_2O ices in the Corona Australis cloud', *Astrophysical Journal*, 430, 779–785.

A.N. Tikhonov, A.V. Goncharski, V.V. Stepanov and I.V. Kochikov (1987), 'Ill-posed image processing problems', *Soviet Physics – Doklady*, 32, 456–458.

D.M. Titterington (1985), 'General structure of regularization procedures in image reconstruction', *Astronomy and Astrophysics*, 144, 381–387.

H. Tong (1990), *Non Linear Time Series*, Clarendon Press, Oxford.

W.S. Torgerson (1958), *Theory and Methods of Scaling*, Wiley, New York.

P.H. Van Cittert (1931), 'Zum Einfluß der Spaltbreite auf die Intensitätsverteilung in Spektrallinien II', *Zeitschrift für Physik*, 69, 298–308.

J.P. Véran, and J.R. Wright (1994), in *Astronomical Data Analysis Software and Systems III*, ASP, San Francisco (anonymous ftp to: uwila.cfht.hawaii.edu:/pub/compfits).

M. Vetterli (1986), 'Filter banks allowing perfect reconstruction', *Signal Processing*, 10, 219–244.

B. Vidaković and P. Müller (1995), 'Wavelets for kids. A tutorial introduction', preprint, available at http://www.isds.duke.edu/~brani/local.html

H. Vorhees and T. Poggio (1987), 'Detecting textons and texture boundaries in natural images', in Proceedings of *First International Conference on Computer Vision*, 250–258.

B.P. Wakker and U.J. Schwarz (1988), 'The multi-resolution CLEAN and its application to the short-spacing problem in interferometry', *Astronomy and Astrophysics*, 200, 312–322.

A.S. Weigend, D.E. Rumelhart and B.A. Huberman (1990), 'Predicting the future: a connectionist approach', *International Journal of Neural Systems*, 1, 195–220.

N. Weir (1991), 'Application of maximum entropy techniques to HST data', in Proceedings of *3rd ESO/ST-ECF Data Analysis Workshop*, European Southern Observatory, Garching.

N. Weir (1992), 'A multi-channel method of maximum entropy image restoration', in: D.M. Worral, C. Biemesderfer and J. Barnes, eds., *Astronomical Data Analysis Software and System 1*, Astronomical Society of the Pacific, San Francisco, 186–190.

J. Weng, T. Huang and N. Ahuja (1989), 'Motion and structure from two perspective views: algorithms, error analysis, and error estimation', *IEEE Transactions on Pattern Analysis and Machine Intelligence*, 11, 451–476.

R. White, M. Postman and M. Lattanzi (1992), in H.T. MacGillivray and E.B. Thompson, eds., *Digitized Optical Sky Surveys*, Kluwer, Dordrecht, 167 (anonymous ftp to stsci.edu:/software/hcompress).

D.C.B. Whittet and J.C. Blades (1980), 'Grain growth in interstellar clouds', *Monthly Notices of the Royal Astronomical Society*, 191, 309–319.

M.V. Wickerhauser (1994), *Adapted Wavelet Analysis from Theory to Practice*, A.K. Peters, Wellesley, MA.

R. W. Wilson, J.E. Baldwin, D.F Busher and P.J Warner (1992), 'High-resolution imaging of Betelgeuse and Mira', *Monthly Notices of the Royal Astronomical Society*, 257, 369–376.

G.U. Yule (1927), 'On a method of investigating periodicities in disturbed series with special reference to Wolfer's sunspot numbers', *Philosophical Transactions of the Royal Society of London Series A*, 26, 247.

Index